TAKING SIDES

Clashing Views on Controversial

Global Issues

THIRD EDITION

TAKING SIDES

Clashing Views on Controversial

Global Issues

THIRD EDITION

Selected, Edited, and with Introductions by

James E. Harf
University of Tampa

and

Mark Owen Lombardi
College of Santa Fe

McGraw-Hill/Dushkin
A Division of The McGraw-Hill Companies

*To my daughter, Marie: May your world
conquer those global issues left unresolved
by my generation. (J.E.H.)*

*For Betty and Marty, who instilled a love of
education and need to explore the world. (M.O.L.)*

Photo Acknowledgment
Cover image: photos.com

Cover Acknowledgment
Maggie Nickles

Manufactured in the United States of America

Third Edition

123456789DOCDOC987654

Library of Congress Cataloging-in-Publication Data
Main entry under title:
Taking sides: clashing views on controversial global issues/selected, edited, and with introductions by James E. Harf and Mark Owen Lombardi.—3rd ed.
Includes bibliographical references and index.
1. Globalization. 2. International relations. 3. Global environmental change. 4. Population. 5. Emigration and immigration. I. Harf, James E., *ed.* II. Lombardi, Mark Owen, *ed.* III. Series.
303
0-07-311163-5
ISSN: 1536-3317

Printed on Recycled Paper

Preface

This volume reflects the changing nature of the international system. The personal reflections of its two editors bring this into sharp focus. We are both products of the separate environments in which we have been raised. Our 20-year age difference suggests that each of us brings to the study of international affairs a particular lens through which two distinct yet interconnected generations view contemporary world issues.

The older editor began his formal schooling a few weeks after the United States dropped two atomic bombs on Japan. His precollegiate experience took place amid the fear that an evil enemy was lurking throughout the world, poised to conquer America and enslave its occupants. During his initial month in college, the Soviet Union launched the first space vehicle, *Sputnik*, forever changing the global landscape. Yet his 1961 international relations textbook assumed that, despite the bomb and the ability to exploit outer space, the driving forces behind national behavior had not changed very much. By the time he accepted his first university teaching position, the world had been locked in a protracted Cold War for two decades. Yet evidence of a new global agenda of problems totally unrelated to the American-Soviet rivalry was beginning to appear. Soon this new set of issues would challenge the traditional agenda for the world's attention and also occupy the older editor's research agenda during much of his career.

When the older editor moved on to a new stage of his long career in the summer 2001, the Cold War existed no more. And the set of problems that had occupied his and other's attention for some time—population growth, resource scarcities, environmental degradation, and the like—were being transformed by the overarching presence of globalization. The Internet and other advances in communication and transportation signal an entirely new environment in which global actors design and carry out their policies. This was brought home in a rather graphic way in December 2000, as the older editor joined others in Kiev, Ukraine, to watch the newspapers debate the pending closure of the remaining operating nuclear power plant at Chernobyl. One month into his new position, he was called from the classroom by his office staff, who informed him that a new heinous form of terrorism had happened. He and his students made it to a television set just in time to see the second plane crash into the World Trade Center, thus ushering in a more widespread and more destructive era of terror. Thus, superimposed on the new agenda and coming at a time when the older editor was assuming a new academic position, the events of September 11, 2001 and the war on terrorism are changing long-held assumptions about war and violence.

This older editor's world has now undergone three major changes during his professional career, from a traditionalist, realist paradigm where military power dominated, to a new global agenda where problems demanded the attention and cooperation of a kaleidoscope of actors, to yet another set of problems and a transformation of existing issues brought on by globalization, where the ease with which people, information, ideas, money and goods cross national boundaries yield both positive and negative consequences for the world at large.

The younger editor began his formal schooling just after the United States and the Soviet Union had settled "comfortably" into the era of mutual assured destruction (MAD). He entered his university studies as increased tensions and U.S.-Soviet rivalry threatened to tear apart the fragile fabric of détente. As he embarked on his teaching career, the communist empire was taking its first steps toward total disintegration. Amid this revolutionary change, deeper and more complex issues of security, technology, development, and ethnic conflict became more pronounced and increasingly present in the consciousness of actors and scholars alike. As he too embarked on a new academic career, September 11 woke all Americans up to the harsh reality that the paradigms of violence and war were changing. Now at the midpoint of his academic life, this younger editor contemplates the speed and force of globalization, and how the issues that this phenomenon has brought forth will affect the lens through which he observes world affairs, influencing his and other research agendas over the next few decades.

The decision of this book's publisher to embark on a text on global issues reflects how the nature of international affairs has changed. Many textbooks on world politics and American foreign policy address longstanding, important concerns of the student of world events, employing for the most part the traditional paradigm of the past century. Simply put, this traditional scholarly view assumed a world where power, particular military might, dominated the world picture. Nation-states, the only really important world actors, focused on pursuing power and then using it in rather conventional ways to become a larger presence on the international scene. The specific problems of the day might have changed, but the fundamental principles that guided the international system remained the same. Most books in these fields reflect this outlook.

But the field of international affairs is ever changing, and textbooks are beginning to reflect this evolution. This new volume takes into account the fact that the age of globalization has accelerated, transforming trends that began over three decades ago. No longer are nation-states the only actors on the global stage. Moreover, their position of dominance is increasingly challenged by an array of other actors—international governmental organizations, private international groups, multi-national corporations, and important individuals—who might be better equipped to address newly emerging issues (or who might also serve as the source of yet other problems). This new agenda took root in the late 1960s where astute observers began to identify disquieting trends: quickening population growth in the poorer sectors of the globe, growing disruptions in the world's ability to feed its population, increasing shortfalls in required resources, and expanding evidence of negative environmental impacts, such as a variety of pollution evils.

An even more recent phenomenon is the unleashing of ethnic pride, which manifests itself in both positive and negative ways. The history of post–Cold War conflict is a history of intrastate, ethnically driven violence that has torn apart those countries that have been unable to deal effectively with the changes brought on by the end of the Cold War. The most insidious manifestation of this emphasis on ethnicity is the emergence of terrorist groups who use religion and other aspects of ethnicity to justify bringing death and destruction on their perceived enemies. As national governments attempt to cope with this latest phenomenon, they too are changing the nature of war and violence. The global agenda's current transformation, brought about by globalization, demands that our attention turn towards the latter's consequences.

The format of *Taking Sides: Clashing Views on Controversial Global Issues*, 3rd edition, follows the successful formula of other books in the Taking Sides series. The book begins with an introduction to the emergence of global issues and the new age of globalization. It then addresses 19 current global issues grouped into four parts. Population takes center stage in Part 1 because it not only represents a global issue by itself, but it also affects the parameters of most other global issues. Part 2 addresses a range of problems associated with global resources and their environmental impact. Parts 3 and 4 feature issues borne out of the emerging agenda of the twenty-first century. The former part examines expanding forces and movements across national boundaries such as the effects of the digital/computer world as well as international financial institutions. Part 4 focuses on new security issues in the post–Cold War and post–September 11 eras such as the debate about preemption as a viable policy option and cultural and ethnic rivalries.

Each issue has two readings, one pro and one con. The readings are preceded by an issue *introduction* that sets the stage for the debate and briefly describes the two readings. Each issue concludes with a *postscript* that summarizes the issues, suggests further avenues for thought, and provides *suggestions for further reading*. At the back of the book is a listing of all the *contributors to this volume* with a brief biographical sketch of each of the prominent figures whose views are debated here.

Changes to this edition   This third edition represents a significant revision. It contains 19 issues. Five of the issues are completely new: "Is Rapid Urbanization in the Developing World a Major Problem?" (Issue 2); "Do Environmentalists Overstate Their Case?" (Issue 4); "Can the Global Community "Win" the Drug War?" (Issue 10); "Do International Financial Institutions and Multinational Corporations Exploit the Developing World?" (Issue 14); and "Is the Transnational Media Hurting Global Society?" (Issue 15). In addition, four issues have been substantially reworked and rephrased: "Is Global Aging in the Developed World a Major Problem?" (Issue 3); "Is the World a Victim of American Cultural Imperialism?" (Issue 13); "Are We Headed Toward a Nuclear 9/11?" (Issue 16); and "Is Preemption a Viable Policy Option in Today's Global Security Environment?" (Issue 17). In several of the remaining issues, one or both selections were replaced to bring the issues up-to-date. Only three of the 19 issues in the book have both original selections.

A word to the instructor   An *Instructor's Manual With Test Questions* (multiple-choice and essay) is available through the publisher for the instructor using *Taking Sides* in the classroom. A general guidebook, *Using Taking Sides in the Classroom,* which discusses methods and techniques for integrating the pro-con approach into any classroom setting, is also available. An online version of *Using Taking Sides in the Classroom* and a correspondence service for *Taking Sides* adopters can be found at http://www.dushkin.com/usingts/.

   Taking Sides: Clashing Views on Controversial Global Issues is only one title in the Taking Sides series. If you are interested in seeing the table of contents for any of the other titles, please visit the Taking Sides Web site at http://www.dushkin.com/takingsides/.

James E. Harf
University of Tampa

Mark Owen Lombardi
College of Santa Fe

Contents In Brief

Contents

The CSIS report, *Meeting the Challenge of Global Aging: A Report to World Leaders from the CSIS Commission on Global Aging,* suggests that the wide range of changes brought on by global aging pose significant challenges to the ability of countries to address problems associated with the elderly directly and those related to the national economy as a whole. Philip Longman takes a more balanced perspective, suggesting that what he calls the coming "baby bust" will yield a variety of positive consequences as well as negative ones.

PART 2   GLOBAL RESOURCES AND THE ENVIRONMENT   63

Environmental journalist Ronald Bailey in his review of the Lomborg book, argues in the subtitle of his critique that "An environmentalist gets it right," suggesting that finally someone has taken the environmental doomsdayers to task for their shoddy use of science. Bioscientist David Pimentel takes to task a controversial book by Bjørn Lomborg, *The Skeptical Environmentalist: Measuring the Real State of the World* (Cambridge University Press, 2001).

The United Nations Environmental Programme presents a comprehensive and gloomy assessment of the global environment at the turn of the millennium. Peter Huber, senior fellow of the Manhattan Institute, contends that humankind is saving the Earth with the technologies that the "soft greens" most passionately oppose.

Hisham Khatib, honorary vice chairman of the World Energy Council and a former Jordanian minister of energy and minister of planning, and his coauthors conclude that reserves of traditional commercial fuels, including oil, "will suffice for decades to come." Lester Brown, former head of the Worldwatch Institute, suggests in this adaptation from his *Plan B: Rescuing a*

Mr. Mayor suggests that drug trafficking and consumption "constitute one of the
most serious threats to our planet" and the world must dry up the demand and
attack the financial power of organized crime. Harry G. Levine, professor of
sociology at Queens College, City University of New York, argues that the
emphasis on drug prohibition should be replaced by a focus on "harm reduction,"
creating mechanisms to address tolerance, regulation, and public health.

Mr. Taylor argues that globalization is not the boogeyman that some scholars
and politicos contend but rather a complex array of economic trends that
provides positive results for the vast majority of people. Mr. Bove contends
that globalization and the capitalist principles that underlie it are misleading and
create inequities and myths regarding progress and wealth creation that are
not supported by an analysis of the facts.

Professor of law Jerry Kang argues that privacy, which is fundamental to the
concept of freedom, is severely threatened by the current love of cyberspace.
He suggest that the U.S. Congress and, by extension, individual states should
take action to ensure individual privacy in cyberspace. Senior researcher
Renata Salecl contends that modern technology does not hinder freedom and
privacy but rather enhances what is only a modern concept.

Julia Galeota, the seventeen-year-old winner of the 2004 Humanist Essay Contest, contends that American cultural imperialism is a reality promoted by commercial images presented through the media and the selling of American products across the globe. Philippe Legrain is a British economist who presents two views of cultural imperialism and argues that the notion of American cultural imperialism "is a myth" and that the spreading of cultures through globalization is a positive, not negative, development.

Nobel Prize winner Joseph Stiglitz argues that when international financial institutions (IFIs) such as the International Monetary Fund (IMF) and multi-national corporations (MNCs) dictate terms of economic exchange for the rest of the world, exploitation occurs. He believes that states must negotiate tough terms that protect their national interests if they are to avoid such exploitation. Economist Jagdish Bhagwati contends that the evidence supports the contention that MNCs do more good than harm in the Third World and that critics are misguided and overly emotional in their condemnations.

Herbet Schiller argues that the globalization of media places image-building and information in the hands of private entities whose interests are not educative or informational but rather designed to create a worldwide consumer society. Benjamin Compaine contends that the image of a global media colossus is overblown and that large private media conglomerates may be a good thing for the public interest as a whole.

Graham Allison and Andrei Kokoshin contend that a U.S.-Russian alliance against terrorism is needed before terrorists acquire nuclear weapons and "launch." They surmise that nuclear terrorism will occur in just a matter of time if these two nuclear powers do not act quickly. Jessica Stern, a lecturer in public policy, argues that Americans are in danger of overestimating terrorist capabilities and thus creating a graver threat than actually exists. She warns that the United States must not overreact in its policy response and that prudent security measures will greatly reduce such threats now and in the future.

Endy Zemenides, a member of the National Strategy Forum Review Editorial Board, writes that international laws against preemption are too strict and ignore the nature of self-defense in today's security world. He contends that such a policy, while tricky, must be part of the optional arsenal of U.S. foreign policy. Neta Crawford, a professor of international affairs at Brown University, contends that preemption is a faulty policy because it creates conflict where none existed and presupposes a level of certainty that no government can adhere to.

David Cole, writing for *The American Prospect*, argues that the Bush administration has restricted civil liberties and begun a process of detention and isolation all in the name of security. Cole contends that these policies are an anathema to the rule of law and hurt our conflict with terrorism, not aid it. Kim R. Holmes, of the Heritage Foundation, and Edwin Meese III, the Ronald Reagan Distinguished Fellow in Public Policy at the Heritage Foundation, argue that the Bush administration's approach to fighting terrorism, including the Patriot Act, strikes a proper balance between liberty and security.

Political scientist Samuel P. Huntington argues that the emerging conflicts of the twenty-first century will be cultural and not ideological. He identifies the key fault lines of conflict and discusses how these conflicts will reshape global policy. Amartya Sen, a Nobel Prize–winning economist, responds that categorizing people into neat cultural containers is artificial and not indicative of the reality of the human experience. He contends that such divisions do not mirror the real world but rather promote division and conflicts, not explain them.

Introduction

Global Issues in the Twenty-First Century

James E. Harf

Mark Owen Lombardi

Threats of the New Millennium

As the new millennium dawned, the world witnessed two very different events whose impacts were deeply felt immediately and that are still being experienced far and wide. One such episode was the tragedy of 9/11, a series of incidents that ushered in a new era of terrorism. It burst upon the international scene with the force of a mega-catastrophe, occupying virtually every waking moment of national and global leaders throughout the world and seizing the attention for months to come of the rest of the planet's citizens who had access to a television set. The focused interest of policymakers was soon transformed into a war on terrorism, while average citizens sought to cope with changes brought on by both the tragic events of September 2001 and the global community's response to them. Both governmental leaders and citizens continue to address the consequences of this first intrusion of the new millennium on a world now far different in many ways since 9/11.

The second event was the creation of a set of millennium development goals by the United Nations. In September 2000 with much fanfare, 189 national governments committed to eight major goals in an initiative known as the UN Millennium Development Goals (MDG): eradicate extreme poverty and hunger; achieve universal primary education; promote gender equality and empower women; reduce child mortality; improve maternal health; combat HIV/AIDS, malaria and other diseases; ensure environmental sustainability, and develop a global partnership for development. This initiative was important not only because the UN was setting an actionable 15-year agenda against a relatively new set of global issues, but also because it signified a major change in how the international community would henceforth address such problems confronting humankind. The new initiative represented a recognition of (1) shared responsibility between rich and poor nations for solving such problems, (2) a link between goals, (3) the paramount role to be played by national governments in the process, and (4) the need for measurable outcome indicators of success. The UN MDG initiative went virtually unnoticed by much of the public, although governmental decision makers involved with the UN understood its significance. Jay Leno, of NBC's *The Tonight Show,* would have a field day questioning passersby on the street about their knowledge of this UN initiative.

These two major events, although vastly different, symbolize the world in which we now find ourselves, a world far more complex and more violent than either the earlier one characterized by the Cold War struggle between the United States and the Soviet Union, or the post–Cold War era of the 1990s, where global and national leaders struggled to identify and then find their proper place in the post-cold war world order. Consider the first event, the 9/11 tragedy. It reminds us all that the use and abuse of military power in pursuit of political goals in earlier centuries is still a viable option for those throughout the world who believe themselves disadvantaged because of various political, economic or social conditions and structures. The only difference is the perpetrators' choice of military hardware and military strategy. Formally declared wars fought by regular national military forces committed (at least on paper) to the tenets of just war theory have now been replaced by a plethora of military actions whose defining characteristics conjure up terrorism, perpetrated by individuals without attachments to a regular military or allegiance to a national government, and who do not hesitate to put ordinary citizens in harm's way.

On the other hand, the second event of the new century, the UN MDG initiative, symbolizes the other side of the global coin, the recognition that the international community is also beset with a number of problems unrelated to military actions or national security, at least in a direct sense. Rather, the past 30-odd years has witnessed the emergence and thrust to prominence of a number of new problems relating to social, economic and environment characteristics of the citizens that inhabit this planet. These problems impact the basic quality of life of global inhabitants in ways very different from the scourges of military violence. Yet they are just as dangerous and just as threatening. At the heart of this global change is a phenomenon called globalization.

The Emergence of the Age of Globalization

The Cold War era, marked by the domination of two superpowers in the decades following the end of World War II, has given way to a new era called globalization. This new epoch is characterized by a dramatic shrinking of the globe in terms of travel and communication, increased participation in global policy making by an expanding array of national and nonstate actors, and an exploding volume of problems with ever-growing consequences. While the tearing down of the Berlin Wall over a decade ago dramatically symbolized the end of the cold war era, the creation of the Internet graphically illustrates the emergence of the globalization era, and the fallen World Trade Center symbolizes a new paradigm for conflict and violence.

Early signs of this transition were manifested during the latter part of the twentieth century by a series of problems that transcended national boundaries. These problems caught the attention of policymakers who demanded comprehensive analyses and solutions. The effects of significant population growth in the developing world, for example, called for multilateral action by leaders in developed and developing nations alike. Acid rain, created by emissions from smokestacks, was the precursor to a host of environmental crises that appeared with increasing frequency, challenging the traditional political order and those that commanded it. Finally, much of the world began to sense that the planet is in reality "Spaceship Earth," with finite resources in danger of being exhausted and of which careful stewardship is necessary.

These global concerns remain today. Some are being addressed success-fully, while others are languishing amid a lack of consensus about the nature of the problems and how to solve them. Also, with the shattering of the cold war system and the advent of the globalization age, new issues have emerged to broaden our conception of the global agenda. In the past 10 years, the freedom of people to move has expanded, and with this freedom are concerns over refu-gees and displaced persons within the developing world. More recently, the world has come to fear an expanded potential for terrorism, as new technolo-gies combined with old but strong hatreds have conspired to make the world far less safe than it had been.

This increase in the movement of information and ideas has ushered in global concerns over cultural imperialism, individual privacy, and freedom. The ability both to retrieve and disseminate information will have an impact in this century as great as, if not greater than, the telephone, radio and television in the twentieth century. The potential for global good or ill is mind-boggling. Finally, traditional notions of great-power security embodied in the cold war rivalry have given way to concerns about terrorism, genocide, nuclear proliferation, cultural conflict and the expansion of international law.

Globalization heightens our awareness of a vast array of global issues that will challenge individuals as well as governmental and nongovernmental actors. Since the demise of the cold war world, analysts and lay-persons have become free to define, examine, and explore solutions to such issues on a truly global scale. This text seeks to identify those issues that are central to the discourse on globalization. The issues in this volume provide a broad overview of the mosaic of global issues that will impact student's daily lives.

What Is a Global Issue?

We begin by addressing what a *global issue* is. (The five characteristics are extra-cted from James E. Harf and B. Thomas Trout, *The Politics of Global Resources,* Duke University Press, 1986, pp. 12–28.) By definition, the word *issue* is charac-terized by disagreement along several related dimensions:

1. whether a problems exists;
2. the characteristics of the problem;
3. the preferred future alternatives or solutions; and/or
4. how preferred futures are to be obtained.

These problems are real, vexing, and controversial because policymakers bring to their analysis different historical experiences, values, goals, and objectives. These differences impede and can even prevent successful problem solving. In short, the key ingredient of an issue is disagreement.

The word *global* in the phrase *global issue* is what makes the set of prob-lems confronting the human race today far different from those that challenged our ancestors. Historically, problems were confined to a village, city, or region. The capacity of the human race to fulfill its daily needs was limited to a much smaller space: the immediate environment. When human feet and the horse were the principal modes of transportation, people pursued their food, fuel and other necessities near their homes. When local resources were exhausted, people moved elsewhere in search of another environment that was better able to supply them. With the invention of the locomotive and the automobile, however, humans were able to travel greater distances on a daily basis. National energy systems and agricultural systems emerged, freeing humans from a reliance on their local community. Problems then became national, as dislocations in the production or distribution system in one part of a country had repercussions in some other part.

With the advent of transoceanic transportation capabilities, national distribution systems gave way to global systems. The "breadbasket" countries of the world—those best able to grow food—expanded production to serve an ever-larger global market. Oil-producing countries with vast reservoirs eagerly increased production to satisfy the needs of an ever-growing worldwide clientele. Countries that were best able to provide other resources to a larger part of the world entered the global trade system as well. These spatial changes were a consequence of significant increases in population and per capita consumption, both of which dramatically heightened the demand for resources.

Additionally, the larger machines of the Industrial Revolution were capable of moving greater quantities of dirt and of emitting greater amounts of pollutants into the atmosphere. With these new production and distribution capabilities came more prevalent manifestations of existing problems as well as new ones. Providers either engaged in greater exploitation of their own environment or searched further a field in order to satisfy additional resource requirements.

The character of these new problems is different from those of earlier eras. First, they transcend national boundaries and impact virtually every corner of the globe. In effect, these issues help make national borders increasingly meaningless. Environmental pollution or poisonous gases do not recognize or respect national borders.

Second, these new issues cannot be resolved by the autonomous action of a single actor, be it a national government, international organization, or multinational corporation. A country cannot guarantee its own energy or food security without participating in a global energy or food system.

Third, these issues are characterized by a wide array of value systems. To a family in the developing world, giving birth to a fifth or sixth child may contribute to the family's immediate economic well-being. But to a research scholar at the UN Population Fund, the consequence of such an action multiplied across the entire developing world leads to expanding poverty and resource depletion.

Fourth, these issues will not go away. They require specific policy action by a consortium of local, national, and international leaders. Simply ignoring the issue cannot eliminate the threat of chemical or biological terrorism, for example. If global warming does exist, it would not disappear unless specific policies are developed and implemented.

These issues are also characterized by their persistence over time. The human race has developed the capacity to manipulate its external environment and, in so doing, has created a host of opportunities and challenges. The accelerating pace of technological change suggests that global issues will proliferate and will continue to challenge human beings throughout the next millennium.

In the final analysis, however, a global issue is defined as such only through mutual agreement by a host of actors within the international community. Some may disagree about the nature, severity, or presence of a given issue, and these concerns become areas of focus only after a significant number of actors (states, international organizations, the UN, and others) begin to focus systematic and organized attention on the issue itself.

The Creation of the Global Issues Agenda

Throughout the first part of the Cold War, the international community found itself more often as an observer of, rather than a player in, the superpower conflict for global domination being waged by the United States and the Soviet Union. Other forces and trends competed for the attention of world leaders during this period. Economic development exploded in many parts of the globe, while other areas floundered in poverty. Resource consumption expanded alongside economic growth throughout the developed world, creating both pollution and scarcity. Population growth rates increased dramatically in societies that were unable to cope with the consequences of a burgeoning population. New forms of violence replaced traditional ways of employing force. And nonstate actors became a larger part of the conflict problem.

As the major participants in the cold war settled into a system of routinized behavior designed to avoid catastrophic nuclear war, other actors in the international system began to worry about the disquieting effects associated with exploding population and consumption. Nation-states, international governmental organizations, and private groups sought to identify these growth-related problems and to seek solutions. The UN soon became the principal force to which those affected by such global behavior turned.

International community members began to join forces under the auspices of the UN to design comprehensive strategies. Such plans became known as international regimes—a set of agreements, structures, and plans of action involving all relevant global actors in addressing specific issues before them.

For example, in response to concerns of the Swedish government in the early 1970s about the discovery of a significant amount of dead fish in its lakes, the UN brought together relevant parties to discuss the problem, which they assumed was caused by acid rain. This 1972 conference, called the United Nations Conference on the Human Environment, represented the first major attempt of the international community to address the global ecological agenda. Representatives from 113 countries appeared in Stockholm, Sweden (the Soviet-bloc countries boycotted the conference for reasons unrelated to the environmental issue). Over 500 international nongovernmental organizations from such disparate interests such as the environment, science, women, religion, and business also sent delegates. Members of the UN's Secretariat and specialized international governmental organizations within the UN, such as the World Bank and the Food and Agricultural Organization (FAO), were present, as well. Experts from all walks of life—scientists, environmentalists and others—attended.

The conference in Stockholm ushered in a new age of thinking about issues in their global context and became the model for future conferences on global issues. Position papers were prepared for delegates, and preparatory committees hammered out draft declarations and plans of action. These served as the starting points for conference deliberations. The final plan of action included the establishment of a formal mechanism, termed the United Nations Environmental Programme (UNEP), for addressing environmental issues in the future.

The most significant outcome of the conference was the creation of awareness among national government officials regarding the severity of ecological issues, the extent to which they cross national boundaries, and the need for international cooperation in order to solve these issues. The era of global issues identification, analysis, and problem solving was born.

Twenty years later, the global community reconvened in Rio de Janeiro, Brazil, at the United Nations Conference on Environment and Development, more popularly known as the Earth Summit. This meeting produced "Agenda 21," a report that outlined a plan of action for achieving development throughout the world while at the same time avoiding environmental degradation.

Another example of the international community's venture into this emerging agenda focused on the global population problem. In the early 1970s, the UN responded to the pleas of a variety of groups to consider the negative consequences of exploding population birth rates in many developing countries. The reason for such concern grew out of the belief that such growth was deterring economic growth in those areas of the world that were the most poverty stricken.

The result was the United Nations World Population Conference held in Bucharest, Romania, in 1974, and attended by delegates from 137 countries as well as by representatives of numerous governmental and nongovernmental organizations. The principal result of the conference was the placement of population on the global agenda and the development of a formal plan of action. The international community has continued its formal attention to the population issue, and has convened every ten years since Bucharest—in Mexico City in 1984 and in Cairo in 1994—to evaluate strategy and assume new tasks to lower growth rates, as well as to address other population-related issues. A decade later, it is currently assessing the strategies agreed upon at Cairo.

These kinds of conferences, where a myriad of experts join a wide range of governments, private groups and individuals, is now characteristic of how the global community is addressing the emerging global agenda. But the international community was far less certain of the wisdom of this strategy back in 1972. However, the success of the United Nations Conference on the Human Environment (now popularly known as the Stockholm Conference) led to similar strategies in a host of other environmental areas and provided the blueprint for the analysis of unforeseen global issues. For most of the issues in the first four parts of this text, the international community is either now formally addressing the problem or is considering its first steps. For the set of issues in the final part, the new global security dilemma, the international community is still coming to grips with how to proceed in the struggle to find appropriate solutions.

The Nexus of Global Issues and Globalization

In 1989, the Berlin Wall fell, and with it a variety of assumptions, attitudes and expectations of the international system. This event did not usher in a utopia, nor did it irrevocably change all existing ideas. It did, however, create a void. As a dominant cold war system entrenched since the end of World War II was broken apart, alternative views, issues, actors and perspectives began to emerge to fill the void left by the collapse of that old system. What helped to bring down the previous era and what has emerged as the new dominant international construct is what we now call globalization.

Throughout the 1990s and into the twentieth-first century, scholars and policymakers have struggled to define this new era. As the early years of the new century ushered in a different and heightened level of violence, a sense of urgency emerged. Some have analyzed the new era in terms of the victory of Western or American ideals, the dominance of global capitalism, and the spread of democracy vs. the have-nots of the world who use religious fanaticism as a ploy to rearrange power within the international system. Others have defined it simply in terms of the multiplicity of actors now performing on the world stage, and how states and their sovereignty have declined in importance and impact vis-à-vis others, such as multinational corporations, and nongovernmental groups like Greenpeace and Amnesty International. Still others have focused on the vital element of technology and its impact on communications, information storage and retrieval, and global exchange.

Whether globalization reflects one, two or all of these characteristics is not as important as the fundamental realization that globalization is the dominant element of a new era in international politics. This new period is characterized by several basic traits that greatly impact the definition, analysis, and solution of global issues. They include the following:

- an emphasis on information technology;
- the increasing speed of information and idea flows;
- a need for greater sophistication and expertise to manage such flows;
- the control and dissemination of technology; and
- the cultural diffusion and interaction that come with information expansion and dissemination.

Each of these areas has helped shape a new emerging global issues agenda. Current issues remain important and, indeed, these factors help us to understand them on a much deeper level. Yet globalization has also created new problems and has brought them to the fore such that significant numbers of actors now recognize their salience in the international system.

For example, the spread of information technology has made ideas, attitudes, and information more available to people throughout the world. Americans in Columbus, Ohio, had the ability to log onto the Internet and speak with their counterparts in Kosovo to discover when NATO bombing had begun and to gauge the accuracy of later news reports on the bombing. Norwegian students can share values and customs directly with their counterparts in South Africa, thereby experiencing cultural attitudes firsthand without the filtering mechanisms of governments or even parents and teachers. Scientific information that is available through computer technology can now be used to build sophisticated biological and chemical weapons of immense destructive capability. Ethnic conflicts and genocide between groups of people are now global news, forcing millions to come to grips with issues of intervention, prevention and punishment. And terrorists in different parts of the globe can communicate readily with one another, transferring plans and even money across national and continental boundaries with ease.

Globalization is an international system and it is also rapidly changing. Because of the fluid nature of this system and the fact that it is both relatively new and largely fueled by the amazing speed of technology, continuing issues are constantly being transformed and new issues are emerging regularly. The nexus of globalization and global issues has now become, in many ways, the defining dynamic of understanding global issues. Whether it is new forms of terrorism, new conceptions of security, expanding international law, solving ethnic conflicts, dealing with mass migration, coping with individual freedom and access to information, or addressing cultural clash and cultural imperialism, the transition from a cold war world to a globalized world helps us understand in part what these issues are and why they are important.

Identifying the New Global Issues Agenda

The analysis of global issues by scholars and policymakers has changed. Our assumptions, ideas, and conceptions of what makes a problem a global issue have expanded. New technologies, new actors, and new strategies within a changing global order have made the study of global issues a broader and more complex undertaking.

The organization of this text reflects these phenomena. Parts 1 and 2 focus on the continuing global agenda of the post-cold war era. The emphasis is on global population and environmental issues and the nexus between these two phenomena. Should the international community continue to address this problem? Is rapid urbanization in the developing world creating a whole new set of problems unique to such urban settings? Is global aging about to unleash a host of problems for governments of the developed world? Do environmentalists overstate their case or is the charge of "crying wolf" by environmental conservatives a misplaced attack? Is environmental degradation worsening or have humans turned the corner is using sustainable practices in fulfilling their resource needs? Should the world continue to rely on oil or should the search for viable alternatives take on a new urgency? Will the world be able to feed itself or provide enough water in the foreseeable future?

Part 3 addresses the consequences of the decline of national boundaries and the resultant increased international flow of information, ideas, money, and material things in this globalization age. Can the global community win the war on drugs? Is this dissemination of knowledge a force for good or a form of indoctrination by one culture over another? And will this greater flow of information promote human empowerment or serve to restrict human freedom? Do the actions of international financial institutions and multinational corporations enhance the capacity of poor countries to develop, or are they simply another form of exploitation perpetrated on the developing world? Are the international media hurting global society?

Part 4 addresses the new global security dilemma. Has the end of the cold war brought a chance for global peace, or has it unleashed a set of forces that will pose greater and more insidious dangers than those faced during the last century? Are we headed for a nuclear 9/11? Is pre-emption a viable policy against terrorism? Are civil liberties being compromised in the war on terrorism? Finally, will cultural and ethnic conflicts be the defining dimensions of the twenty-first century?

The revolutionary changes of the last few decades present us with serious challenges unlike any others in human history. However, as in all periods of historic change, we possess significant opportunities to overcome problems. The task ahead is to define these issues, explore their context, and develop solutions that are comprehensive in scope and effect. The role of all researchers in this field, or any other, is to analyze objectively such problems and search for workable solutions. As students of global issues, your task is to educate yourselves about these issues and become part of solution.

Population Reference Bureau

The Population Reference Bureau provides current information on international population trends and their implications from an objective viewpoint. The PopNet section of this Web site offers maps with regional and country-specific population information as well as information divided by selected topics.

http://www.prb.org

United Nations Population Fund (UNFPA)

The United Nations Population Fund (UNFPA) was established in 1969 and was originally called the United Nations Fund for Population Activities. This organization works with developing countries to educate people about reproductive and sexual health as well as about family planning. The UNFPA also supports population and development strategies that will benefit developing countries and advocates for the resources needed to accomplish these tasks. Explore this site to learn more about the UNFPA, including the organization's support of individuals' right to decide the number and spacing of their children without outside coercion.

http://www.unfpa.org

Population Connection

Population Connection (formerly Zero Population Growth) is a national, nonprofit organization working to slow population growth and to achieve a sustainable balance between Earth's people and its resources. The organization seeks to protect the environment and to ensure a high quality of life for present and future generations. Population Connection's education and advocacy programs aim to influence public policies, attitudes, and behavior on national and global population issues and related concerns.

http://www.populationconnection.org

The CSIS Global Aging Initiative

The Center for Strategic and International Studies (CSIS) is a public policy research institution that approaches the issue of the aging population in developed countries in a bipartisan manner. The CSIS is involved in a two-year project to explore the global implications of aging in developed nations and to seek strategies on dealing with this issue. This site includes a list of publications that were presented at previous events.

http://www.csis.org/gai/

The Population Council

The Population Council is an international, nonprofit organization that conducts research on population matters from biological, social science, and public health perspectives. It was established in 1952 by John D. Rockefeller, III.

http://www.popcouncil.org

PART 1

Global Population

*I*t is not a coincidence that many of the global issues in this book emerged at about the same time as world population growth exploded. No matter what the issue, the presence of a large and fast-growing population alongside it exacerbates the issue and transforms its basic characteristics. And the rapid growth within urban areas of the developing world pose a different set of problems. Finally, the emergence of a graying population throughout the globe but particularly in the developed world has the potential for significant impact as well. The ability of the global community to respond to any given issue is diminished by certain population conditions, be it an extremely young consuming population in a poor country in need of producers, an expanding urban population whose local public officials are unable to provide an appropriate infrastructure, a large working-age group in a nation without sufficient jobs, or an ever-growing senior population for whom additional services are needed.

Thus we begin this text with a series of issues directly related to various aspects of world population. It serves as both a separate global agenda and as a context within which other issues are examined.

- Should the International Community Attempt to Curb Population Growth in the Developing World?

- Is Rapid Urbanization in the Developing World a Major Problem?

- Is Global Aging in the Developed World a Major Problem?

ISSUE 1

Should the International Community Attempt to Curb Population Growth in the Developing World?

YES: Robert S. McNamara, from "The Population Explosion," *The Futurist* (November/December 1992)

NO: Steven W. Mosher, from "McNamara's Folly: Bankrolling Family Planning," *PRI Review* (March–April 2003)

ISSUE SUMMARY

YES: Robert McNamara, former president of the World Bank, argues in this piece written during his presidency that the developed countries of the world and international organizations should help the countries of the developing world reduce their population growth rates.

NO: Steven W. Mosher, president of the Population Research Institute, an organization dedicated to debunking the idea that the world is overpopulated, argues that McNamara's World Bank and other international financial lending agencies have served for over a decade as "loan sharks" for those groups and individuals who were pressuring developing countries to adopt fertility reduction programs for self-interest reasons.

The history of the international community's efforts to lower birth rates throughout the developing world goes back to the late 1960s, when the annual growth rate hovered around 2.35 percent. At that time, selected individuals in international governmental organizations, including the United Nations, were persuaded by a number of wealthy national governments as well as by international nongovernmental population agencies that a problem of potentially massive proportions had recently emerged. Quite simply, demographers had observed a pattern of population growth in the poorer regions of the world quite unlike that which had occurred in the richer countries during the previous 150–200 years.

Population growth in the developed countries of the globe had followed a rather persistent pattern during the last two centuries. Prior to the Industrial Revolution, these countries typically experienced both high birth rates and death rates. As industrialization took hold and advances in the quality of life for citizens of these countries occurred, death rates fell, resulting in a period of time when the size of the population rose. Later, birth rates also began to decline, in large part because the newly industrialized societies were better suited to families with fewer children. After awhile, both birth and death rates leveled off at a much lower level than during preindustrial times.

This earlier transition throughout the developed world differed, however, from the newer growth pattern in the poorer regions of the globe observed by demographers in the late 1960s. First, the transition in the developed world occurred over a long period of time, allowing the population to deal more readily with such growth. On the other hand, post-1960s' growth in the developing world had taken off at a much faster pace, far outstripping the capacity of these societies to cope with the changes accompanying such growth.

Second, the earlier growth in the developed world began with a much smaller population base and a much larger resource base than did the developing world, again allowing the richer societies to cope more easily with such growth. The developing world of the 1960s, however, found percentages of increase based on a much higher base. Coping under the latter scenario proved much more difficult.

Finally, industrialization accompanied population change in the developed world, again allowing for those societies to address resultant problems more easily. Today's developing world has no such luxury. New jobs are not available, expanded educational facilities are non-existent, unsatisfactory health services remain unchanged, and modern infrastructures have not been created.

The international community formally placed the population issue—defined primarily as excessive birth rates in the developing world—on the global agenda in 1974 with the first major global conference on population, held in Bucharest, Romania. There was much debate over the motives of both sides. Both rich and poor countries eventually pledged to work together.

Finally, each side bought into the assumption that "the best contraceptive was economic development," but until development was achieved, national family planning programs would help lower growth rates. By 1994 when nations of the world reconvened in Cairo to assess progress, considerable success had been achieved in getting developing countries to accept such programs. As the new millennium occurred, some analysts even called for an ending to such ventures, declaring that the growth problem had diminished.

In the first selection, Robert McNamara argues that high population growth is exacerbating an already dire set of conditions in the developing world and that the industrialized countries of the globe should embark on a massive assistance program to help the "have-not" countries reduce fertility. In the second selection, Steven Mosher views the efforts of organizations such as the United Nations Population Fund much differently. In his view, these organizations have always sought to impose birth-control methods on the developing world in the misguided name of "virtuous and humanitarian motives," while attacking the motives of their opponents as self-serving or worse.

Robert S. McNamara  **YES**

The Population Explosion

For thousands of years, the world's human population grew at a snail's pace. It took over a million years to reach 1 billion people at the beginning of the last century. But then the pace quickened. The second billion was added in 130 years, the third in 30, and the fourth in 15. The current total is some 5.4 billion people.

Although population growth rates are declining, they are still extraordinarily high. During this decade, about 100 million people per year will be added to the planet. Over 90% of this growth is taking place in the developing world. Where will it end?

The World Bank's latest projection indicates that the plateau level will not be less than 12.4 billion. And Nafis Sadik, director of the United Nations Population Fund, has stated that "the world could be headed toward an eventual total of 14 billion."

What would such population levels mean in terms of alleviating poverty, improving the status of women and children, and attaining sustainable economic development? To what degree are we consuming today the very capital required to achieve decent standards of living for future generations?

More People, Consuming More

To determine whether the world—or a particular country—is on a path of sustainable development, one must relate future population levels and future consumption patterns to their impact on the environment.

Put very simply, environmental stress is a function of three factors: increases in population, increases in consumption per capita, and changes in technology that may tend to reduce environmental stress per unit of consumption.

Were population to rise to the figure referred to by Sadik—14 billion—there would be a 2.6-fold increase in world population. If consumption per capita were to increase at 2% per annum—about two-thirds the rate realized during the past 25 years—it would double in 35 years and quadruple in 70 years. By the end of the next century, consumption per capita would be eight times greater than it is today.

Some may say it is unreasonable to consider such a large increase in the per capita incomes of the peoples in the developing countries. But per capita income in the United States rose at least that much in this century, starting from a much higher base. And today, billions of human beings across the globe are now living in intolerable conditions that can only be relieved by increases in consumption.

A 2.6-fold increase in world population and an eightfold increase in consumption per capita by 2100 would cause the globe's production output to be 20 times greater than today. Likewise, the impact on non-renewable and renewable resources would be 20 times greater, assuming no change in environmental stress per unit of production.

On the assumptions I have made, the question becomes: Can a 20-fold increase in the consumption of physical resources be sustained? The answer is almost certainly "No." If not, can substantial reductions in environmental stress—environmental damage—per unit of production be achieved? Here, the answer is clearly "Yes."

Reducing Environmental Damage

Environmental damage per unit of production can— and will— be cut drastically. There is much evidence that the environment is being stressed today. But there are equally strong indications that we can drastically reduce the resources consumed and waste generated per unit of "human advance."

With each passing year, we are learning more about the environmental damage that is caused by present population levels and present consumption patterns. The superficial signs are clearly visible. Our water and air are being polluted, whether we live in Los Angeles, Mexico City, or Lagos. Disposal of both toxic and nontoxic wastes is a worldwide problem. And the ozone layer, which protects us all against skin cancer, is being destroyed by the concentration of chlorofluorocarbons in the upper atmosphere.

But for each of these problems, there are known remedies—at least for today's population levels and current consumption patterns. The remedies are costly, politically difficult to implement, and require years to become effective, but they can be put in place.

The impact, however, of huge increases in population and consumption on such basic resources and ecosystems as land and water, forests, photosynthesis, and climate is far more difficult to appraise. Changes in complex systems such as these are what the scientists describe as nonlinear and subject to discontinuities. Therefore, they are very difficult to predict.

A Hungrier Planet?

Let's examine the effect of population growth on natural resources in terms of agriculture. Can the world's land and water resources produce the food required to feed 14 billion people at acceptable nutritional levels? To do so would require a four-fold increase in food output.

Modern agricultural techniques have greatly increased crop yields per unit of land and have kept food production ahead of population growth for several decades. But the costs are proving to be high: widespread acceleration of erosion and nutrient depletion of soils, pollution of surface waters, overuse and contamination of groundwater resources, and desertification of overculti-vated or overgrazed lands.

The early gains of the Green Revolution have nearly run their course. Since the mid-1980s, increases in worldwide food production have lagged behind popu-lation growth. In sub-Saharan Africa and Latin America, per capita food produc-tion has been declining for a decade or more.

What, then, of the future? Some authorities are pessimistic, arguing that maximum global food output will support no more than 7.5 billion people. Others are somewhat more optimistic. They conclude that if a variety of actions were taken, beginning with a substantial increase in agricultural research, the world's agricultural system could meet food requirements for at least the next 40-50 years.

However, it seems clear that the actions required to realize that capacity are not now being taken. As a result, there will be severe regional shortfalls (e.g., in sub-Saharan Africa), and as world population continues to increase, the likeli-hood of meeting global food requirements will become ever more doubtful.

Similar comments could be made in regard to other natural resources and ecosystems. More and more biologists are warning that there are indeed biologi-cal limits to the number of people that the globe can support at acceptable stan-dards of living. They say, in effect, "We don't know where those limits are, but they clearly exist."

Sustainability Limits

How much might population grow and production increase without going beyond sustainable levels—levels that are compatible with the globe's capacity for waste disposal and that do not deplete essential resources?

Jim MacNeil, Peter Winsemaus, and Taizo Yakushiji have tried to answer that question in *Beyond Interdependence,* a study prepared recently for the Trilateral Commission. They begin by stating: "Even at present levels of economic activity, there is growing evidence that certain critical global thresholds are being approached, perhaps even passed."

They then estimate that, if "human numbers double, a five- to ten-fold increase in economic activity would be required to enable them to meet [even] their basic needs and minimal aspirations." They ask, "Is there, in fact, any way to multiply economic activity a further five to ten times, without it undermining itself and compromising the future completely?" They clearly believe that the answer is "No."

Similar questions and doubts exist in the minds of many other experts in the field. In July 1991, Nobel laureate and Cal Tech physicist Murray Gell-Mann and his associates initiated a multiyear project to try to understand how "humanity can make the shift to sustainability." They point out that "such a change, if it could be achieved, would require a series of transitions in fields rang-ing from technology to social and economic organization and ideology."

The implication of their statement is not that we should assume the outlook for sustainable development is hopeless, but rather that each nation individually, and all nations collectively, should begin now to identify and introduce the changes necessary to achieve it if we are to avoid costly—and possibly coercive—action in the future.

One change that would enhance the prospects for sustainable, development across the globe would be a reduction in population growth rates.

Population and Poverty

The developing world has made enormous economic progress over the past three decades. But at the same time, the number of human beings living in "absolute poverty" has risen sharply.

When I coined the term "absolute poverty" in the late 1960s, I did so to distinguish a particular segment of the poor in the developing world from the billions of others who would be classified as poor in Western terms. The "absolute poor" are those living, literally, on the margin of life. Their lives are so characterized by malnutrition, illiteracy, and disease as to be beneath any reasonable definition of human dignity.

Today, their number approaches 1 billion. And the World Bank estimates that it is likely to increase further—by nearly 100 million—in this decade.

A major concern raised by poverty of this magnitude lies in the possibility of so many children's physical and intellectual impairment. Surveys have shown that millions of children in low-income families receive insufficient protein and calories to permit optimal development of their brains, thereby limiting their capacity to learn and to lead fully productive lives. Additional millions die each year, before the age of five, from debilitating disease caused by nutritional deficiencies.

High population growth is not the only factor contributing to these problems; political organization, macroeconomic policies, institutional structures, and economic growth in the industrial nations all affect economic and social advance in developing countries. But intuitively we recognize that the immediate effects of high population growth are adverse.

Our intuition is supported by facts: In Latin America during the 1970s, when the school-age population expanded dramatically, public spending per primary-school student fell by 45% in real terms. In Mexico, life expectancy for the poorest 10% of the population is 20 years less than for the richest 10%.

Based on such analyses, the World Bank has stated: "The evidence points overwhelmingly to the conclusion that population growth at the rates common in most of the developing world slows development. . . . Policies to reduce population growth can make an important contribution to [social advance]."

A Lower Plateau for World Population?

Any one of the adverse consequences of the high population growth rates—environmentally unsustainable development, the worsening of poverty, and the negative impact on the status and welfare of women and children—would be reason enough for developing nations across the globe to move more quickly to reduce fertility rates. Taken together, they make an overwhelming case.

Should not every developing country, therefore, formulate long-term population objectives—objectives that will maximize the welfare of both present and future generations? They should be constrained only by the maximum feasible rate at which the use of contraception could be increased in the particular nation.

If this were done, I estimate that country family-planning goals might lead to national population-stabilization levels that would total 9.7 billion people for the globe. That is an 80% increase over today's population, but it's also 4.3 billion fewer people than the 14 billion toward which we may be heading. At the consumption levels I have assumed, those additional 4.3 billion people could require a production output several times greater than the world's total output today.

Reducing Fertility Rates

Assuming that nations wish to reduce fertility rates to replacement levels at the fastest possible pace, what should be done?

The Bucharest Population Conference in 1974 emphasized that high fertility is in part a function of slow economic and social development. Experience has indeed shown that as economic growth occurs, particularly when it is accompanied by broadly based social advance, birth rates do tend to decline. But it is also generally recognized today that not all economic growth leads to immediate fertility reductions, and in any event, fertility reduction can be accelerated by direct action to increase the use of contraceptives.

It follows, therefore, that any campaign to accelerate reductions in fertility should focus on two components: (1) increasing the pace of economic and social advance, with particular emphasis on enhancing the status of women and on reducing infant mortality, and (2) introducing or expanding comprehensive family-planning programs.

Much has been learned in recent years about how to raise rates of economic and social advance in developing countries. I won't try to summarize those lessons here. I do wish to emphasize, however, the magnitude of the increases required in family planning if individual countries are to hold population growth rates to levels that maximize economic and social advance.

The number of women of childbearing age in developing countries is projected to increase by about 22% from 1990 to 2000. If contraception use were to increase from 50% in 1990 to 65% in 2000, the number of women using contraception must rise by over 200 million.

That appears to be an unattainable objective, considering that the number of women using contraception rose by only 175 million in the past *two* decades, but it is not. The task for certain countries and regions—for example, India, Pakistan, and almost all of sub-Saharan Africa—will indeed be difficult, but other nations have done as much or more. Thailand, Indonesia, Bangladesh, and Mexico all increased use of contraceptives at least as rapidly. The actions they took are known, and their experience can be exported. It is available to all who ask.

Financing Population Programs

A global family-planning program of the size I am proposing for 2000 would cost approximately $8 billion, with $3.5 billion coming from the developed nations (up from $800 million spent in 1990). While the additional funding appears large, it is very, very small in relation to the gross national products and overseas development assistance projected for the industrialized countries.

Clearly, it is within the capabilities of the industrialized nations and the multilateral financial institutions to help developing countries finance expanded family-planning programs. The World Bank has already started on such a path, doubling its financing of population projects in the current year. Others should follow its lead. The funds required are so small, and the benefits to both families and nations so large, that money should not be allowed to stand in the way of reducing fertility rates as rapidly as is desired by the developing countries.

The developed nations should also initiate a discussion of how their citizens, who consume seven times as much per capita as do those of the developing countries, may both adjust their consumption patterns and reduce the environmental impact of each unit of consumption. They can thereby help ensure a sustainable path of economic advance for all the inhabitants of our planet.

Steven W. Mosher   ↩ **NO**

McNamara's Folly: Bankrolling Family Planning

At the same time that Reimert Ravenholt was setting up his "powerful population program," the nations of Western Europe, along with Japan, were being encouraged by the administration of President Lyndon B. Johnson to make family planning a priority of their own aid programs. International organizations, primarily the UN and its affiliated agencies, were also being leveraged on board. Together, they helped to create and maintain the illusion that the international community was solidly behind population control programs. (It wasn't, and isn't, as we shall see.) But it was the World Bank and its billions that was the real prize for the anti-natalists. And they captured it when one of their own, Robert McNamara, was appointed as President in 1968.[1]

McNamara Moves In

McNamara came to the World Bank from the post of Secretary of Defense, where he had unsuccessfully prosecuted the Vietnam War by focusing on "kill ratios" and the "pacification of the natives" instead of victory. A former automobile executive, he was prone to cost-cutting measures which sometimes proved to be false economies, as when he decreed that a new class of ship—the fleet frigate—should have only one screw instead of the customary two. This saved the expense of a second turbine and drive train, but the frigate—known to the Navy as McNamara's Folly—lacked speed, was hard to berth, and had to be retired early.[2] The population policies he was to advocate suffered from similar defects.

When the Boards of the World Bank and the International Monetary Fund convened on October 1 of that year, President Johnson made a surprise appearance.[3] Technology in the underdeveloped nations, he said, had "bought time for family planning policies to become effective. But the fate of development hinges on how vigorously that time is used."

No More People

The stage was now set for McNamara to get up and attack the "population explosion," saying that it was "one of the greatest barriers to the economic growth and social well-being of our member states." The World Bank would no longer stand idly by in the face of this threat, McNamara said, but would:

> Let the developing nations know the extent to which rapid population growth slows down their potential development, and that, in consequence, the optimum employment of the world's scarce development funds requires attention to this problem. Seek opportunities to finance facilities required by our member countries to carry out family planning programs. Join with others in programs of research to determine the most effective methods of family planning and of national administration of population control programs.[4]

It quickly became evident that "the optimum employment of the world's scarce development funds" meant in practice that the World Bank, the International Monetary Fund (IMF), and its network of regional development banks would act as loan sharks for the anti-natalists, pressuring sovereign nations into accepting family planning programs on pain of forfeiting vital short-term, long-term, and soft loans.[5] This practice is well known in the developing world, as when a Dhaka daily, *The New Nation*, headline read, "WB [World Bank] Conditions Aid to Population Control."[6]

McNamara also began providing loans for population and family planning projects, including those which involved abortion (both surgical and through abortifacient chemicals). By 1976 the National Security Council (NSC) was able to praise the World Bank for being "the principal international financial institution providing population programs."[7] Details are hard to come by, however. The World Bank is one of the most secretive organizations in the world, besides being effectively accountable to no one. It is known that there is a carefully segregated population division, which reportedly employs approximately 500 people. But those who work on conventional development projects are not privy to what goes on in this division, which is off-limits to all but those who work there.[8]

Fewer People, More Money

A rare inside look at the organization's activities in this area is provided by a recent World Bank report, entitled *Improving Reproductive Health: The Role of the World Bank*. Written in a distinctly self-congratulatory tone, the document reveals that the Bank has spent over $2.5 billion over the last twenty-five years to support 130 reproductive health projects in over 70 countries. Indonesia and Lesotho, for example, have been the site of "'information, education and communication' campaigns about sex and reproductive health." India has been the beneficiary of several different programs, which the report claims have "helped bring India two-thirds of the way towards her goal of replacement level fertility." No mention is made of the fact that the Indian campaigns have been notorious for their coercive tactics. Or that McNamara visited India at the height of the compulsory sterilization campaign in 1976 to congratulate the government for its "political will and determination" in the campaign and, one would suspect, to offer new loans.[9]

The World Bank also promotes abortion. *Improving Reproductive Health* openly admits that, since the 1994 Cairo Conference on Population and Development, the first of the World Bank's goals in the area of reproductive health has been "providing access and *choice* in family planning." [italics added] Except for its candor, this promotion of abortion should come as no surprise. In Burkina Faso, for example, we are told that World Bank projects have included "mobilizing public awareness and political support" [that is, lobbying] for abortion and other reproductive health services.

The Bank has long been accused of pressuring nations, such as Nigeria, into legalizing abortion. In 1988, for example, abortion was virtually unthinkable as an official family planning practice in Nigeria. As recently as 1990, the Planned Parenthood Federation of Nigeria was forced to defend itself against allegations that it promoted the sale and use of "contraceptives" that were abortifacient in character. A year later—and two months after approval of a $78 million World Bank population loan—the government announced proposals for allowing abortion under certain circumstances.[10]

Population control loans skyrocketed after the Cairo conference. The Bank reported that, in the two years that followed, it had "lent almost $1 billion in support of population and reproductive health objectives."[11] And the numbers have been climbing since then. But even this is just the tip of the iceberg. As Jacqueline Kasun notes, "Given the conditions which the bank imposes on its lending, the entire $20 billion of its annual disbursements is properly regarded as part of the world population control effort."[12]

No More Reform

Despite his predilection for population control, McNamara never abandoned more conventional aid modalities, roads, dams, power plants, and the like. Not so James Wolfensohn, who became the head of the Bank in 1995. Asked at the 1996 World Food Summit in Rome how the World Bank understood its mission towards the developing world, Wolfensohn replied that there was a "new paradigm" at the Bank. "From now on," Wolfensohn said, "the business of the World Bank will not be primarily economic reform, or governmental reform. The business of the World Bank will primarily be social reform." The Bank has learned, he added, that attempting to reform a nation's economics or government without first reforming the society "usually means failure."

The benefits to nations who are willing to fall into line in the "civil society" will be immediate and intensely attractive. "The World Bank will be willing to look favorably on any reasonable plan for debt reduction—and even debt forgiveness," Wolfensohn told the assembled reporters, "provided that the nation in question is willing to follow a sensible social policy." Wolfensohn went on to tell reporters that population control activities are a *sine qua non* for any social policy to be considered "sensible."[13]

The World Bank is also, according to Wolfensohn, prepared to begin "directly funding—not through loans" certain NGOs in the countries involved, to further ensure that governments adopt "sensible social policies." Thus fueled with money from the World Bank, the heat these favored NGOs will be able to generate on their governments to adopt, say, population control programs, including legalized abortion, will be considerable.[14] Of course, other international organizations, not to mention USAID and European aid agencies, have been using this tactic for many years with great effect. Recalcitrant governments (who may innocently believe that they do not have a population "problem") are thus sandwiched between the demands of international lenders and aid givers on the one hand, and the demands of "local" NGOs—loud, persistent and extremely well-funded—on the other.

Rapid Spread of Programs

With the U.S., international organizations, and an increasing number of developed countries now working in tandem to strong-arm developing countries into compliance, anti-natalist programs spread with startling rapidity. Bernard Berelson, the head of the Population Council, happily reported in 1970 that:

> In 1960 only three countries had anti-natalist population policies (all on paper), only one government was offering assistance [that is, funding population control programs overseas], no international organizations was working on family planning. In 1970 nearly 25 countries on all three developing continents, with 67 percent of the total population, have policies and programs; and another 15 or so, with 12 percent of the population, provide support in the absence of an explicitly formulated policy . . . five to ten governments now offer external support (though only two in any magnitude); and the international assistance system is formally on board (the U.N. Population Division, the UNDP, WHO, UNESCO, FAO, ILO, OECD, the World Bank).[15]

The recklessness with which Ravenholt, McNamara and others forced crude anti-natal programs upon the developing world dismayed many even within the movement. Ronald Freedman, a leading sociologist/demographer, complained in 1975 that, "If reducing the birth rate is that important and urgent, then the results of the expanded research during the 1960s are still *pathetically inadequate. There are serious proposals for social programs on a vast scale to change reproductive institutions and values that have been central to human society for millennia.*"[16] [italics added] This was social engineering with a vengeance, Freedman was saying, and *we don't know what we are doing.*

With even committed controllers saying "Slow down!" one might think that the anti-natalists would hesitate. But their army had already been assembled and its generals had sounded the advance; it could not be halted now. Even Freedman, rhetorically throwing up his hands, conceded that "many people . . . are eager for knowledge that can be used in action programs aimed at accelerating fertility decline," and that the programs would have to proceed by "a process of trial and error." The *trials* of course would be funded by the developed world; while the *errors*, murderous and costly, would be borne by poor women and families in the developed world.

What justification was offered for this massive investment of U.S. prestige and capital in these programs? Stripped of its later accretions—protecting the environment, promoting economic development, advancing the rights of women—at the outset it was mostly blatant self-interest. McNamara, who headed an organization ostensibly devoted to the welfare of the developing countries, had told the World Bank's Board of Governors in 1968 that "population growth slows down their potential development." But he told the *Christian Science Monitor* some years later that continued population growth would lead to "poverty, hunger, stress, crowding, and frustration," which would threaten social, economic and military stability. This would not be "a world that anyone wants," he declared.[17] It was certainly not the world that many in the security establishment wanted, as secret National Security Council deliberations would soon make starkly clear.

Cold War Against Population

As the populations of developing world countries began to grow after World War Two, the U.S. national security establishment—the Pentagon, the Central intelligence Agency, the National Security Agency, and the National Security Council—became concerned. Population was an important element of national power, and countries with growing populations would almost inevitably increase in geopolitical weight. This was obviously a concern in the case of countries opposed to U.S. interests, such as the Soviet Union and China. But even allies might prove less pliable as their populations and economies grew. Most worrisome of all was the possibility that the rapidly multiplying peoples of Asia, Africa and Latin America would turn to communism in their search for independence and economic advancement *unless their birth rate was reduced*. Thus did population control become a weapon in the Cold War. . . .

Earth First (People Last): Environmental Movement Signs On

Every sorcerer deserves an apprentice. Hugh Moore, grand wizard of the population explosion, got his in the person of a young Stanford University entomologist by the name of Paul Ehrlich. In the very first sentence of his very first book Ehrlich proved beyond all doubt that he had already mastered Moore's panic-driven style. "The battle to feed all of humanity is over," he wrote. "In the 1970s the world will undergo famines—hundreds of million of people will starve to death in spite of any crash programs embarked upon now."[18]

In fact, he had gone Moore one better, as overzealous acolytes are prone to do. His book should have been named *The Population Explosion*, instead of *The Population Bomb*, for according to Ehrlich the "bomb" had already gone off and there was nothing to do now but wait for the inevitable human die-back. "Too many people" were chasing "too little food."[19] The most optimistic of Ehrlich's "scenarios" involved the immediate imposition of a harsh regimen of population control and resource conservation around the world, with the goal of reducing the number of people to 1.5 billion (about a fourth of its current level) over the next century or two. Even so, about a fifth of the world's population would still starve to death in the immediate future.

Such a prediction took pluck, for when the book appeared in 1968 there was no hint of massive famine on the horizon. The days of Indian food shortages were past. (We wouldn't learn about China's man-made calamity until a decade later.) The Green Revolution was starting to pay off in increased crop yields. And experts like Dr. Karl Brandt, the Director of the Stanford Food Research Institute, rebuked Ehrlich, saying that "Many nations need more people, not less, to cultivate food products and build a sound agricultural economy . . . every country that makes the effort can produce all the food it needs."[20]

But it wasn't his forecast of a massive human die-off that catapulted Ehrlich into the front rank of environmental prophets. (In a motif that has since become familiar, the book left readers with the impression that this might not be such a bad thing.) Rather it was his startling claim that our reckless breeding had jeopardized earth's ability to support life. All life, not just human life. Our planet was literally dying. Not only were the Children of Earth killing ourselves, we were going to take Mother with us as well.

The Population Bomb

Heavily promoted by the Sierra Club, *The Population Bomb* sold over a million copies. Ehrlich became an instant celebrity, becoming as much of a fixture on the "Tonight Show" as Johnny Carson's sidekick Ed McMahon. He command[ed] hefty lecture fees wherever he went (and he went everywhere), and always drew a crowd. People found it entertaining to hear about the end of the world. Likening the earth to an overloaded spaceship or sinking lifeboat, issuing apocalyptic warnings about the imminent "standing room only" problem, he captured the popular imagination. His prescriptions were always the same: "Join the environmental movement, stop having children, and save the planet."[21]

While Ehrlich fiddled his apocalyptic tunes, Moore burned to commit the growing environmental movement firmly to a policy of population control. His ad campaign, still ongoing, began suggesting that the best kind of environmental protection was population control. "Whatever Your Cause, It's a Lost Cause Unless We Control Population," one ad read. "Warning: The Water You are Drinking May be Polluted," read another, whose text went on to equate more people with more pollution. A third, addressed to "Dear President Nixon," claimed that "We can't lick the environment problem without considering this little fellow." It featured a picture of a newborn baby.

Birth of Earth Day

Moore went all out for the first Earth Day in 1970, printing a third of a million leaflets, folders, and pamphlets for campus distribution. College newspapers received free cartoons highlighting the population crisis and college radio stations a free taped show (featuring Paul Ehrlich). With his genius for marketing, Moore even announced a contest with cash awards for the best slogans relating environmental problems to what he called "popullution" [population pollution]. Students on over 200 campuses participated. The winner, not surprisingly, was "People Pollute."[22]

By 1971 most of the leading environmental groups had signed on to the anti-natal agenda, having been convinced that reducing the human birth rate would greatly benefit the environment. Perhaps it was their interest in "managing" populations of other species—salmon, condors, whales, etc.—that predisposed them to impose technical solutions on their own species. In any event, many of them were population hawks, who believed that simply making abortion, sterilization and contraception widely available was not enough. "Voluntarism is a farce," wrote Richard Bowers of Zero Population Growth as early as 1969. "The private sector effort has failed . . . [even the expenditure] of billions of dollars will not limit growth." Coercive measures were required. He proposed enacting "criminal laws to limit population, if the earth is to survive."[23]

Those who held such views were not content to merely stop people from multiplying, they demanded radical reductions in human numbers. The group Negative Population Growth wanted to cut the-then U.S. population of 200 million by more than half, to 90 million.[24] Celebrated oceanographer Jacques Cousteau told the *UNESCO Courier* in 1991, "In order to stabilize world populations, we must eliminate 350,000 people per day." Garrett Hardin of "The Tragedy of the Commons" fame opined that the "carrying capacity" of the planet was 100 million and that our numbers should be reduced accordingly. (Do we pick the lucky 100 million by lottery?) To carry out these decimations, Malthusian solutions are proposed, as when novelist William T. Vollman stated that, "there are too many people in the world and maybe something like AIDS or something like war may be a good thing on that level."[25] And lest we have compunctions about resorting to such measures, we should bear in mind, as Earth First! Founder Dave Foreman wrote, "We humans have become a disease, the Humanpox."

The Feminist Dilemma

The most radical of the feminists had a different definition of disease. Why should women be "subject to the species gnawing at their vitals," as Simone de Beauvoir so memorably wrote in her feminist classic *The Second Sex?* Why endure pregnancy at all, if contraception, sterilization and, especially, abortion, could be made widely available? With the legalization of abortion in the U.S. in 1973, feminists increasingly looked overseas, eager to extend their newfound rights to "women of color" elsewhere in the world. They had read their Ehrlich as well as their Beauvior, and knew that the world had too many people, or soon would. But family planning, especially abortion, provided a way out. "Let us bestow upon all the women of the world the blessing that we women in the privileged West have received—freedom from fear of pregnancy," the feminists said to themselves. "We will, at the same time and by the same means, solve the problem of too many babies. For surely impoverished Third World women do not actually want all those children they are bearing. Patriarchy has made them into breeding machines, but we will set them free."

Abortion "Needs" Appear

At the time, the population control movement remained ambivalent over the question of abortion. Hugh Moore had long wanted it as a population control measure, but Frank Notestein was still arguing in the early 1970s that the Population Council should "consistently and firmly take the anti-abortion stance and use every occasion to point out that the need for abortions is the proof of program failure in the field of family planning and public health education."[26]

But the women's movement would not be put off with the promise of a perfect contraceptive. They knew, better than anyone (and often from painful personal experience) that contraception, because of the inevitable failures, *always* led to abortion. As Sharon Camp of the Population Crisis Committee wrote "both abortion and contraception are presently on the rise in most developing countries."[27]

Abortion was, in the end, accepted by most controllers because it came to be seen as a necessary part of the anti-natal arsenal. The Rockefeller Commission, established by President Nixon, wrote that "We are impressed that induced abortion has a demographic effect wherever legalized" and on these grounds went on to call for "abortion on demand."[28] The Population Council followed the Commission in endorsing abortion as a means of population control by 1975.

In the end, feminist advocacy of abortion had proven decisive. The feminists had given the population control movement an additional weapon, abortion, to use in its drive to reduce human fecundity, and encouraged its aggressive use.

Third World Women

At the same time, it was soon apparent to many feminists that birth control was not an unmixed blessing for Third World women, who continued to be targets of ever-more aggressive programs in places like Indonesia, India, and Bangladesh. They began to demand further changes in the way programs were carried out, starting with male contraceptives and more vasectomies. Frank Notestein wrote of the feminists that, "As second-generation suffragists they were not at all disposed to allow the brutish male to be in charge of contraception. Women must have their own methods!" But more recent feminists "complain violently that the men are trying to saddle the women with all the contraceptive work. You can't please them if you do, and can't please them if you don't."[29]

Although expressed somewhat crudely, Notestein's comment pointed out the dilemma faced by feminists. On the one hand, they sought to impose a radically pro-abortion agenda on population control programs, whose general purpose—fertility reduction—they applauded. On the other they tried to protect women from the abuses that invariably accompanied such programs. But with the exception of the condom, other methods of contraception all put the burden on women. Vasectomies could easily be performed on men, but it was usually the woman who went under the knife to have her tubes tied. And abortions could only be performed on women. So, as a practical matter, the burden of fertility reduction was placed disproportionately on women. And when programs took a turn towards the coercive, as they were invariably prone to do in the Third World, it was overwhelmingly women who paid the price.

Feminist complaints did lead to some changes, but these were mostly cosmetic. Population controllers did learn, over time, to speak a different language or, rather, several different languages, to disguise the true, anti-natal purpose of their efforts. When Western feminists need to be convinced of the importance of supporting the programs, reproductive rights rhetoric is the order of the day. Thus we hear Nafis Sadik telling Western reporters on the eve of the 1994 U.N. Conference on Population and Development that the heart of the discussion "is the recognition that the low status of women is a root cause of inadequate reproductive health care." Such language would ring strange in the ears of Third World women, who are instead the object of soothing lectures about "child-spacing" and "maternal health." Population control programs were originally unpopular in many Middle Eastern countries and sub-Sarahan countries until they were redesigned, with feminist input, as programs to "help" women. As Peter Donaldson, the head of the Population Reference Bureau, writes, "The idea of limiting the number of births was so culturally unacceptable [in the Middle East and sub-Saharan Africa] that family planning programs were introduced as a means for promoting better maternal and child health by helping women space their births."[30] James Grant of the U.N. Children's Fund (UNICEF), in an address to the World Bank, was even more blunt: "Children and women are to be the Trojan Horse for dramatically slowing population growth."[31]

Corrupted Feminist Movement

The feminists did not imagine, when they signed onto the population control movement, that they would merely be marketing consultants. It is telling that many Third World feminists have refused to endorse population control programs at all, arguing instead that these programs violate the rights of women while ignoring their real needs. It must be painful for Western feminists to contemplate, but their own movement has been used or, to use Betsy Hartmann's term, "co-opted," by another movement for whom humanity as a whole, and women in particular, remain a faceless mass of numbers to be contracepted, sterilized, and aborted. For, despite the feminist rhetoric, the basic character of the programs hasn't changed. They are a numbers-driven, technical solution to the "problem of overpopulation"—which is, in truth, a problem of poverty—and they overwhelmingly target women.

This is, in many respects, an inevitable outcome. To accept the premise that the world is overpopulated and then seek to make the resulting birth control programs "women-friendly," as many feminists have, is a fateful compromise. For it means that concern for the real needs of women is neither the starting point of these programs, nor their ultimate goal, but merely a consideration along the way. Typical of the views of feminists actively involved in the population movement are those of Sharon Camp, who writes, "There is still time to avoid another population doubling, but only if the world community acts very quickly to make family planning universally available and to invest in other social programs, like education for girls, which can help accelerate fertility declines."[32] Here we see the population crisis mentality in an uneasy alliance with programs for women which, however, are justified chiefly because they "help accelerate fertility declines."

The alliance between the feminists and population controllers has been an awkward affair. But the third of the three most anti-natalist movements in history gave the population controllers new resources, new constituencies, new political allies, a new rhetoric, and remains a staunch supporter even today.

Population Firm Funding

Over the past decade the Population Firm has become more powerful than ever. Like a highly organized cartel, working through an alphabet soup of United Nations agencies and "nongovernmental organizations," its tentacles reach into nearly every developing country. It receives sustenance from feeding tubes attached to the legislatures of most developed countries, and further support through the government-financed population research industry, with its hundreds of professors and thousand of students. But unlike any other firm in human history, its purpose is not to produce anything, but rather to destroy— to destroy fertility, to prevent babies from being conceived and born. It diminishes, one might say, the oversupply of people. It does this for the highest of motives—to protect all of us from "popullution." Those who do not subscribe to its ideology it bribes and browbeats, bringing the combined weight of the world's industrial powers to bear on those in countries which are poor.

In 1991, the U.N. estimated that a yearly sum of $4.5 to $5 billion was being directed to population programs in developing countries. This figure, which has grown tremendously in the last 10 years, includes contributions from bilateral donors such as the U.S., the European nations and Japan, from international agencies like those associated with the UN, and from multilateral lending institutions, including the World Bank and the various regional development banks. It includes grants from foundations, like Ted Turner's U.N. Foundation, and wealthy individuals like Warren Buffet.

Moreover, a vast amount of money not explicitly designated as "population" finance is used to further the family planning effort. As Elizabeth Liagin notes, "During the 1980s, the diversion of funds from government non-population budgets to fertility-reduction measures soared, especially in the U.S., where literally hundreds of millions from the Economic Support Funds program, regional development accounts, and other non-population budgets were redirected to "strengthen" population planning abroad."[33]

More Money Spent

An almost unlimited variety of other "development" efforts—health, education, energy, commodity imports, infrastructure, and debt relief, for example—are also used by governments and other international agencies such as the World Bank, to promote population control policies, either through requiring recipient nations to incorporate family planning into another program or by holding funds or loans hostage to the development of a national commitment to tackle the "over-population" problem.

In its insatiable effort to locate additional funds for its insatiable population control programs, USAID has even attempted to redirect "blocked assets"—profits generated by international corporations operating in developing nations that prohibit the transfer of money outside the country—into population control efforts. In September 1992, USAID signed a $36.4 million contract and "statement of work" with the accounting firm Deloitte and Touche to act as a mediator with global corporations and to negotiate deals that would help turn the estimated

$200 *billion* in blocked assets into "private" contributions for family planning in host countries. The corporations would in return get to claim a deduction on their U.S. tax return for this "charitable contribution." The Profit initiative, as it is fittingly called, is not limited to applying its funds directly to family planning "services," but is also encouraged to "work for the removal" of "trade barriers for contraceptive commodities" and "assist in the development of a regulatory framework that permits the expansion of private sector family planning services." This reads like a bureaucratic mandate to lobby for the elimination of local laws which in any way interfere with efforts to drive down the birth rate, such as laws restricting abortion or sterilization.[34]

U.S.'s Real Foreign Aid Policy

Throughout the nineties, the idea of the population controllers that people in their numbers were somehow the enemy of all that is good reigned supreme. J. Brian Atwood, who administered the U.S. Agency for International Development in the early days of the Clinton administration, put it this way: "If we aren't able to find and promote ways of curbing population growth, we are going to fail in *all* of our foreign policy initiatives." [italics added] (Atwood also went on to announce that the U.S. "also plans to resume funding in January [1994] to the U.N. Fund for Population Activities (UNFPA).")[35] Secretary of State Warren Christopher offered a similar but even more detailed defense of population programs the following year. "Population and sustainable development are back where they belong in the mainstream of American foreign policy and diplomacy." He went on to say, in a line that comes rights out of U.S. National Security Study Memo 200, that population pressure "ultimately jeopardizes America's security interests." But that's just the beginning. Repeating the now familiar litany, he claimed of population growth that, "It strains resources. It stunts economic growth. It generates disease. It spawns huge refugee flows, and ultimately it threatens our stability. . . . We want to continue working with the other donors to meet the rather ambitious funding goals that were set up in Cairo."[36]

The movement was never more powerful than it was in 2000 in terms of money, other resources, and political clout.

Losing Momentum

Like a wave which crests only seconds before it crashes upon the shore, this appearance of strength may be deceiving. There are signs that the anti-natal movement has peaked, and may before long collapse of its own overreaching. U.S. spending on coercive population control and abortion overseas have long been banned. In 1998 the U.S. Congress, in response to a flood of reports about human rights abuses, for the first time set limits on what can be done to people in the name of "voluntary family planning."[37] Developing countries are regularly denouncing what they see as foreign interference into their domestic affairs, as the Peruvian Congress did in 2002. Despite strenuous efforts to co-opt them, the opposition of feminists to population control programs (which target women) seems to be growing.[38] Many other groups—libertarians, Catholics, Christians of other denominations, the majority of economists, and those who define themselves as pro-life—have long been opposed.

As population control falls into increasing disrepute worldwide, the controllers are attempting to reinvent themselves, much the same way that the Communists in the old Soviet Union reemerged as "social democrats" following its collapse. Organizations working in this area have found it wise to disguise their agenda by adopting less revealing names. Thus Zero Population Growth in June 2002 became Population Connection, and the Association for Voluntary Surgical Contraception the year before changed its name to Engender Health. Similarly, the U.N., in documents prepared for public consumption, has recently found it expedient to cloak its plans in language about the "empowerment of women," "sustainable development," "safe motherhood," and "reproductive health." Yet the old anti-natal zeal continues to come through in internal discussions, as when Thoraya Obaid affirmed to her new bosses on the U.N. Commission on Population and Development her commitment to "slow and eventually stabilize population growth." "And today I want to make one thing very clear," she went on to say. "The slowdown in population growth does not mean we can slow down efforts for population and reproductive health—quite the contrary. If we want real progress and if we want the projections to come true, we must step up efforts . . . while population growth is slowing, it is still growing by 77 million people every year."[39] And so on.

Such efforts to wear a more pleasing face for public consumption will in the end avail them nothing. For, as we will see, their central idea—the Malthusian notion that you can eliminate poverty, hunger, disease, and pollution by eliminating the poor—is increasingly bankrupt.

Reducing the numbers of babies born has not and will not solve political, societal and economic problems. It is like trying to kill a gnat with a sledgehammer, missing the gnat entirely, and ruining your furniture beyond repair. It is like trying to protect yourself from a hurricane with a bus ticket. Such programs come with massive costs, largely hidden from the view of well-meaning Westerners who have been propagandized into supporting them. And their "benefits" have proven ephemeral or worse. These programs, as in China, have done actual harm to real people in the areas of human rights, health care, democracy, and so forth. And, with falling birth rates everywhere, they are demographic nonsense. Where population control programs are concerned, these costs have been largely ignored (as the cost of doing business) while the benefits to people, the environment, and to the economy, have been greatly exaggerated, as we will see. Women in the developing world are the principal victims.

Notes

1. The World Bank is to a large degree under the control of the United States, which provides the largest amount of funding. This is why the head of the World Bank is always an American. The activities of the Bank are monitored by the National Advisory Council on International Monetary and Financial Policies—called the NAC for short, of the Treasury Department. The 1988 annual report of the NAC states that "the council [NAC] seeks to ensure that . . . the . . . operations [of the World Bank and other international financial institutions] are conducted

in a manner consistent with U.S. policies and objectives . . ." International Finance: National Advisory Council on International Monetary and Financial Policies, Annual Report to the President and to the Congress for the Fiscal year 1988, (Wash., D.C., Department of the Treasury), Appendix A, p. 31. Quoted in Jean Guilfoyle, "World Bank Population Policy: Remote Control," *PRI Review* 1:4 (July/August 1991), p. 8.

2. I served on board a ship of this class, the USS Lockwood, from 1974–76. As the Main Propulsion Assistant—the officer in charge of the engine room—I can personally attest that this fleet frigate, as it was called, was anything but fleet. On picket duty, it could not keep up with the big flattops that it was intended to protect from submarine attacks.

3. The 1968 meeting was 23rd joint annual meeting of the Boards of Governors of the World Bank and the International Monetary Fund. The two organizations always hold their annual meetings in tandem, underscoring their collaboration on all matters of importance.

4. McNamara moderated his anti-natal rhetoric on this formal occasion. More often, he sounded like Hugh Moore, as when he wrote: "the greatest single obstacle to the economic and social advancement of the majority of the peoples in the underdeveloped world is rampant population growth. . . . The threat of unmanageable population pressures is very much like the threat of nuclear war. . . . Both threats can and will have catastrophic consequences unless they are dealt with rapidly." *One Hundred Countries, Two Billion People* (London: Pall Mall Press, 1973), pp. 45–46. Quoted in Michael Cromartie, ed., *The 9 Lives of Population Control*, (Washington: Ethics and Public Policy Center, 1995), p. 62. McNamara never expressed any public doubts about the importance of population control, although he did once confide in Bernard Berelson that "many of our friends see family planning as being 'too simple, too narrow, and too coercive.'" As indeed it was—and is. Quote is from Donald Crichtlow, *Intended Consequences: Birth Control, Abortion, and the Federal Government in Modern America*, (Oxford: Oxford University Press, 1999), p. 178.

5. See Fred T. Sai and Lauren A. Chester, "The Role of the World Bank in Shaping Third World Population Policy," in Godfrey Roberst, ed., *Population Policy: Contemporary Issues* (New York: Praeger, 1990). Cited in Jacqueline Kasun, *The War Against Population*, Revised Ed. (San Fransisco: Ignatius Press, 1999), p. 104.

6. 7 September 1994, p. 1. Cited in Kasun, p. 104.

7. U.S. International Population Policy: First Annual report, prepared by the Interagency Task Force on Population Policy, (Wash., D.C., U.S. National Archives, May 1976). Quoted in Jean Guilfoyle, "World Bank Population Policy: Remote Control," *PRI Review* 1:4 (July/August 1991), p. 8.

8. Personal Communication with the author from a retired World Bank executive who worries that, if his identity is revealed, his pension may be in jeopardy.

9. Peter T. Bauer and Basil S. Yamey, "The Third World and the West: An Economic Perspective," in W. Scott Thompson, ed., *The Third World: Premises of U.S. Policy* (San Francisco: Institute for Contemporary Studies, 1978), p. 302. Quoted in Kasun, p. 105.

10. See Elizabeth Liagin, "Money for Lies," *PRI Review*, July/October 1998, for the definitive history of the imposition of population control on Nigeria; The Nigerian case is also discussed by Barbara Akapo, "When family planning meets population control," *Gender Review*, June 1994, pp. 8–9.

11. Word Bank, 1995 Annual Report, p. 18. Quoted in *PRI Review* 5:6 (November/December 1995), p. 7.

12. Kasun, p. 277.

13. David Morrison, "Weaving a Wider Net: U.N. Move to Consolidate its Anti-Natalist Gains," *PRI Review* 7:1 (January–February 1997), p. 7.

14. Ibid.

15. Bernard Berelson, "Where Do We Stand," paper prepared for Conference on Technological change and Population Growth, California Institute of Technology, May 1970, p. 1. Quoted in Ronald Freedman, *The Sociology of Human Fertility: An Annotated Bibliography* (New York: Irvington Publishers, 1975), p. 3. It's worth noting that Freedman's book was a subsidized product of the institution Berelson then headed. As Freedman notes in his "Preface," the "staff at the Population Council were very helpful in reading proof, editorial review, and making detailed arrangements for publication. Financial support was provided by the Population Council." (p. 2.)

16. Freedman, p. 4.

17. *Christian Science Monitor*, 5 July 1977. He went on to say that, if present methods of population control "fail, and population pressures become too great, nations will be driven to more coercive methods."

18. Paul R. Ehrlich, *The Population Bomb* (New York: Ballantine books, 1968).

19. The first three sections of Ehrlich's book were called, "Too many people," "Too little Food," "A Dying Planet."

20. *Is there a Population Explosion?*, Daniel Lyons, (Catholic Polls: New York, 1970), p. 5.

21. Ehrlich has continued on the present day, writing one book after another, each one chock full of predictions of imminent disasters that fail to materialize. People wonder why Ehrlich doesn't learn from his experiences? The answer, I think, is that he has learned very well. He has learned that writing about "overpopulation and environmental disaster" sells books, *lots* of books. He has learned that there is no price to pay for being wrong, as long as he doesn't admit his mistakes *in print* and glibly moves on to the next disaster. In one sense, he has far outdone Hugh Moore in this regard. For unlike Moore, who had to spend his own money to publish the original *The Population Bomb*, Ehrlich was able to hype the population scare *and* make money by doing so. He is thus the archetype of a figure familiar to those who follow the anti-natal movement: the population hustler.

22. Lawrence Lader, *Breeding Ourselves to Death*, pp. 79–81.

23. Richard M. Bowers to ZPG members, 30 September 1969, Population Council (unprocessed), RZ. Quoted in Critchlow, p. 156.

24. In later years, as U.S. population continued to grow, NPG has gradually increased its estimate of a "sustainable" U.S. population to 150 million. See Donald Mann, "A No-Growth Steady-State Economy Must Be Our Goal," NPG Position Paper, June 2002.

25. Quoted in David Boaz, "Pro-Life," *Cato Policy Report* (July/August 2002), p. .2.

26. Frank Notestein to Bernard Berelson, 8 February 1971, Rockefeller Brother Fund Papers, Box 210, RA. Quoted in Critchlow, p. 177. These concerns, while real enough to Notestein, apparently did not cause him to reflect on the fact that he was a major player in a movement that "detracted from the value of human life" by suggesting that there was simply too much of that life, and working for its selective elimination.

27. Population Action International, "Expanding Access to Safe Abortion: Key Policy Issues," Population Policy Information Kit 8 (September 1993). Quoted in Sharon Camp, "The Politics of U.S. Population Assistance," in Mazur, *Beyond the Numbers*, p. 126.

28. Critchlow, p. 165.

29. Frank Notestein to Bernard Berelson, April 27, 1971, Notestein Papers, Box 8, Princeton University.

30. Peter Donaldson and Amy Ong Tsui, "The International Family Planning Movement," in Laurie Ann Mazur, ed., *Beyond the Numbers* (Island Press, Washington,

D.C., 1994), p. 118. Donaldson was, at the time, president of the Population Reference Bureau, and Tsui was deputy director of the Carolina Population Center.

31.   World Bank 1993 International Development Conference, Washington, D.C., 11 January 1993, p. 3. Also quoted in *PRI Review* (September–October 1994), p. 9.

32.   Sharon Camp, "Politics of U.S. Population Assistance," in *Beyond the Numbers*, pp. 130–1. Camp for many years worked at the Population Crisis Committee, founded by Hugh Moore in the early sixties. It has recently, perhaps in recognition of falling fertility rates worldwide, renamed itself Population Action International.

33.   Quoted from Elizabeth Liagin, "Profit or Loss: Cooking the Books at USAID," *PRI Review* 6:3 (November/December 1996), p. 1.

34.   Ibid., p. 11.

35.   John M. Goshko, "Planned Parenthood gets AID grant . . . ," *Washington Post*, 23 November 1993, A12-13.

36.   Reuters, "Christopher defends U.S. population programs," Washington, D.C, 19 December 1994.

37.   The Tiahrt Amendment.

38.   See Betsy Hartmann, *Reproductive Rights and Wrongs* (Boston: South End Press, 1995).

39.   Thoraya Ahmed Obaid, "Reproductive Health and Reproductive Rights With Special Reference to HIV/AIDS," Statement to the U.N. Commission on Population and Development, 1 April 2002.

POSTSCRIPT

Should the International Community Continue to Curb Population Growth in the Developing World?

There are at least two basic dimensions to this issue. First, should the international community involve itself in reducing fertility throughout the developing world? That is, is it a violation of either national sovereignty (a country should be free from extreme outside influence) or human rights (an individual has the right to make fertility decisions unencumbered by outside pressure, particularly those from another culture)? And second, if so, what should its motives be? To put it another way, is advocacy of population control really a form of ethnic or national genocide?

McNamara answers these questions with the same set of arguments. His belief that the international community has an obligation to get involved is based on his assessment of the resultant damage to both the developing world where such high levels of growth are found and the rest of the world that must compete with the poorer countries for increasingly scare resources. For McNamara, a 2.6-fold increase in population and an 8-fold increase in consumption per capita (by the end of the twenty-first century) would result in a 20-fold impact on nonrenewable and renewable resources. He cites the projected agricultural needs to demonstrate that the Earth cannot sustain such consumption levels. Additionally, McNamara suggests that other consequences include environmentally unsustainable development, worsening of poverty, and negative impacts on the welfare of women and children. They appear to make an overwhelming case for fertility reduction. And the cost of policy intervention is small, in his judgment, compared to both gross national products and overseas general development assistance. And finally, given the projections, the developing world should embrace such policy intervention for its own good.

Those who oppose such intervention point to several different reasons. The first, originally articulated at the 1974 Bucharest conference, argues that economic development is the best contraceptive. The demographic transition worked in the developed world, and there was no reason to assume that it will not work in the developing world. This was the dominant view of the developing countries at Bucharest, who feared international policy intervention, and at the same time wanted and needed external capital to develop. They soon came to realize that characteristics particular to the developing world in the latter part of the twentieth century meant that foreign aid alone would not be enough. It had to be accompanied by fertility reduction programs, wherever the origin.

Second, some who oppose intervention do not see the extreme negative environmental consequences of expanding populations. For them, the pronatalists (those favoring fertility reduction programs) engage in inflammatory discourse, exaggerated arguments, and scare tactics devoid of much scientific evidence.

Mosher goes further in the second selection, accusing McNamara, who served as U.S. Secretary of Defense during the Vietnam era, of using his position as one of the most important leaders in the international financial lending world to help prevent continuing population growth in the developing world because of its potential threat to the national security of countries like the United States. He also concludes that the interventionists are engaged in social engineering, not economic development.

And finally, cultural constraints can counteract any attempt to impose fertility reduction values from the outside. Children are valued in most societies. They, particularly male heirs, serve as the social security system for most families. They become producers at an early age, contributing to the family income.

McNamara is correct if one assumes that no changes—technological, social, and organizational—accompany predicted population growth. But others would argue that society has always found a way to use technology to address newly emerging problems successfully. The question remains, though: Are the current problems and the world in which they function too complex to allow for a simple solution?

The bottom line is that each has a valid point. Environmental stress is a fact of life, and increased consumption does play a role. If the latter originates because of more people, than fertility reduction is a viable solution. Left to its own devices, the developing world is unable to provide both the rationale for such action and the resources to accomplish it. The barriers are too great. While the international community must guard against behavior that is or appears to be either genocidal in nature or violates human rights, it must nonetheless move forward.

An excellent account of the 1994 Cairo population conference's answer to how the international community ought to respond is found in Lori S. Ashford's *New Perspectives on Population: Lessons from Cairo* (1995). Other sources that address the question of the need for international action include William Hollingworth's *Ending the Explosion: Population Policies and Ethics for a Humane Future* (1996), Elizabeth Liagin's *Excessive Force: Power, Politics, and Population* (1996), and Julian Simon's *The Ultimate Resource 2* (1996).

There are both United Nations and external analyses of the Cairo conference and assessments of progress made in implementing its Plan of Action. The United Nations Population Fund's Web site (www.unfpa.org) has a section entitled *ICPD and Key Actions*. There you will find numerous in-house analyses of both the conference and post-1994 action. In the broader UN Web site (www.un.org/popin), one finds a major study, *Progress Made in Achieving the Goals and Objectives of the Program of Action of the International Conference on Population and Development.* An external assessment is the Population Institute's report entitled *The Hague Forum: Measuring ICPD Progress Since Cairo* (1999). See its Web site at www.populationinstitute.org.

ISSUE 2

Is Rapid Urbanization in the Developing World a Major Problem?

YES: Barbara Boyle Torrey, from "Urbanization: An Environmental Force to Be Reckoned With," *Population Reference Bureau* (April 2004)

NO: Gordon McGranahan and David Satterthwaite, from "Urban Centers: An Assessment of Sustainability," *Annual Review of Energy and the Environment* (vol. 28, 2003)

ISSUE SUMMARY

YES: Barbara Boyle Torrey, a member of the Population Reference Bureau's board of trustees and a writer/consultant, argues that extremely high urban growth rates are resulting in and will continue to create a range of negative environmental problems.

NO: Gordon McGranahan and David Satterthwaite, members of London's International Institute for Environment and Development, suggest that there is little research about urban centers and their ability to provide sustainable development.

In 1950, 55 percent of the population of the developed world resided in urban areas, compared to only 18 percent in the developing world. By 2000, 76 percent of those in the developed world were urbanized, and it is expected, according to UN projections, to reach 82 percent by 2025. But because there will be low population growth throughout the developed world in the coming decades, the impact will not be substantial.

The story is different in the developing world. In 2000, the level of urbanization had risen to almost 40 percent, and it is projected to be 54 percent in 2025. The percentages tell only part of the story, however, as these percentages are not based on a stable national population level but will occur in the context of substantial increases in the national population level. To illustrate the dual implication of urban growth as the consequence of both migration to the cities and increased births to those already living in urban areas, the UN projects that the urban population in the developing world will nearly double in size between 2000 and 2025, from just under 2 billion to more than 3.5 billion. And urban growth will not stop in 2025.

The UN reported in UN Chronicle (no. 3, 2002) a number of other conclusions relating to urban growth:

- Almost all population increase in the next three decades will occur in the urban areas of the developed world.
- This growth in the developed world's urban centers will continue at a rate of at least 3 percent.
- There are marked differences in the level and pace of urbanization among major geographical areas of the developing world.
- The proportion of the world's population living in megacities is small, as only 3.7 percent reside in cities of at least 10 million inhabitants.

There are two ways to examine rapid urbanization. One can study the ability of society to provide services to the urban population. A second approach is to examine the adverse impacts of the urbanized area on the environment. The best place to begin a discussion of urbanization's effects is found in "An Urbanizing World" (Martin P. Brockerhoff, *Population Bulletin*, Population Reference Bureau, 2000). To Brockerhoff, increasing urbanization in the poor countries can be seen "as a welcome or as an alarming trend." He suggests that cities have been the "engines of economic development and the centers of industry and commerce." The diffusion of ideas is best found in cities around the world. And Brockerhoff observes the governmental cost savings of delivering goods and services to those in more densely populated environments.

The current problem is not only that cities of the developing world are growing but that they are expanding at a rapid pace. This calls into question the ability of both government and the private sector to determine what is necessary for urbanites to not only survive but to thrive. Many researchers believe that poverty and health problems (both physical and mental) are consequences of urbanization. Brockerhoff also alludes to the potential for greater harm to residents of cities from natural disasters and environmental hazards.

There is a more recent concern emerging among researchers about urbanization's impact, this time on biodiversity. One source has coined the phrase "heavy ecological footprints" ("Impact on the Environment," *Population Reports*, 2002) to describe the adverse effects. One study concludes, for example, that urban sprawl in the United States endangers more species than any other human behavior (Michael L. McKinney, "Urbanization, Biodiversity, and Conservation," *Bioscience*, 2002).

The two selections for this issue address the question of the consequences of urbanization. In the first selection, Torrey asserts that environmental degradation is a consequence of rapid urbanization. McGranahan and Satterthwaite argue not so fast. Instead they suggest that controversies exist among scholars regarding how well cities are able to promote sustainable development.

YES ↵

Urbanization: An Environmental Force to Be Reckoned With

Human beings have become an increasingly powerful environmental force over the last 10,000 years. With the advent of agriculture 8,000 years ago, we began to change the land.[1] And with the industrial revolution, we began to affect our atmosphere. The recent increase in the world's population has magnified the effects of our agricultural and economic activities. But the growth in world population has masked what may be an even more important human-environmental interaction: While the world's population is doubling, the world's urban population is tripling. Within the next few years, more than half the world's population will be living in urban areas (see Figure 1).[2]

The level and growth of urbanization differ considerably by region (see Figure 2). Among developing countries, Latin American countries have the highest proportion of their population living in urban areas. But East and South Asia are likely to have the fastest growth rates in the next 30 years. Almost all of future world population growth will be in towns and cities. Both the increase in and the redistribution of the earth's population are likely to affect the natural systems of the earth and the interactions between the urban environments and populations.

The best data on global urbanization trends come from the United Nations Population Division and the World Bank.[3] The UN, however, cautions users that the data are often imprecise because the definition of urban varies country by country. Past projections of urbanization have also often overestimated future rates of growth. Therefore, it is important to be careful in using urbanization data to draw definitive conclusions.

The Dynamics of Urbanization

In 1800 only about 2 percent of the world's population lived in urban areas. That was small wonder: Until a century ago, urban areas were some of the unhealthiest places for people to live. The increased density of populations in urban areas led to the rapid spread of infectious diseases. Consequently, death rates in urban areas historically were higher than in rural areas. The only way urban areas maintained their existence until recently was by the continual in-migration of rural people.[4]

From *Population Reference Bureau* by Barbara Boyle Torrey. Copyright © 2004 by Population Reference Bureau. Reprinted by permission.

Figure 1

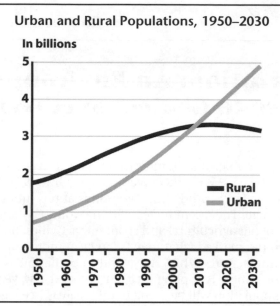

Urban and Rural Populations, 1950–2030

Source: UN, *World Urbanization Prospects: The 2003 Revision* (2004).

Figure 2

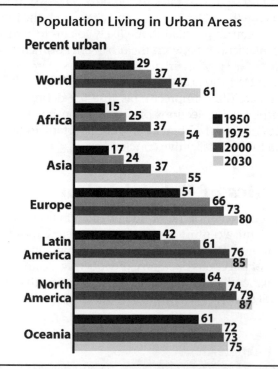

Population Living in Urban Areas

Source: UN, *World Urbanization Prospects: The 2003 Revision* (2004).

In only 200 years, the world's urban population has grown from 2 percent to nearly 50 percent of all people. The most striking examples of the urbanization of the world are the megacities of 10 million or more people. In 1975 only four megacities existed; in 2000 there were 18. And by 2015 the UN estimates that there will be 22.[5] Much of the future growth, however, will not be in these huge agglomerations, but in the small to medium-size cities around the world.[6]

The growth in urban areas comes from both the increase in migration to the cities and the fertility of urban populations. Much of urban migration is driven by rural populations' desire for the advantages that urban areas offer. Urban advantages include greater opportunities to receive education, health care, and services such as entertainment. The urban poor have less opportunity for education than the urban nonpoor, but still they have more chance than rural populations.[7]

Urban fertility rates, though lower than rural fertility rates in every region of the world, contribute to the growth of urban areas. Within urban areas, women who migrated from rural areas have more children than those born in urban areas.[8] Of course, the rural migrants to urban areas are not a random selection of the rural population; they are more likely to have wanted fewer children even if they had stayed in the countryside. So the difference between the fertility of urban migrants and rural women probably exaggerates the impact of urban migration on fertility.

In sub-Saharan Africa, the urban fertility rates are about 1.5 children less than in rural areas; in Latin America the differences are almost two children.[9] Therefore, the urbanization of the world is likely to slow population growth. It is also likely to concentrate some environmental effects geographically.

Environmental Effects of Urbanization

Urban populations interact with their environment. Urban people change their environment through their consumption of food, energy, water, and land. And in turn, the polluted urban environment affects the health and quality of life of the urban population.

People who live in urban areas have very different consumption patterns than residents in rural areas.[10] For example, urban populations consume much more food, energy, and durable goods than rural populations. In China during the 1970s, the urban populations consumed more than twice as much pork as the rural populations who were raising the pigs.[11] With economic development, the difference in consumption declined as the rural populations ate better diets. But even a decade later, urban populations had 60 percent more pork in their diets than rural populations. The increasing consumption of meat is a sign of growing affluence in Beijing; in India where many urban residents are vegetarians, greater prosperity is seen in higher consumption of milk.

Urban populations not only consume more food, but they also consume more durable goods. In the early 1990s, Chinese households in urban areas were two times more likely to have a TV, eight times more likely to have a washing machine, and 25 times more likely to have a refrigerator than rural households.[12] This increased consumption is a function of urban labor markets, wages, and household structure.

Energy consumption for electricity, transportation, cooking, and heating is much higher in urban areas than in rural villages. For example, urban populations have many more cars than rural populations per capita. Almost all of the cars in the world in the 1930s were in the United States. Today we have a car for every two people in the United States. If that became the norm, in 2050 there would be 5.3 billion cars in the world, all using energy.[13]

In China the per capita consumption of coal in towns and cities is over three times the consumption in rural areas.[14] Comparisons of changes in world energy consumption per capita and GNP show that the two are positively correlated but may not change at the same rate.[15] As countries move from using noncommercial forms of energy to commercial forms, the relative price of energy increases. Economies, therefore, often become more efficient as they develop because of advances in technology and changes in consumption behavior. The urbanization of the world's populations, however, will increase aggregate energy use, despite efficiencies and new technologies. And the increased consumption of energy is likely to have deleterious environmental effects.

Urban consumption of energy helps create heat islands that can change local weather patterns and weather downwind from the heat islands. The heat island phenomenon is created because cities radiate heat back into the atmosphere at a rate 15 percent to 30 percent less than rural areas. The combination of the increased energy consumption and difference in albedo (radiation) means that cities are warmer than rural areas (0.6 to 1.3 C).[16] And these heat islands become traps for atmospheric pollutants. Cloudiness and fog occur with greater frequency. Precipitation is 5 percent to 10 percent higher in cities; thunderstorms and hailstorms are much more frequent, but snow days in cities are less common.

Urbanization also affects the broader regional environments. Regions downwind from large industrial complexes also see increases in the amount of precipitation, air pollution, and the number of days with thunderstorms.[17] Urban areas affect not only the weather patterns, but also the runoff patterns for water. Urban areas generally generate more rain, but they reduce the infiltration of water and lower the water tables. This means that runoff occurs more rapidly with greater peak flows. Flood volumes increase, as do floods and water pollution downstream.

Many of the effects of urban areas on the environment are not necessarily linear. Bigger urban areas do not always create more environmental problems. And small urban areas can cause large problems. Much of what determines the extent of the environmental impacts is how the urban populations behave— their consumption and living patterns—not just how large they are.

Health Effects of Environmental Degradation

The urban environment is an important factor in determining the quality of life in urban areas and the impact of the urban area on the broader environment. Some urban environmental problems include inadequate water and sanitation, lack of rubbish disposal, and industrial pollution.[18] Unfortunately, reducing the problems and ameliorating their effects on the urban population are expensive.

The health implications of these environmental problems include respiratory infections and other infectious and parasitic diseases. Capital costs for building improved environmental infrastructure—for example, investments in a cleaner public transportation system such as a subway—and for building more hospitals and clinics are higher in cities, where wages exceed those paid in rural areas. And urban land prices are much higher because of the competition for space. But not all urban areas have the same kinds of environmental conditions or health problems. Some research suggests that indicators of health problems, such as rates of infant mortality, are higher in cities that are growing rapidly than in those where growth is slower.[19]

Urban Environmental Policy Challenges

Since the 1950s, many cities in developed countries have met urban environmental challenges. Los Angeles has dramatically reduced air pollution. Many towns that grew up near rivers have succeeded in cleaning up the waters they befouled with industrial development. But cities at the beginning of their development generally have less wealth to devote to the mitigation of urban environmental impacts. And if the lack of resources is accompanied by inefficient government, a growing city may need many years for mitigation. Strong urban governance is critical to making progress. But it is often the resource in shortest supply.[20] Overlapping jurisdictions for water, air, roads, housing, and industrial development frustrate efficient governance of these vital environmental resources. The lack of good geographic information systems means that many public servants are operating with cataracts. The lack of good statistics means that many urban indicators that would inform careful environmental decisionmaking are missing.[21]

When strong urban governance is lacking, public-private partnerships can become more important.[22] These kinds of partnerships can help set priorities that are shared broadly, and therefore, implemented. Some of these public-private partnerships have advocated tackling the environmental threats to human health first. "Reducing soot, dust, lead, and microbial disease presents opportunities to achieve tangible progress at relatively low cost over relatively short periods," concluded conferees at a 1994 World Bank gathering on environmentally sustainable development.[23] But ultimately there are many other urban environmental priorities that produce chronic problems for both people and the environment over the long term that also have to be addressed.

Much of the research that needs to be done on the environmental impacts of urban areas has not been done because of a lack of data and funding. Most of the data that exist are at a national level. But national research is too coarse for the environmental improvement of urban areas. Therefore, data and research at the local level need to be developed to provide the local governments with the information they need to make decisions. Certainly the members of the next generation, the majority of whom will be living in urban areas, will judge us by whether we were asking the right questions today about their urban environments. They will want to know whether we funded the right research to address those questions. And they will also want to know whether we used the research findings wisely.

References

1. M. Gordon Wolman, "Population, Land Use, and Environment: A Long History," in *Population and Land Use in Developing Countries*, ed. Carole L. Jolly and Barbara Boyle Torrey, Committee on Population, Commission on Behavioral and Social Sciences and Education, National Research Council (Washington, DC: National Academies Press, 1993).

2. United Nations, *World Urbanization Prospects: The 2003 Revision* (New York: UN, 2004).

3. World Bank, *World Development Report 2002: Building Institutions for Markets* (New York: Oxford University Press for the World Bank, 2002).

4. Nathan Keyfitz, "Impact of Trends in Resources, Environment and Development on Demographic Prospects," in *Population and Resources in a Changing World*, ed. Kingsley Davis et al. Stanford, CA: Morrison Institute for Population and Resource Studies, 1989).

5. United Nations, *World Urbanization Prospects*.

6. National Research Council, *Cities Transformed: Demographic Change and Its Implications in the Developing World*, ed. Mark R. Montgomery et al., Panel on Urban Population Dynamics, Committee on Population, Commission on Behavioral and Social Sciences and Education, National Research Council (Washington, DC: National Academies Press, 2003).

7. United Nations, *World Urbanization Prospects:* 193.

8. Martin Brockerhoff, "Fertility and Family Planning in African Cities: The Impact of Female Migration," *Journal of Biosocial Science* 27, no. 3 (1995): 347–58; and Robert Gardner and Richard Blackburn, "People Who Move: New Reproductive Health Focus," *Population Reports* Vol. 24, no. 3 (Baltimore, MD: Johns Hopkins School of Public Health, Population Information Program, November 1996).

9. Estimates calculated from 90 Demographic and Health Surveys as reported in National Research Council, *Cities Transformed: Demographic Change and Its Implications in the Developing World*.

10. Jyoti K. Parikh et al., Indira Gandhi Institute of Development Research, "Consumption Patterns: The Driving Force of Environmental Stress" (presented at the United Nations Conference on Environment and Development, August 1991).

11. Jeffrey R. Taylor and Karen A. Hardee, *Consumer Demand in China: A Statistical Factbook* (Boulder, CO: Westview Press, 1986): 112.

12. Taylor and Hardee, *Consumer Demand in China:* 148.

13. U.S. Census Bureau, *Statistical Abstract of the United States: 2003* (Washington, DC: Government Printing Office, 2003).

14. Taylor and Hardee, *Consumer Demand in China:* 125.

15. Gretchen Kolsrud and Barbara Boyle Torrey, "The Importance of Population Growth in Future Commercial Energy Consumption," in *Global Climate Change: Linking Energy, Environment, Economy and Equity,* ed. James C. White (New York: Plenum Press, 1992): 127–42.

16. Andrew S. Goudie, *The Human Impact on the Natural Environment*, 2d ed. (Cambridge, MA: MIT Press, 1987): 263.

17. Goudie, *The Human Impact on the Natural Environment:* 265.

18. Kolsrud and Torrey, "The Importance of Population Growth in Future Commercial Energy Consumption": 268.

19. Martin Brockerhoff and Ellen Brennan, "The Poverty of Cities in Developing Regions," *Population and Development Review* 24, no. 1 (March 1998): 75–114.

20. Eugene Linden, "The Exploding Cities of the Developing World," *Foreign Affairs* 75, no. 1 (1996): 52–65.

21. Organisation of Economic Co-operation and Development (OECD), *Better Understanding Our Cities, The Role of Urban Indicators* (Paris: OECD, 1997).

22. Ismail Serageldin, Richard Barrett, and Joan Martin-Brown, "The Business of Sustainable Cities," *Environmentally Sustainable Development Proceedings Series*, no. 7 (Washington, DC: The World Bank, 1994).

23. Serageldin, Barrett, and Martin-Brown, "The Business of Sustainable Cities": 33.

Gordon McGranahan and
David Satterthwaite

 **NO**

Urban Centers: An Assessment of Sustainability

Introduction

At its core, the concept of sustainable development is about reconciling "development" and "environment." Development, i.e., the meeting of people's needs, requires use of resources and implies generation of wastes. The environment has finite limits to the use of many resources and on the capacity of ecosystems to absorb or breakdown wastes or render them harmless at local, regional, and global scales. If development implies extending to all current and future populations the levels of resource use and waste generation that are the norm among middle-income groups in high-income nations, it is likely to conflict with local or global systems with finite resources and capacities to assimilate wastes. There is good evidence that such conflicts are occurring in more and more localities and also that the richest localities have overcome local constraints by drawing on the resources and sinks of other localities to the point where some resources and ecological processes are threatened both in these localities and globally (1). This implies that sustainable development requires a commitment to ensuring everyone's needs are met with modes of development that are less rooted in high-consumption, high-waste lifestyles.

Urban centers have particular relevance to any discussion of sustainable development for three reasons. The first is an increasingly urbanized world; today close to half the world's population live in urban centers, and the proportion is likely to continue growing as an increasing amount of the world's economic activities concentrate in urban centers (2). As described below, a significant proportion of the world's population with unmet needs lives in urban areas. The second reason is that urban centers concentrate most of the world's economic activities, including most industrial production, and thus are the sites that concentrate most demands for the natural resources used in such production and the sites that generate most industrial wastes. The third is that much of the world's middle- and upper-income groups live and work in urban centers, and it is their demands for goods and services that underpin most of the (rural and urban) resource demands and waste outputs from production worldwide. As a result of urban centers' concentration of middle- and upper-income groups and of the world's nonagricultural production, inevitably they are also the sites for the generation of a high proportion of greenhouse gases. In addition, many of the greenhouse gases generated outside urban centers are linked to urban-based demands—as in the greenhouse gases generated by fossil-fuel power stations, oil wells, and farms that are outside urban boundaries but where the electricity, oil, and farm products are destined for urban producers or consumers. Thus, the quality of environmental management within urban centers, which include measures to increase resource use efficiency and reduce waste generation, heavily influences not only the quality of life for the urban population but also populations that may be far outside of urban areas. As a result, the scale of resource use and waste generation arising from production and consumption located in urban centers has major implications for broad ecological sustainability. Thus, the key ecological issue for urban centers is not sustainable cities but cities and smaller urban centers that have production systems and inhabitants with patterns of consumption that are compatible with sustainable development within their region (encompassing both rural and urban areas) and globally.

An Urbanizing World

By 2003, the world's urban population reached 3 billion people, around the same size as the world's total population in 1960 (3). Close to half of the world's population now lives in urban centers, compared to less than 15% in 1900 (4). During the twentieth century, the world's urban population increased more than tenfold, and many aspects of urban change over the last fifty years are unprecedented. These include not only the world's level of urbanization and the size of its urban population but also the number of countries becoming more urbanized and the size and number of very large cities. Many cities have had populations that grew ten to twentyfold in the last 30 years (2). Most of the world's largest cities are now in Asia, not in Europe or North America. . . .

Most of the world's urban population is now outside Europe and North America. Asia alone contains almost half the world's urban population, even if more than three fifths of its people still live in rural areas. Africa, which is generally perceived as overwhelmingly rural, now has close to two fifths of its population in urban areas and a larger total urban population than North America. The urban population of Africa, Asia, Latin America, and the Caribbean is now nearly three times the size of the urban population of the rest of the world. United Nations' (U.N.) projections suggest that urban populations are growing so much faster than rural populations that 85% of the growth in the world's population between 2000 and 2010 will be in urban areas, and virtually all of this growth will be in Africa, Asia, and Latin America (3). . . .

Ecological Footprints of Cities

Consideration of the use of resources and sinks by producers and consumers located in any urban center must also take into account the ecological impacts of this use in distant regions. In general, the larger and wealthier the urban center, the larger the area from which resources are drawn. Many cities now draw on the freshwater resources of distant ecosystems, as their demand for freshwater has long exceeded local capacities and often destroyed local capacities by overexploiting groundwater and polluting surface water (7). Dakar, the capital of Senegal, now needs to draw water from a lake 200 km away (29); Mexico City has to supplement local supplies with water drawn from neighboring basins that has to be pumped over 1000 meters in height and drawn from up to 150 km away (30). Prosperous cities also draw heavily on the resources and waste-assimilation capacities of "distant elsewheres"—as highlighted by the concept developed by William Rees of cities' ecological footprints (31, 32). A city may perform extremely well in environmental terms in regard to having a safe, stimulating environment for its inhabitants and a very well-managed region around it (with good watershed management, many forests, and careful protection of sites of particular scientific interest), yet it can have a very high environmental impact on distant regional systems and on global systems because of the high demand from its population and businesses for goods whose fabrication and transport to that city required high resource inputs and generated high levels of pollution and greenhouse gas emissions. This is part of a more general tendency for cities to pass on environmental burdens to other people, other ecosystems, or to global systems as they become more prosperous and larger (33). The lower Fraser Valley, in which the city of Vancouver is located, has an ecological footprint that is about 20 times its actual area, to produce the food and forestry products its inhabitants and businesses use and to grow vegetation to absorb the carbon dioxide that they create (31). London's ecological footprint is estimated to be 125 times its actual size, based on similar criteria (34). Small relatively poor urban centers generally have ecological footprints that are relatively small and very local. Most of the largest more wealthy cities draw resources from huge areas; many extend beyond their nations' borders and have high levels of greenhouse gas emissions as discussed in detail below. It can be misleading to compare the ecological footprints of cities in that there are large disparities within any city population in regard to a household's ecological footprint. For instance, the poorest 20 percent of Canada's population have average ecological footprints that are less than a quarter of those of the wealthiest 20 percent of the population (32). The disparities between the ecological footprints of high-income and low-income households in cities in low- and middle-income nations is likely to be much larger than this.

Nonrenewable Resources and Sinks

There are also dramatic contrasts between high-income nations and low-income nations in regard to the use of nonrenewable resources and sinks. Most high-income nations have levels of commercial energy consumption per person that are 20 to 30 times that in many low-income nations (35). Comparable contrasts exist between per capita consumption in rich and poor nations for most nonrenewable resources. There are fewer figures comparing city populations' nonrenewable resource consumption, but those that do exist also show very large differences. Gasoline use per capita in cities, such as Houston, Detroit, and Los Angeles, with among the world's highest consumption levels, are 100 times or more those of most cities in low-income nations (36). Average waste generation per inhabitant in urban centers can vary more than 20-fold when comparing urban citizens in high-income nations (who may generate 1000 kg or more of waste per person per year) to those in some of the lowest-income nations (who may generate less than 50 kg per person per year) (5, 6). The disparities in terms of the amount of nonrenewable resources thrown away in the garbage (especially metals) are much higher because of the higher proportion of metals discarded in the household wastes in cities in high-income nations. Indeed, many low-income urban dwellers in Africa and Asia hardly throw away any nonrenewable resources because they cannot afford goods made from such resources; many do the opposite (in effect they create nonrenewable resources) because their livelihoods are based on finding and reclaiming items from waste streams for reuse or recycling (5).

This comparison of nonrenewable resource consumption between nations or between cities can be misleading in that it is the middle- and upper-income groups that account for most resource use and most generation of household wastes. This becomes a high-income:low-income country issue (or a North: South issue) because most of the world's middle- and upper-income people with high consumption lifestyles live in Europe, North America, and Japan, and because international politics reflect this imbalance. High-income households in cities such as Lagos, Sao Paulo, and Bangkok may have levels of nonrenewable resource use comparable to high-income households in Los Angeles or Houston; it is primarily the fact that there are fewer of them within the city population that keeps city averages much lower. . . .

What Sustainable Development Implies for Urban Authorities

A commitment to sustainable development by city or municipal authorities means adding new goals to those that are their traditional concerns. Meeting development goals has long been among their main responsibilities. These generally include attracting new investment, better social conditions (and fewer social problems), ensuring basic services and adequate housing, and (more recently) better environmental standards within their jurisdiction. This does not imply that city and municipal authorities need be major providers of housing and basic services, but they can act as supervisors and/or supporters of private or community provision.

A concern for sustainable development retains these conventional concerns and adds two more. The first is a concern for the environmental impact of urban-based production and consumption on the needs of all people, not just those within their jurisdiction. The second is an understanding of the finite nature of many natural resources (or the ecosystems from which they are drawn) and of the capacities of natural systems in the wider regional, national, and international context to absorb or break down wastes. Historically, these have not been considered within the purview of city authorities. Indeed, many cities in high-income nations have only made considerable progress in achieving sustainable development goals within their own boundaries (i.e., reducing poverty, ensuring high quality living environments, protecting local ecosystems, and developing more representative and accountable government) by drawing heavily on the environmental capital of other regions or nations and on the waste absorption capacity of "the global commons" (31, 33). But in the long term, all cities will suffer if the aggregate impact of all cities' production and their inhabitants' consumption draws on global resources at unsustainable rates and deposits wastes in global sinks at levels that disrupt the functioning of ecosystems and global cycles. A large proportion of the world's urban population are particularly at risk from the growth in the number and intensity of extreme weather events that are likely to occur if greenhouse gas emissions are not controlled (42).

Adding a concern for ecological sustainability onto existing development concerns means setting limits on the rights of city enterprises or consumers to use scarce resources (wherever they come from) and to generate nonbiodegradable wastes. Such limits can be implemented through local authorities' guidelines and regulations in planning and regulating the built environment, e.g., building material production, construction, building design and performance, site and settlement planning, and efficiency standards for appliances and fixtures. Goals relating to local or global ecological sustainability can be incorporated into the norms, codes, and regulations that influence the built environment and its heating or cooling. But city authorities need national guidelines and often national encouragement. In most political systems, national governments have the primary role in developing guidelines and supporting innovation allied to regional or global conventions or guidelines where international agreement is reached on setting such limits. . . .

Implementing Sustainable Development in Cities

The regulations and incentives needed to support the achievement of development goals, within a framework that promotes local and global ecological sustainability, is relatively easy to conceive as an abstract exercise. The poverty suffered by the minority of urban dwellers in richer nations and the majority in poorer nations can be drastically reduced without a large expansion in resource use (and waste generation). The economic and ecological costs of providing safe and sufficient water supplies, provision for sanitation, garbage removal, and health care and ensuring safe, secure shelters are often overstated. The quality of life of wealthy (generally high-consumption) individuals and households need not diminish and in certain aspects may indeed improve within a long-term program to cut their draw on the world's natural capital (44).

But translating this into reality within nations and globally is another issue. Powerful vested interests oppose most of the needed policies and priorities. For instance, reducing the resource use and greenhouse gas emissions of middle- and upper-income groups implies less profits for many companies and their politically influential coalitions, and the fact that it may also mean more profits for as yet unformed coalitions is largely irrelevant politically. Economic vested interests are particularly strong in the richest nations, but it is in these nations where changes are also most needed, for at least three reasons:

1. These are the nations with the highest current and historic contributions to greenhouse gases.
2. These nations have no moral basis for demanding more resource conserving (less greenhouse gas emitting) patterns of development among lower income nations unless they (and their wealthiest citizens) set an example of how to combine high quality lives with much lower resource use and waste generation.
3. Low-income nations that need to develop a stronger and more prosperous economic base will generally need to increase their greenhouse gas emissions, and only by reducing emissions per person in the richer nations can this be possible, within a commitment to restrict global greenhouse gas emissions.

Large cuts in greenhouse gas emissions within the richest countries will bring higher costs, especially to those who currently consume most. Technological change can help limit the rise in costs. Examples include moderating the impact of rising gasoline prices through the relatively rapid introduction of increasingly fuel-efficient automobiles and moderating the impact of higher electricity prices (especially where these are generated by fossil-fueled power stations) through more efficient electrical appliances and better designed and managed buildings that restrict the need for space heating or cooling. Many industries can also limit the impact of higher fossil fuel prices or water scarcity by increasing the efficiency with which fuel or water is used. In addition, a steady increase in the price of resources increases the economic incentives to replace them with renewable sources or improvements in efficiency (and technological change helps reduce the cost of tapping alternative resources). The scope for using renewable energy resources for space heating and cooling is also much increased in energy-efficient buildings. But if combating atmospheric warming does demand a rapid reduction in greenhouse gas emissions in high-income nations, this may require limitations in middle- and upper-income groups' right to use inefficient private automobiles, have unlimited air travel, and use inefficient space heating or cooling, which cannot be met by new technologies and alternative (renewable) fuels—at least at costs which will prove politically possible. In addition, so many existing commercial, industrial, and residential buildings and urban forms (for instance low density suburban developments and out-of-town shopping malls) have high levels of energy use built into them, and these are not easily or rapidly changed.

At the same time, in Africa, Asia, and Latin America, the achievement of development goals that minimize the call on local and global natural capital demands a competence and capacity to act by city and municipal government that is currently rarely present. There is a widely shared recognition that too little attention has been paid by most governments in low-income nations and by most international agencies to developing the competence, capacity, and accountability of urban governments (5, 6). As noted by Stren, one of the most experienced specialists on issues of urban governance, in a review of African cities: "ultimately, solutions to problems of urban finance, housing, public transport, the siting and standards of urban infrastructure, public health and public cleansing services, water, electricity and numerous other urban amenities must be formulated locally, by local people, on the basis of local experience and information" (55). It is difficult to see much success in the achievement of both the development and the ecological sustainability components of sustainable development in urban areas in low- and middle-income nations without more competent, effective, accountable local governments. Local governments cannot take on these roles without a stronger financial base, the support of national government, and an appropriate legislative, regulatory, and incentive structure (5, 6). There is also the need for mechanisms to allow resource transfers between local governments otherwise only local governments in more prosperous areas will have the resources to address development and sustainability goals.

There is also a recognition that the capacity of local government to work in partnership with community organizations, nongovernment organizations, nonprofit foundations, and private sector enterprises is central to the achievement of development goals, and this is especially true when economic circumstances limit the investment capacity of local government. This stress on the importance of such partnerships is evident in Agenda 21 that came out of the U.N. Earth Summit in 1992 and also in the recommendations that came out of the 2002 U.N. Summit on Sustainable Development. A stress on enablement at local level is to provide the support and advice that will encourage community initiatives and multiply many-fold the number that start and succeed. Such policy directions imply the need for new kinds of enabling institutions widely distributed within each nation to provide funding and technical advice (56). . . .

References

1. Rees WE. 1995. Achieving sustainability: reform or transformation? *J. Plan. Lit.* 9(4):343–61

2. Satterthwaite D. 2002. *Coping With Rapid Urban Growth*. London: R. Inst. Chart. Surv. pp. 35

3. UN Popul. Div. 2002. *World Urbanization Prospects: The 2001 Revision*. New York: UN Dep. Econ. Soc. Aff., ST/ESA/SER.A/216. pp. 321

4. Graumann JV. 1977. Orders of magnitude of the world's urban and rural population in history. *UN Popul. Bull.* 8:16–33

5. UN Cent. Hum. Settl. (Habitat). 1996. *An Urbanizing World: Global Report on Human Settlements, 1996*. Oxford: Oxford Univ. Press. pp. 559

6. Hardoy JE, Mitlin D, Satterthwaite D. 2001. *Environmental Problems in an Urbanizing World: Finding Solutions for Cities in Africa, Asia and Latin America*. London: Earthscan. pp. 470

7. UN Habitat. 2003. *Water and Sanitation in the World's Cities*. London: Earthscan. pp. 274

8. Garza G. 2002. *Urbanisation of Mexico during the twentieth century*. Urban Change Work. Pap. 7, Int. Inst. Environ. Dev., London

9. UN Popul. Div. 1975. *Trends and Prospects in the Population of Urban Agglomerations as assessed in 1973–75*. ESA/P/WP.58 New York: UN Dep. Int. Econ. Soc. Aff.

10. Brown L. 1974. *In the Human Interest*. New York: Norton

11. Potts D. 2001. *Urban growth and urban economies in eastern and southern Africa: an overview*. Presented at Workshop Afr. Urban Econ.: Viability Vitality Vitiation Major Cities East South. Afr., Nov. Neth.

12. Sassen S. 2002. Locating cities on global circuits. *Environ. Urban.* 14(1):13–30

13. World Bank. 2003. *Global Development Finance 2003*. Table A 19, Stat. Annex, p. 198. Washington, DC: World Bank

14. World Comm. Environ. Dev. 1987. *Our Common Future*. Oxford: Oxford Univ. Press. pp. 383

15. Newman P. 1996. Reducing automobile dependence *Environ. Urban.* 8(1):67–92

16. World Health Organ: 1999. Creating healthy cities in the 21st century. Chapter 6. In *The Earthscan Reader on Sustainable Cities,* ed. D Satterthwaite, pp. 137–72. London: Earthscan

17. Wratten E. 1995. Conceptualizing urban poverty. *Environ. Urban.* 7(1):11–36

18. World Health Organ. 1992. *Our planet our health*. Rep. Comm. Health Environ. Geneva pp. 282

19. Hardoy JE, Satterthwaite D. 1989. *Squatter Citizen: Life in the Urban Third World*. London: Earthscan. pp. 388

20. Satterthwaite D, Taneja B. 2003. *Agriculture and urban development*. Presented at World Bank, Washington, DC, 28 pp. http://www.worldbank.org/urban/urbanruralseminar/

21. Hardoy JE, Satterthwaite D. 1988. Small and intermediate urban centres in the Third World; what role for government? *Third World Plan. Rev.* 10(1):5–26

22. Tacoli T. 1998. *Bridging the divide: rural-urban interactions and livelihood strategies*. Gatekeeper Ser. 77. Int. Inst. Environ. Dev., Sustain. Agric. Rural Livelihoods Programme, London. pp. 17

23. Eaton D, Hilhorst T. 2003. Opportunities for managing solid waste flows in the peri-urban interface of Bamako and Ouagadougou. *Environ. Urban.* 15(1):53–64

24. Smit J, Ratta A, Nasr J. 1996. *Urban Agriculture: Food Jobs and Sustainable Cities*. Publ. Ser. Habitat II, Vol. 1. New York: UN Dev. Programme. pp. 302

25. Mayers J, Bass S. 1999. *Policy That Works for Forests and People*. London: Int. Inst. Environ. Dev. pp. 324

26. Rietbergen S. 1989. Africa. In *No Timber Without Trees*, ed. D Poore, pp. 40–73. London: Earthscan

27. Bhatt CP. 1990. The Chipko Andolan: forest conservation based on people's power. *Environ. Urban.* 2(1):7–18

28. Leach G, Mearns R. 1989. *Beyond the Woodfuel Crisis: People, Land and Trees in Africa*. London: Earthscan. pp. 309

29. White R. 1992. The international transfer of urban technology: Does the North have anything to offer for the global environmental crisis? *Environ. Urban.* 4(2):109–20

30. Connolly P. 1999. Mexico City: our common future? *Environ. Urban.* 11(1):53–78

31. Rees WE. 1992. Ecological footprints and appropriated carrying capacity: what urban economics leaves out. *Environ. Urban.* 4(2):121–30

32. Wackernagel M, Rees WE. 1995. *Our Ecological Footprint: Reducing Human Impact on the Earth*. Gabriola Isl., Can.: New Society. pp. 176

33. McGranahan G, Jacobi P, Songsore J, Surjadi C, Kjellén M. 2001. *Citizens at Risk: From Urban Sanitation to Sustainable Cities*. London: Earthscan

34. Jopling J, Girardet H. 1996. *Creating a Sustainable London*. London: Sustain. London Trust. pp. 45

35. World Bank. 1999. *Entering the 21st Century: World Development Report 1999/2000*. Oxford/New York: Oxford Univ. Press. pp. 300

36. Newman P, Kenworthy J. 1999. *Sustainability and Cities: Overcoming Automobile Dependence*. Washington, DC: Island. pp. 442

37. World Resourc. Inst. 2000. *World Resources 2000–2001: People and Ecosystems: the Fraying Web of Life*. Table AC. 1. Washington, DC: World Resourc. Inst.

38. Nishioka S, Noriguchi Y, Yamamura S. 1990. Megalopolis and climate change: the case of Tokyo. In *Cities and Global Climate Change*, ed. J McCulloch, pp. 108–33. Washington, DC: Clim. Inst.

39. Agarwal A, Narain S, Sen S, eds. 1999. *State of India's Environment: The Citizens' Fifth Report*. New Delhi: Cent. Sci. Environ. pp. 300

40. UN Environ. Programme. 1991. *Environmental Data Report, 1991–2*, GEMS Monit. Assess. Res. Cent. Oxford: Blackwell. pp. 408

41. Satterthwaite D. 1997. Sustainable cities or cities that contribute to sustainable development. *Urban Stud.* 34(10): 1667–91

42. Scott M, Gupta S, Jáuregui E, Nwafor J, Satterthwaite D, et al. 2001. Human settlements, energy and industry. In *Climate Change 2001; Impacts, Adaptation, and Vulnerability*, ed. JJ McCarthy, OF Canziani, NA Leary, DJ Dokken, KS White, pp. 381–416. Cambridge: Cambridge Univ. Press

43. Haughton G. 1998. Environmental justice and the sustainable city. *J. Plan. Edu. Res.* 18(3):233–43

44. Von Weizsäcker E, Lovins AB, Lovins LH. 1997. *Factor Four: Doubling Wealth, Halving Resource Use*. London: Earthscan. pp. 322

45. Velasquez LS. 1998. Agenda 21; a form of joint environmental management in Manizales, Colombia. *Environ. Urban.* 10(2):9–36

46. Miranda L, Hordijk M. 1998. Let us build cities for life: the national campaign of local Agenda 21s in Peru. *Environ. Urban.* 10(2):69–102

47. Menegat R. 2002. Participatory democracy and sustainable development: integrated urban environmental management in Porto Alegre, Brazil. *Environ. Urban.* 14(2): 181–206

48. Rabinovitch J. 1992. Curitiba: towards sustainable urban development. *Environ. Urban.* 4(2):62–77

49. Gleick PH. 2003. Water use. *Annu. Rev. Environ. Resour* 28:275–314

50. McGranahan G, Satterthwaite D. 2000. Environmental health or ecological sustainability? Reconciling the brown and green agendas in urban development. In *Sustainable Cities in Developing Countries*, ed. Cedric Pugh, pp. 73–90. London: Earthscan

51. McGranahan G. 2002. *Demand-Side Water Strategies and the Urban Poor*. PIE Ser. 4 London: Int. Inst. Environ. Dev. pp. 67

52. Hasan A. 1997. *Working with Government: The Story of the Orangi Pilot Project's Collaboration with State Agencies for Replicating its Low Cost Sanitation Programme*. Karachi: City, Press. pp. 269

53. Mangin W. 1967. Latin American squatter settlements; a problem and a solution. *Latin Am. Res. Rev.* 2(3):65–98

54. Turner JFC. 1969. Uncontrolled urban settlements: problems and policies. In *The City in Newly Developed Countries*, ed. G Breese, pp. 507–34. New Jersey: Prentice Hall

55. Stren RE. 1989. Administration of urban services. In *African Cities in Crisis*, ed. RE Stren, RR White, pp. 37–67. Boulder Westview.

56. Satterthwaite D. 2001. Reducing urban poverty: constraints on the effectiveness of aid agencies and development banks and some suggestions for change. *Environ. Urban.* 13(1):137–57

POSTSCRIPT

Is Rapid Urbanization in the Developing World a Major Problem?

It appears self-evident that rapidly growing urbanized areas, particularly in the developing world, create special circumstances. Our visual image of such places accents this fact. First-world travelers to the developing world are likely to take away from that experience a litany of pictures that paint a bleak image of life there. The critical question, though, is whether such environments really do create major problems for those who live within such areas and policy makers who must provide goods and services. Conventional wisdom that such problems must exist is found everywhere throughout the urbanization literature. We easily could have selected any one of a dozen or more articles that asserted with much conviction.

Perhaps, however, those problems observed by urban visitors might, in fact, be a consequence of some other situation unrelated to urbanization. Moreover, others allude to the advantages of urbanization for modernization. These individuals tend to bring a historical perspective to their analysis, but therein lies a potential fatal flaw. Urbanization occurred first in the current developed world, where cities grew more slowly and where governments and the capacity and the inclination to provide a better quality of life for urban dwellers.

Torrey's article carries the force of its publisher, the prestigious Population Reference Bureau, well-respected for its balanced and scholarly approach to population analysis. Her short article is well documented and presents a convincing case for the adverse consequences of rapid urbanization in the developing world. While acknowledging that more research is needed, she nonetheless puts forth a strong case.

McGranahan and Satterthwaite do acknowledge that ecologically more sustainable patterns of urban development are needed. While they admit the existence of certain environmental impacts of urbanization, they are more concerned about how cities will handle these impacts in the future rather than concentrating on present deficiencies.

The most definitive source of information about urbanization is found in the United Nations' *World Urbanization Prospects: The 2003 Revision* (U.N., 2004). Another important work is the National Research Council's *Cities Transformed: Demographic Change and Its Implications in the Developing World* (National Academies Press, 2003). The United Nations Human Settlements Programme has presented a comprehensive picture of the poorer sections of the planet's urban centers in *The Challenge of Slums: Global Report on Human Settlements* (2003). An interesting approach to the same subject is found in Mike Davis' "Planet of Slums" (*Harper's,*

June 2004). The Population Reference Bureau has produced a comprehensive monogram called "An Urbanizing World" by Martin P. Brockerhoff (*Population Bulletin*, September 2004). A sophisticated analysis is found in a 2003 article in *Sociological Perspectives* by John M. Shandra, Bruce London, and John B. Williamson ("Environmental Degradation, Environmental Sustainability, and Overurbanization in the Developing World: A Quantitative, Cross-National Analysis," 2003). The *UN Chronicle* compares urban changes in different continents for the years 1985, 2000 and 2015 in "Our Urban World" (vol. 38, issue I, March–May 2001).

An excellent account of the effect of urbanization on habitat loss and species extinction is found in Michael L. McKinney's "Urbanization, Biodiversity, and Conservation" (*Bioscience*, vol. 52, issue 10, October 2002). The effect of urbanization on the health of children is well summarized in M. Gracey's "Child Health in an Urbanizing World" (*Acta Paediatrica*, vol. 91, issue 1, January 2002). The role of urbanization in allergies is described in G. D'Amato's "Environmental Urban Factors" (*Allergy*, vol. 57, issue 6, June 2002).

Is Global Aging in the Developed World a Major Problem?

YES: The Center for Strategic and International Studies (CSIS), from *Meeting the Challenge of Global Aging: A Report to World Leaders from the CSIS Commission on Global Aging* (CSIS Press, 2002)

NO: Phillip Longman, from "The Global Baby Bust," *Foreign Affairs* (May/June 2004)

ISSUE SUMMARY

YES: The CSIS report, *Meeting the Challenge of Global Aging: A Report to World Leaders from the CSIS Commission on Global Aging,* suggests that the wide range of changes brought on by global aging pose significant challenges to the ability of countries to address problems associated with the elderly directly and those related to the national economy as a whole.

NO: Philip Longman takes a more balanced perspective, suggesting that what he calls the coming "baby bust" will yield a variety of positive consequences as well as negative ones.

While the poorer nations of the world have endured the consequences of over 30 years of huge birthrates, the richer sectors of the globe have witnessed the opposite phenomenon for even a longer period. The demographic transition began far earlier in the then newly industrial countries of Western Europe and the United States. And for almost half a century, these nations have witnessed the third and final stage of this transition—low birth and death rates. The drop in death rates in these countries has been a function of several factors. One major reason has been the dramatic decline in both infant mortality (within the first year of birth) and child mortality (within the first five years) due to women being healthier during pregnancy and nursing periods, and to the virtually universal inoculation of children against principal childhood diseases. A second factor is that once people reach adulthood, they are living longer, in large part because of medical advances against key adult illnesses, such as cancer and heart disease.

Declining mortality rates yield an aging population in need of a variety of services—heath care, housing, and guards against inflation, for example—provided, in large part, by the tax dollars of the younger, producing sector of society. As the "gray" numbers of society grow, the labor force is increasingly called upon to provide more help for this class. Moreover, the decline of infant and child mortality suggests that a potentially larger group could proceed to adulthood and become part of the producing class, with the potential to produce more children who will continue the cycle. This possible outcome is counteracted, however, by declining birth rates throughout the developed world.

Declining birth and death rates in developed countries means that significantly more services will be needed to provide for the aging populations of these countries, while at the same time, fewer individuals will be joining the workforce to provide the resources to pay for these services. However, some experts say that the new workforce will be able to take advantage of the skills of the more aged, unlike previous eras. In order for national economies to grow in the information age, an expanding workforce may not be as an important prerequisite as it once was. Expanding minds, not bodies, may be the key to expanding economies and increased abilities to provide public services.

However, the elderly and the young are not randomly distributed throughout society, which is likely to create a growing set of regional problems. In the United States, for example, the educated young are likely to leave the "gray belt" of the north for the Sun Belt of the south, southwest, and west. Who will be left in the older, established sectors of the country that were originally at the forefront of the industrial age to care for the disproportionately elderly population? Peter G. Peterson introduces the phrase "the Floridization of the developed world," where retirees continue to flock in unprecedented numbers, in order to capture the essence of the problems associated with the changing age composition in industrial societies. What will happen 30 and 40 years from now, when the respective sizes of the young and the elderly populations throughout the developed world will yield a much larger population at the twilight of their existence? Although the trend is most evident in the richer part of the globe, people are also living longer in the developing world, primarily because of the diffusion of modern medical practices. But unless society can accommodate their skills of later years, they may become an even bigger burden in the future for their national governments.

In the first selection, The CSIS Commission on Global Aging suggest that the inevitable changing demographics of a much higher percentage of population in the developed world falling into the elderly category will create "significant challenges" to society to maintain and improve the quality of life for seniors as well as for the entire population of the developed world. In the second section, Phillip Longman first describes the genesis of this transition, then turns to an assessment of its likely consequences. For him, there are substantial positive possibilities as well as negative consequences. And with respect to the latter, he offers potential solutions to these emerging adverse effects.

YES ⬅

The Center for Strategic and International Studies (CSIS)

Foreword

$\mathbf{T}$ he advent of modern old age represents one of the great achievements of the democratic, market economies of the developed world. Advances in medicine and public health have led to a substantial increase in life spans, producing nearly a threefold increase in over-65 populations among the OECD countries since 1950. Meanwhile, pension income from public and private sources now provides the kind of income security in old age that was only a dream a century ago.

Although these advances have greatly improved the quality of life for the elderly, demographic developments beginning later this decade and continuing for the next 50 years will create significant challenges to this achievement that must be addressed in a timely manner. The surge of baby-boomer retirements and further gains in longevity, combined with continued below-replacement fertility rates, will lead to a significant increase in old-age dependency ratios in all developed nations. The portion of elderly in the populations of developed countries is expected to double from current levels, even as working populations grow more slowly or decline. This, in turn, will bring substantial fiscal pressures and, absent offsetting trends, is likely to constrain economic growth. In sum, a demographic transition is occurring in all the developed nations, albeit at varying degrees from one nation to another, making it a global aging phenomenon—one that requires the attention of world leaders, the academic and business communities, electorates, and individuals.

This challenge prompted the Center for Strategic and International Studies in 1999 to launch its two-year Global Aging Initiative and convene the Commission on Global Aging. The commission is a panel of 85 leading voices in politics, government, business, academia, and the nongovernmental sector from three continents.

Following more than a year of intensive research and discussion, CSIS surveyed the commission members during the spring of 2001 to determine their views on the fiscal, economic, financial, and international political challenges presented by global aging. Many were contacted by telephone and otherwise for more extensive input. The responses of 73 of these commissioners are presented and analyzed in this report.

From *Meeting the Challenge of Global Aging* by Ryutaro Hashimoto, Walter F. Mondale, and Karl Otto Pohl, pp. vii–xi. Copyright © 2002 by Center for Strategic and International Studies. Reprinted by permission.

Summary and Highlights

With a high degree of consensus, the commission found that the changes wrought by global aging are fundamental and unprecedented. They pose significant challenges to the ability of nations to sustain current benefits for the elderly as well as to sustain economic growth rates and the recent historic rise in living standards.

The commission found that pessimism is not warranted if the nations most directly affected act promptly to manage their aging transitions. The commission also found, however, that postponing reforms will lead to greater sacrifice later on. Rising dependency ratios will require industrial nations to examine carefully their pay-as-you-go old-age guarantees. In some cases these guarantees are not sustainable in their present form at current tax rates and will require significant reforms this decade. In other cases, the challenges facing pay-as-you-go pensions are less daunting if addressed in a timely manner.

The commission called on countries to undertake a multifaceted program of policy reforms to social protection schemes, private pensions, labor law, financial services, family policy, immigration, civil society, and international diplomacy in order to manage the aging transition and ensure sustainable old-age pension and health care systems. Although there are many ways that nations can prepare for the aging transition, no single policy reform by itself will be sufficient.

The growing interdependence of the major economies with one another, and with the larger world economy, means that the interests at stake transcend national boundaries. Although retirement policies must reflect unique national issues such as demographics and culture, the commission recommended that nations develop a framework of international consultation and monitoring. This will enable nations to share best practices and, where appropriate, coordinate policies.

Several major themes emerged from the commission's deliberations:

- **The fiscal challenge.** The commission was unanimous in its finding that demographic trends throughout the developed world pose a significant challenge to the sustainability of current social insurance pension and health guarantees for the aged. This remains true in spite of more than a decade of reforms around the world. The commission found that this fiscal challenge could make it difficult for nations to afford other important spending priorities, including infrastructure, defense, and education. Further, it found that indebted nations with large social insurance guarantees would be vulnerable to economic shocks that would reduce revenue and lead to large, potentially destabilizing budget deficits. Such instability could adversely affect the global economy.
- **The economic challenge.** The commission found that a rising portion of the retired populations in the developed world combined with slow growing—or, in several cases, sharply declining—numbers of labor force entrants will lead to sluggish economic growth and slower gains in standards of living. These forces may be compounded as total populations decline in Japan and most European Union member states. To the extent that nations can find ways to boost capital and labor productivity, they

will be able to partially offset this effect. This can be done by, among other things, labor law reform and economic and sectoral restructuring.

- **Aged-society industry as a boost for economy.** Population aging will result in large-scale sectoral shifts that could provide a stimulus to economic growth. The commission found that although some economic sectors, such as real estate, may decline owing to population aging, the expansion of aged-society-industry sectors will provide substantial growth opportunities. These sectors include not only medical and nursing care but also supply of the products and services to meet the needs of the aged, including barrier-free and universal-design products, housework assistance, and leisure activities.

- **Prefunding of public pensions.** The commission called on nations to undertake a gradual transition to market financing of public pension systems, provided that such reforms retain a public guarantee of an adequate retirement income beginning at a specified eligibility age. Such financing could take the form of the investment of government trust funds in equities and other private assets or the creation of individual accounts. The course any country takes would depend on its unique political and technical circumstances. Because funded systems are not directly affected by changes in the old-age dependency ratio, commissioners view funded systems as more sustainable in the long term. Funded systems also can help to insulate retirement security from adverse national economic trends through cross-border diversification. However, funded systems are not completely immune to demographic vulnerabilities. In particular, the commission found that the trend to dissaving after 2020 could reduce equity return in some countries. Funded pension systems also are vulnerable to fluctuations in economic performance.

- **Funded supplement to pay-as-you-go systems.** In some nations, pay-as-you-go systems are generous enough to sustain preretirement living standards without the need for saving. The commission recommended gradual reductions in benefit levels in these nations and in other nations where it is deemed appropriate owing to fiscal pressures and high and rising government debt. To make up for those benefit reductions, the commissioners also recommend that nations adopt supplemental funded pension systems designed to attract widespread participation. The combination of pay-as-you go and funded pensions should reduce risk by diversifying retiree income sources.

- **Older populations are becoming fitter.** The commissioners found that advances in medicine have brought—and will continue to bring—longer, healthier, more productive lives with declining rates of disability for the elderly. Further, the commission found that some countries may be significantly underestimating future gains in longevity. To the extent that people who have reached pension eligibility age are willing to continue working, current and future medical advances could enable more of the elderly to work past normal retirement age. Increasing full-time and part-time opportunities for older people who wish to work offers one of the best potential prescriptions to the challenges presented by population aging. The commissioners also unanimously agreed that increased labor participation by the elderly could produce significant social benefit.

- **Lengthening work lives.** To alleviate what is expected to be a severe tightening of labor markets in coming years, the commission recommended that nations enact reforms designed to reduce programmatic incentives for early retirement and to make it possible for those who wish to work after the statutory retirement age to do so. The commission also endorses raising statutory retirement ages under social security systems in correspondence with increases in longevity. Some commissioners, particularly in the United States where the retirement age is already relatively high, strongly believe that raising the eligibility age may not be necessary, particularly if prudent actions are taken soon to ensure long-term sustainability of social security systems. Thus, although the commission endorsed raising eligibility ages, the application of this measure would need to take into account the unique circumstances of each country. The private sector should likewise prepare for a world where employers will need older workers.
- **Opportunities for women.** Nations where the labor force participation rates of women are low may want to pursue policies that allow more women who wish to do so to participate in the labor force. Employers are encouraged to provide more flexible work opportunities that make it easier for women to both work and have a family. The commission strongly endorses the concept of phased retirement together with rules that forbid age and gender discrimination.
- **Financial markets.** The commission found that a rise in the old-age dependency ratio may depress saving rates and the value of equities. Demand for public sector borrowing to meet pension obligations could increase at the same time that large retired populations are spending down their retirement savings and also when defined pension plans are paying out more in benefits than they are receiving in employer contributions. The combined effect may put upward pressure on interest rates and downward pressure on equity and other financial asset values. To help ensure adequate global savings and stable financial markets after 2020, the commission recommended that nations without employer-sponsored pension schemes establish them. It also recommended that developed nations encourage the establishment of funded social insurance schemes in both the developed and developing worlds. The commission recommended that funded pensions be managed in accord with prudent-expert fiduciary standards and a minimum of cross-border investment restrictions and that financial markets be well regulated and transparent.
- **The role of globalization.** The commission found that economic integration with the "youthful" developing world could produce several benefits for developed and developing societies. It could enable the developed nations to restructure their economies around high-value-added activities, thereby raising labor productivity. It also could allow pension fund managers and individuals to invest a portion of their retirement assets in the developing world, both enhancing investment returns and spurring global growth. Finally, immigration from the developing world could increase the labor supply and support economic specialization. But it should also reflect the needs of the developing countries. In support of this objective, the commission recommended that the developed nations take steps to promote tolerance of nonnative

races or cultures, and to make it easier for nonnative residents to achieve citizenship or permanent residency.

- **Mechanisms for international consultation.** The commission recommended that the challenges of global aging be a permanent item on the agenda of the Group of Eight nations. They also recommend that a mechanism be created for senior-level consultations among officials and experts around the world on an ongoing basis to better develop and coordinate beneficial cooperative strategies. These consultations should include the emerging market economies of the developing world.

In conclusion, we wish to thank our fellow commissioners and the Global Aging Initiative staff of the Center for Strategic and International Studies for their many contributions to the analyses and ideas reflected in this broad and far-reaching report. Like all important research, our work raised as many questions as it has answered. Inevitably, these uncertainties gave rise to disagreement among the commissioners. Yet the breadth and depth of our consensus reflects the reality that global aging will usher in a profoundly new era in which everything from the way we plan our lives to the role of nations in the global system will change.

We commend this report to the world's leaders and the international public in the hope that you will join us in meeting the fundamental challenge of global aging.

Ryutaro Hashimoto
Former Prime Minister of Japan

Walter F. Mondale
Former Vice President of the United States

Karl Otto Pöhl
Former President of Deutsche Bundesbank

January 7, 2002

 **NO**

The Global Baby Bust

The Wrong Reading

You awaken to news of a morning traffic jam. Leaving home early for a doctor's appointment, you nonetheless arrive too late to find parking. After waiting two hours for a 15-minute consultation, you wait again to have your prescription filled. All the while, you worry about the work you've missed because so many other people would line up to take your job. Returning home to the evening news, you watch throngs of youths throwing stones somewhere in the Middle East, and a feature on disappearing farmland in the Midwest. A telemarketer calls for the third time, telling you, "We need your help to save the rain forest." As you set the alarm clock for the morning, one neighbor's car alarm goes off and another's air conditioner starts to whine.

So goes a day in the life of an average American. It is thus hardly surprising that many Americans think overpopulation is one of the world's most pressing problems. To be sure, the typical Westerner enjoys an unprecedented amount of private space. Compared to their parents, most now live in larger homes occupied by fewer children. They drive ever-larger automobiles, in which they can eat, smoke, or listen to the radio in splendid isolation. Food is so abundant that obesity has become a leading cause of death.

Still, both day-to-day experience and the media frequently suggest that the quality of life enjoyed in the United States and Europe is under threat by population growth. Sprawling suburban development is making traffic worse, driving taxes up, and reducing opportunities to enjoy nature. Televised images of developing-world famine, war, and environmental degradation prompt some to wonder, "Why do these people have so many kids?" Immigrants and other people's children wind up competing for jobs, access to health care, parking spaces, favorite fishing holes, hiking paths, and spots at the beach. No wonder that, when asked how long it will take for world population to double, nearly half of all Americans say 20 years or less.

Yet a closer look at demographic trends shows that the rate of world population growth has fallen by more than 40 percent since the late 1960s. And forecasts by the UN and other organizations show that, even in the absence of major wars or pandemics, the number of human beings on the planet could well start to decline within the lifetime of today's children. Demographers at

From *Foreign Affairs* by Phillip Longman, vol. 83, no, 3, May/June 2004, pp. 64–79. Copyright © 2004 by Foreign Affairs. Reprinted by permission.

the International Institute for Applied Systems Analysis predict that human population will peak (at 9 billion) by 2070 and then start to contract. Long before then, many nations will shrink in absolute size, and the average age of the world's citizens will shoot up dramatically. Moreover, the populations that will age fastest are in the Middle East and other underdeveloped regions. During the remainder of this century, even sub-Saharan Africa will likely grow older than Europe is today.

Free Falling

The root cause of these trends is falling birthrates. Today, the average woman in the world bears half as many children as did her counterpart in 1972. No industrialized country still produces enough children to sustain its population over time, or to prevent rapid population aging. Germany could easily lose the equivalent of the current population of what was once East Germany over the next half-century. Russia's population is already contracting by three-quarters of a million a year. Japan's population, meanwhile, is expected to peak as early as 2005, and then to fall by as much as one-third over the next 50 years—a decline equivalent, the demographer Hideo Ibe has noted, to that experienced in medieval Europe during the plague.

Although many factors are at work, the changing economics of family life is the prime factor in discouraging childbearing. In nations rich and poor, under all forms of government, as more and more of the world's population moves to urban areas in which children offer little or no economic reward to their parents, and as women acquire economic opportunities and reproductive control, the social and financial costs of childbearing continue to rise.

In the United States, the direct cost of raising a middle-class child born this year through age 18, according to the Department of Agriculture, exceeds $200,000—not including college. And the cost in forgone wages can easily exceed $1 million, even for families with modest earning power. Meanwhile, although Social Security and private pension plans depend critically on the human capital created by parents, they offer the same benefits, and often more, to those who avoid the burdens of raising a family.

Now the developing world, as it becomes more urban and industrialized, is experiencing the same demographic transition, but at a faster pace. Today, when Americans think of Mexico, for example, they think of televised images of desperate, unemployed youths swimming the Rio Grande or slipping through border fences. Yet because Mexican fertility rates have dropped so dramatically, the country is now aging five times faster than is the United States. It took 50 years for the American median age to rise just five years, from 30 to 35. By contrast, between 2000 and 2050, Mexico's median age, according to UN projections, will increase by 20 years, leaving half the population over 42. Meanwhile, the median American age in 2050 is expected to be 39.7.

Those televised images of desperate, unemployed youth broadcast from the Middle East create a similarly misleading impression. Fertility rates are falling faster in the Middle East than anywhere else on earth, and as a result, the region's population is aging at an unprecedented rate. For example, by mid-century, Algeria will see its median age increase from 21.7 to 40, according to UN projections. Postrevolutionary Iran has seen its fertility rate plummet by nearly two-thirds and will accordingly have more seniors than children by 2030.

Countries such as France and Japan at least got a chance to grow rich before they grew old. Today, most developing countries are growing old before they get rich. China's low fertility means that its labor force will start shrinking by 2020, and 30 percent of China's population could be over 60 by mid-century. More worrisome, China's social security system, which covers only a fraction of the population, already has debts exceeding 145 percent of its GDP. Making demographics there even worse, the spreading use of ultrasound and other techniques for determining the sex of fetuses is, as in India and many other parts of the world, leading to much higher abortion rates for females than for males. In China, the ratio of male to female births is now 117 to 100—which implies that roughly one out of six males in today's new generation will not succeed in reproducing.

All told, some 59 countries, comprising roughly 44 percent of the world's total population, are currently not producing enough children to avoid population decline, and the phenomenon continues to spread. By 2045, according to the latest UN projections, the world's fertility rate as a whole will have fallen below replacement levels.

Repaying the Demographic Dividend

What impact will these trends have on the global economy and balance of power? Consider the positive possibilities. Slower world population growth offers many benefits, some of which have already been realized. Many economists believe, for example, that falling birthrates made possible the great economic boom that occurred in Japan and then in many other Asian nations beginning in the 1960s. As the relative number of children declined, so did the burden of their dependency, thereby freeing up more resources for investment and adult consumption. In East Asia, the working-age population grew nearly four times faster than its dependent population between 1965 and 1990, freeing up a huge reserve of female labor and other social resources that would otherwise have been committed to raising children. Similarly, China's rapid industrialization today is being aided by a dramatic decline in the relative number of dependent children.

Over the next decade, the Middle East could benefit from a similar "demographic dividend." Birthrates fell in every single Middle Eastern country during the 1990s, often dramatically. The resulting "middle aging" of the region will lower the overall dependency ratio over the next 10 to 20 years, freeing up more resources for infrastructure and industrial development. The appeal of radicalism could also diminish as young adults make up less of the population and Middle Eastern societies become increasingly dominated by middle-aged people concerned with such practical issues as health care and retirement savings. Just as population aging in the West during the 1980s was accompanied by the disappearance of youthful indigenous terrorist groups such as the Red Brigades and the Weather Underground, falling birthrates in the Middle East could well produce societies far less prone to political violence.

Declining fertility rates at first bring a "demographic dividend." That dividend has to be repaid, however, if the trend continues. Although at first the fact that there are fewer children to feed, clothe, and educate leaves more for adults to enjoy, soon enough, if fertility falls beneath replacement levels, the number of productive workers drops as well, and the number of dependent elderly increase. And these older citizens consume far more resources than children do. Even after considering the cost of education, a typical child in the United States consumes 28 percent less than the typical working-age adult, whereas elders consume 27 percent more, mostly in health-related expenses.

Largely because of this imbalance, population aging, once it begins creating more seniors than workers, puts severe strains on government budgets. In Germany, for example, public spending on pensions, even after accounting for a reduction in future benefits written into current law, is expected to swell from an already staggering 10.3 percent of GDP to 15.4 percent by 2040—even as the number of workers available to support each retiree shrinks from 2.6 to 1.4. Meanwhile, the cost of government health-care benefits for the elderly is expected to rise from today's 3.8 percent of GDP to 8.4 percent by 2040.

Population aging also depresses the growth of government revenues. Population growth is a major source of economic growth: more people create more demand for the products capitalists sell, and more supply of the labor capitalists buy. Economists may be able to construct models of how economies could grow amid a shrinking population, but in the real world, it has never happened. A nation's GDP is literally the sum of its labor force times average output per worker. Thus a decline in the number of workers implies a decline in an economy's growth potential. When the size of the work force falls, economic growth can occur only if productivity increases enough to compensate. And these increases would have to be substantial to offset the impact of aging. Italy, for example, expects its working-age population to plunge 41 percent by 2050—meaning that output per worker would have to increase by at least that amount just to keep Italy's economic growth rate from falling below zero. With a shrinking labor supply, Europe's future economic growth will therefore depend entirely on getting more out of each remaining worker (many of them unskilled, recently arrived immigrants), even as it has to tax them at higher and higher rates to pay for old-age pensions and health care. . . .

Global Aging and Global Power

Current population trends are likely to have another major impact: they will make military actions increasingly difficult for most nations. One reason for this change will be psychological. In countries where parents generally have only one or two children, every soldier becomes a "Private Ryan"—a soldier whose loss would mean overwhelming devastation to his or her family. In the later years of the Soviet Union, for example, collapsing birthrates in the Russian core meant that by 1990, the number of Russians aged 15–24 had shrunk by 5.2 million from 25 years before. Given their few sons, it is hardly surprising that Russian mothers for the first time in the nation's history organized an antiwar movement, and that Soviet society decided that its casualties in Afghanistan were unacceptable.

Another reason for the shift will be financial. Today, Americans consider the United States as the world's sole remaining superpower, which it is. As the cost of pensions and health care consume more and more of the nation's wealth, however, and as the labor force stops growing, it will become more and more difficult for Washington to sustain current levels of military spending or the number of men and women in uniform. Even within the U.S. military budget, the competition between guns and canes is already intense. The Pentagon today spends 84 cents on pensions for every dollar it spends on basic pay. Indeed, except during wartime, pensions are already one of the Pentagon's largest budget categories. In 2000, the cost of military pensions amounted to 12 times what the military spent on ammunition, nearly 5 times what the Navy spent on new ships, and more than 5 times what the Air Force spent on new planes and missiles.

Of course, the U.S. military is also more technically sophisticated than ever before, meaning that national power today is much less dependent on the ability to raise large armies. But the technologies the United States currently uses to project its power—laser-guided bombs, stealth aircraft, navigation assisted by the space-based Global Positioning System, nuclear aircraft carriers—are all products of the sort of expensive research and development that the United States will have difficulty affording if the cost of old-age entitlements continues to rise.

The same point applies to the U.S. ability to sustain, or increase, its levels of foreign aid. Although the United States faces less population aging than any other industrialized nation, the extremely high cost of its health care system, combined with its underfunded pension system, means that it still faces staggering liabilities. According to the International Monetary Fund (IMF), the imbalance between what the U.S. federal government will collect in future taxes under current law and what it has promised to pay in future benefits now exceeds 500 percent of GDP. To close that gap, the IMF warns, "would require an immediate and permanent 60 percent hike in the federal income tax yield, or a 50 percent cut in Social Security and Medicare benefits." Neither is likely. Accordingly, in another 20 years, the United States will be no more able to afford the role of world policeman than Europe or Japan can today. Nor will China be able to assume the job, since it will soon start to suffer from the kind of hyper-aging that Japan is already experiencing. . . .

POSTSCRIPT

Is Global Aging in the Developed World a Major Problem?

The issue of the changing age composition in the developed world was fore-seen a few decades ago, but its heightened visibility is relatively recent. This visibility culminated in a UN-sponsored conference on aging in Madrid in April 2002. Its plan of action commits governments to address the problem of aging and provided them with a set of 117 specific recommendations covering three basic areas: older individuals and development, advancing health and well-being into old age, and ensuring enabling and supportive environments.

With the successful demographic transition in the industrial world, the percentage of those above the age of 60 is on the rise, while the labor force percentage is decreasing. In 1998, 19 percent of the first world fell into the post-60 category (10 percent worldwide). Children under age 15 also make up 19 percent of the developed world's population while the labor force is at 62 percent. With birth rates hovering around 1 percent or less, and life expectancy increasing, the percentages will likely continue to grow toward the upper end of the scale.

The aforementioned Paul Peterson has argued that the costs of global aging will not only outweigh the benefits, the capacity of the developed world to pay for these costs is questionable at best. The economic burden on the labor force will be "unprecedented, he suggests," and offers a number of solutions ("Gray Dawn: The Global Aging Crisis," *Foreign Affairs* (January/February 1999).

A particularly outspoken opponent of the "gloom" viewpoint is Phil Mullan. His book, *The Imaginary Time Bomb; Why an Ageing Population Is Not a Social Problem* (I.B. Tauris, New York, 2000), criticizes how the idea of an aging developed world has become "a kind of mantra for opponents of the welfare state and for a collection of alarmists."

The most succinct presentation of the effects of global aging can be found in John Hawksworth's "Seven Key Effects of Global Aging" (Pricewaterhouse-Coopers' Web site, www.pwcglobal.com). This is "must reading" for those who want a concise objective description of the potential consequence of the changing demographics associated with age distribution. His presentation focuses on the effects on economic growth, pensions systems, working lives, equity and bond markets, international capital flows, migration, and business strategies.

A good source for the effect of declining population in Europe is "Population Policy Dilemmas in Europe at the Dawn of the Twenty-First Century" (*Population and Development Review*, The Population Council, Inc., March 2003).

For some, such as Leon F. Bouvier and Jane T. Bertrand (*World Population: Challenges for the 21st Century*, Seven Locks Press, 1999), there seems to be a potential silver lining on the horizon. Although future increases in immigration will counterbalance the decline of the indigenous population, they assert, the real advance will be the decoupling of productivity expansion and workforce increases. The information age is knowledge-intensive, and becoming more so, not labor-intensive.

One author who accepts Bouvier and Bertrand's thesis is the noted scholar of management, Peter Drucker. In "The Future That Has Already Happened," *The Futurist* (November 1998), Drucker predicts that retirement age in the developed world will soon rise to 75 primarily because their greatest skill, knowledge, will become even more of an asset. He maintains that knowledge resources will become the most important commodity. Another author who sees global aging in a positive manner is Lincoln H. Day (*The Future of Low-Birth-Rate Populations*, Routledge, 1992).

An important book on the fiscal problems facing the developed world because of aging can be found in Robert Stowe England's *The Fiscal Challenge of an Aging Industrial World* (CSIS *Significant Issues* Series, November 2001). An earlier report from CSIS is *Global Aging: The Challenge of the New Millennium (CSIS and Watson Wyatt Worldwide)*. This document's presentation of the raw data is particularly useful.

The Center for Strategic and International Studies (CSIS) in Washington is at the forefront of research and policy advocacy on the issue of global aging in the developed world. Its Global Aging Initiative (GAI) explores the international economic, financial, political, and security implications of aging and depopulation. One can find numerous speeches and other short papers on the issue on its Web site, including the most extensive bibliography on the effects of global aging in the developed world (www.csis.org).

Other Web sites include LinkAge 2000: Policy Implications of Global Aging (library.thinkquest.org/), The International Center on Global Aging (www.globalaging.org/resources), and The Environmental Literacy Council (www.enviroliteracy.org).

Finally, the Second World Assembly on Aging in Madrid in April 2002 produced a large number of documents. Its Plan of Action and other reports can be found at www.un.org/ageing/coverage.

UNEP World Conservation Monitoring Centre

The United Nations Environment Programme s World Conservation Monitoring Centre Web site contains information on conservation and sustainable use of the globe s natural resources. The center provides information to policymakers concerning global trends in conservation, biodiversity, loss of species and habitats, and more. This site includes a list of publications and environmental links.

http://www.unep-wcmc.org

The International Institute for Sustainable Development

This nonprofit organization based in Canada provides a number of reporting services on a range of environmental issues, with special emphasis on policy initiatives associated with sustainable development.

http://www.iisd.ca

The Hunger Project

The Hunger Project is a nonprofit organization that seeks to end global hunger. This organization asserts that society-wide actions are needed to eliminate hunger and that global security depends on ensuring that everyone's basic needs are fulfilled. Included on this site is an outline of principles that guide the organization, information on why ending hunger is so important, and a list of programs sponsored by the Hunger Project in 11 developing countries across South Asia, Latin America, and Africa.

http://www.thp.org

Global Warming Central

The Global Warming Central Web site provides information on the global warming debate. Links to the best global warming debate sites as well as key documents and reports are included. Explore the recent news section to find the latest articles on the subject.

http://www.law.pace.edu/env/energy/
globalwarming.html

International Association for Environmental Hydrology

The International Association for Environmental Hydrology (IAEH) is a worldwide association of environmental hydrologists dedicated to the protection and cleanup of freshwater resources. The IAEH's mission is to provide a place to share technical information and exchange ideas and to provide a source of inexpensive tools for the environmental hydrologist, especially hydrologists and water resource engineers in developing countries.

http://www.hydroweb.com

PART 2

Global Resources and the Environment

*T*he availability of resources and the manner in which the planet's inhabitants use them characterize another major component of the global agenda. Many believe that environmentalists overstate their case because of ideology, not science. Many others state that renewable resources are being consumed at a pace that is too fast to allow for replenishment, while non-renewable resources are being consumed at a pace that is faster than our ability to find suitable replacements.

The production, distribution, and consumption of these resources also leave their marks on the planet. A basic set of issues relates to whether these impacts are permanent, too degrading to the planet, too damaging to one's quality of life, or simply beyond a threshold of acceptability.

- Do Environmentalists Overstate Their Case?

- Is Environmental Degradation Worsening?

- Should the World Continue to Rely on Oil as a Major Source of Energy?

- Will the World Be Able to Feed Itself in the Foreseeable Future?

- Is the Threat of Global Warming Real?

- Is the Threat of a Global Water Shortage Real?

ISSUE 4

Do Environmentalists Overstate Their Case?

YES: Ronald Bailey, from "Debunking Green Myths," *Reason* (February 2002)

NO: David Pimentel, from "Skeptical of the Skeptical Environmentalist," *Skeptic* (vol. 9, no. 2, 2002)

ISSUE SUMMARY

YES: Environmental journalist Ronald Bailey in his review of the Lomborg book, argues in the subtitle of his critique that "An environmentalist gets it right," suggesting that finally someone has taken the environmental doomsdayers to task for their shoddy use of science.

NO: Bioscientist David Pimentel takes to task a controversial book by Bjørn Lomborg, *The Skeptical Environmentalist: Measuring the Real State of the World* (Cambridge University Press, 2001).

For a few decades, those skeptics of the claims of many environmentalists that the world was in danger of ecological collapse and in the not so distant future, looked to Julian Simon for guidance. And Simon did not disappoint, as he constantly questioned these researchers' motives and methodology—their models, data, and data analysis techniques. Two seminal works, *The Ultimate Resource* and *The Ultimate Resource 2*, in particular, attempted to demonstrate that much research was really bad science. With his death in 1998, such criticism lost its focus, or at least a central rallying cry.

Bjørn Lomborg, a young Danish political scientist, changed all of that with the 2001 publication of his The *Skeptical Environmentalist: Measuring the Real State of the World* (a take-off on the annual State of the World Series produced by Lester Brown and the Worldwatch Institute). Lomborg's central thesis is that statistical analyses of principal environmental indicators reveal that environmental problems have been overstated by most in the environmental movement.

What distinguished Simon's body of work that earned him the unofficial title of "doomslayer" from Lomborg's book was that the latter received much greater and more widespread attention, both by the popular media and by those in academic and scientific circles. In effect, it has become the most popular anti-environmental book ever, prompting a huge backlash by those vested in the scientific community. Because the popular press appeared to accept Lomborg's assertions with an uncritical eye, the scientific community began a comprehensive counter-attack against *The Skeptical Environmentalist. Scientific American*, in January 2002, published almost a dozen pages of critiques of the book by four experts and concluded that the book's purpose of showing the real state of the world was a failure.

The attention paid Lomborg's book by *Scientific American* was typical of the responses found in every far corner of the scientific community. Not only was Lomborg's analyses attacked, but his credentials were as well. Researchers scurried to discredit both him and his work, with a passion unseen heretofore in the debate over the potential for global environmental catastrophe. The Danish Committees on Scientific Dishonesty was called upon to investigate the work. The Danish Ministry of Science, Technology and Innovation found serious flaws in Lomborg's critique. One reviewer concluded by observing that he wished he could find that the book had some scientific merit but he could not. The BBC (British Broadcasting Company) devoted a three-part series to Lomborg's claims. One critique was titled "No Friend of the Earth."

These examples illustrate the debate put forth in this issue—namely, do environmentalists overstate the case for environmental decay and potential catastrophe? Ronald Bailey, probably the unofficial successor to Julian Simon's role as principal critic of environmentalist ideology, provides one of the few positive critiques of *The Skeptical Environmentalist.* His initial statement places the genesis of modern environmentalism in the radical movements of the 1960s, suggesting that their aim is to demonstrate that "the world is going to hell in a handbasket." Calling environmentalism an ideology, Bailey argues that like Marxists, environmentalists "have had to force the facts to fit their ideology." In sum, he suggests that the book deals a major blow to environmentalist ideology "by superbly documenting a response to environmental doomsaying." David Pimentel, a professor of insect ecology and agricultural sciences, argues that those who contend that the environment is not threatened are using data selectively while ignoring much evidence to the contrary.

Ronald Bailey

Debunking Green Myths: An Environmentalist Gets It Right

Modern environmentalism, born of the radical movements of the 1960s, has often made recourse to science to press its claims that the world is going to hell in a handbasket. But this environmentalism has never really been a matter of objectively describing the world and calling for the particular social policies that the description implies.

Environmentalism is an ideology, very much like Marxism, which pretended to base its social critique on a "scientific" theory of economic relations. Like Marxists, environmentalists have had to force the facts to fit their theory. Environmentalism is an ideology in crisis: The massive, accumulating contradictions between its pretensions and the actual state of the world can no longer be easily explained away.

The publication of *The Skeptical Environmentalist,* a magnificent and important book by a former member of Greenpeace, deals a major blow to that ideology by superbly documenting a response to environmental doomsaying. The author, Bjorn Lomborg, is an associate professor of statistics at the University of Aarhus in Denmark. On a trip to the United States a few years ago, Lomborg picked up a copy of Wired that included an article about the late "doomslayer" Julian Simon.

Simon, a professor of business administration at the University of Maryland, claimed that by most measures, the lot of humanity is improving and the world's natural environment was not critically imperiled. Lomborg, thinking it would be an amusing and instructive exercise to debunk a "right-wing" anti-environmentalist American, assigned his students the project of finding the "real" data that would contradict Simon's outrageous claims.

Lomborg and his students discovered that Simon was essentially right, and that the most famous environmental alarmists (Stanford biologist Paul Ehrlich, Worldwatch Institute founder Lester Brown, former Vice President Al Gore, Silent Spring author Rachel Carson) and the leading environmentalist lobbying groups (Greenpeace, the World Wildlife Fund, Friends of the Earth) were wrong. It turns out that the natural environment is in good shape, and the prospects of humanity are actually quite good.

Lomborg begins with "the Litany" of environmentalist doom, writing: "We are all familiar with the Litany. . . . Our resources are running out. The

From *Reason* by Ronald Bailey, February 2002, pp. 396–403, 406–409, 416–420. Copyright © 2002 by Reason Foundation. Reprinted by permission.

population is ever growing, leaving less and less to eat. The air and water are becoming ever more polluted. The planet's species are becoming extinct in vast numbers. . . . The world's ecosystem is breaking down. . . . We all know the Litany and have heard it so often that yet another repetition is, well, almost reassuring." Lomborg notes that there is just one problem with the Litany: "It does not seem to be backed up by the available evidence."

Lomborg then proceeds to demolish the Litany. He shows how, time and again, ideological environmentalists misuse, distort, and ignore the vast reams of data that contradict their dour visions. In the course of The Skeptical Environmentalist, Lomborg demonstrates that the environmentalist lobby is just that, a collection of interest groups that must hype doom in order to survive monetarily and politically.

Lomborg notes, "As the industry and farming organizations have an obvious interest in portraying the environment as just-fine and no-need-to-do-anything, the environmental organizations also have a clear interest in telling us that the environment is in a bad state, and that we need to act now. And the worse they can make this state appear, the easier it is for them to convince us we need to spend more money on the environment rather than on hospitals, kindergartens, etc. Of course, if we were equally skeptical of both sorts of organization there would be less of a problem. But since we tend to treat environmental organizations with much less skepticism, this might cause a grave bias in our understanding of the state of the world." Lomborg's book amply shows that our understanding of the state of the world is indeed biased.

So what is the real state of humanity and the planet?

Human life expectancy in the developing world has more than doubled in the past century, from 31 years to 65. Since 1960, the average amount of food per person in the developing countries has increased by 38 percent, and although world population has doubled, the percentage of malnourished poor people has fallen globally from 35 percent to 18 percent, and will likely fall further over the next decade, to 12 percent. In real terms, food costs a third of what it did in the 1960s. Lomborg points out that increasing food production trends show no sign of slackening in the future.

What about air pollution? Completely uncontroversial data show that concentrations of sulfur dioxide are down 80 percent in the U.S. since 1962, carbon monoxide levels are down 75 percent since 1970, nitrogen oxides are down 38 percent since 1975, and ground level ozone is down 30 percent since 1977. These trends are mirrored in all developed countries.

Lomborg shows that claims of rapid deforestation are vastly exaggerated. One United Nations Food and Agriculture survey found that globally, forest cover has been reduced by a minuscule 0.44 percent since 1961. The World Wildlife Fund claims that two-thirds of the world's forests have been lost since the dawn of agriculture; the reality is that the world still has 80 percent of its forests. What about the Brazilian rainforests? Eighty-six percent remain uncut, and the rate of clearing is falling. Lomborg also debunks the widely circulated claim that the world will soon lose up to half of its species. In fact, the best evidence indicates that 0.7 percent of species might be lost in the next 50 years if nothing is done. And of course, it is unlikely that nothing will be done.

Finally, Lomborg shows that global warming caused by burning fossil fuels is unlikely to be a catastrophe. Why? First, because actual measured temperatures aren't increasing nearly as fast as the computer climate models say they should be—in fact, any increase is likely to be at the low end of the predictions, and no one thinks that would be a disaster. Second, even in the unlikely event that temperatures were to increase substantially, it will be far less costly and more environmentally sound to adapt to the changes rather than institute draconian cuts in fossil fuel use.

The best calculations show that adapting to global warming would cost $5 trillion over the next century. By comparison, substantially cutting back on fossil fuel emissions in the manner suggested by the Kyoto Protocol would cost between $107 and $274 trillion over the same period. (Keep in mind that the current yearly U.S. gross domestic product is $10 trillion.) Such costs would mean that people living in developing countries would lose over 75 percent of their expected increases in income over the next century. That would be not only a human tragedy, but an environmental one as well, since poor people generally have little time for environmental concerns.

Where does Lomborg fall short? He clearly understands that increasing prosperity is the key to improving human and environmental health, but he often takes for granted the institutions of property and markets that make progress and prosperity possible. His analysis, as good as it is, fails to identify the chief cause of most environmental problems. In most cases, imperiled resources such as fisheries and airsheds are in open-access commons where the incentive is for people to take as much as possible of the resource before someone else beats them to it. Since they don't own the resource, they have no incentive to protect and conserve it.

Clearly, regulation has worked to improve the state of many open-access commons in developed countries such as the U.S. Our air and streams are much cleaner than they were 30 years ago, in large part due to things like installing catalytic converters on automobiles and building more municipal sewage treatment plants. Yet there is good evidence that assigning private property rights to these resources would have resulted in a faster and cheaper cleanup. Lomborg's analysis would have been even stronger had he more directly taken on ideological environmentalism's bias against markets. But perhaps that is asking for too much in an already superb book.

"Things are better now," writes Lomborg, "but they are still not good enough." He's right. Only continued economic growth will enable the 800 million people who are still malnourished to get the food they need; only continued economic growth will let the 1.2 billion who don't have access to clean water and sanitation obtain those amenities. It turns out that ideological environmentalism, with its hostility to economic growth and technological progress, is the biggest threat to the natural environment and to the hopes of the poorest people in the world for achieving better lives.

"The very message of the book," Lomborg concludes, is that "children born today—in both the industrialized world and the developing countries—will live longer and be healthier, they will get more food, a better education, a higher standard of living, more leisure time and far more possibilities—without the global environment being destroyed. And that is a beautiful world."

NO ↵

David Pimentel

Skeptical of the Skeptical Environmentalist

Bjørn Lomborg discusses a wide range of topics in his book and implies, through his title, that he will inform readers exactly what the real state of world is. In this effort, he criticizes countless world economists, agriculturists, water specialists, and environmentalists, and furthermore, accuses them of misquoting and/or organizing published data to mislead the public concerning the status of world population, food supplies, malnutrition, disease, and pollution. Lomborg bases his optimistic opinion on his selective use of data. Some of Lomborg's assertions will be examined in this review, and where differing information is presented, extensive documentation will be provided.

Lomborg reports that "we now have more food per person than we used to."[1] In contrast, the Food and Agricultural Organization (FAO) of the United Nations reports that food per capita has been declining since 1984, based on available cereal grains.[2] Cereal grains make up about 80% of the world's food. Although grain yields per hectare (abbreviated ha) in both developed and developing countries are still increasing, these increases are slowing while the world population continues to escalate.[3] Specifically from 1950 to 1980, U.S. grains yields increased at about 3% per year, but after 1980 the rate of increase for corn and other grains has declined to only about 1%.

Obviously fertile cropland is an essential resource for the production of foods but Lomborg has chosen not to address this subject directly. Currently, the U.S. has available nearly 0.5 ha of prime cropland per capita, but it will not have this much land if the population continues to grow at its current rapid rate.[4] Worldwide the average cropland available for food production is only 0.25 ha per person.[5] Each person added to the U.S. population requires nearly 0.4 ha (1 acre) of land for urbanization and transportation.[6] One example of the impact of population growth and development is occurring in California where an average of 156,000 ha of agricultural land is being lost each year.[7] At this rate it will not be long before California ceases to be the number one state in U.S. agricultural production.

In addition to the quantity of agricultural land, soil quality and fertility is vital for food production. The productivity of the soil is reduced when it is eroded by rainfall and wind.[8] Soil erosion is not a problem, according to Lomborg,

From *Skeptic* by David Pimentel, vol. 9 no. 2, 2002, pp. 90–93. Copyright © 2002 by David Pimentel. Reprinted by permission of the author.

especially in the U.S. where soil erosion has declined during the past decade. Yes, as Lomborg states, instead of losing an average of 17 metric tons per hectare per year on cropland, the U.S. cropland is now losing an average of 13 t/ha/yr.[9] However, this average loss is 13 times the sustainability rate of soil replacement.[10] Exceptions occur, as during the 1995–96 winter in Kansas, when it was relatively dry and windy, and some agricultural lands lost as much as 65 t/ha of productive soil. This loss is 65 times the natural soil replacement in agriculture.[11]

Worldwide soil erosion is more damaging than in the United States. For instance, in India soil is being lost at 30 to 40 times its sustainability.[12] Rate of soil loss in Africa is increasing not only because of livestock overgrazing but also because of the burning of crop residues due to the shortages of wood fuel.[13] During the summer of 2000, NASA published a satellite image of a cloud of soil from Africa being blown across the Atlantic Ocean, further attesting to the massive soil erosion problem in Africa. Worldwide evidence concerning soil loss is substantiated and it is difficult to ignore its effect on sustainable agricultural production.

Contrary to Lomborg's belief, crop yields cannot continue to increase in response to the increased applications of more fertilizers and pesticides. In fact, field tests have demonstrated that applying excessive amounts of nitrogen fertilizer stresses the crop plants, resulting in declining yields.[14] The optimum amount of nitrogen for corn, one of the crops that require heavy use of nitrogen, is approximately 120 kg/ha.[15]

Although U.S. farmers frequently apply significantly more nitrogen fertilizer than 120 kg/ha, the extra is a waste and pollutant. The corn crop can only utilize about one-third of the nitrogen applied, while the remainder leaches either into the ground or surface waters.[16] This pollution of aquatic ecosystems in agricultural areas results in the high levels of nitrogen and pesticides occurring in many U.S. water bodies.[17] For example, nitrogen fertilizer has found its way into 97% of the well-water supplies in some regions, like North Carolina.[18] The concentrations of nitrate are above the U.S. Environmental Protection Agency drinking-water standard of 10 milligrams per liter (nitrogen) and are a toxic threat to young children and young livestock.[19] In the last 30 years, the nitrate content has tripled in the Gulf of Mexico,[20] where it is reducing the Gulf fishery.[21]

In an undocumented statement Lomborg reports that pesticides cause very little cancer.[22] Further, he provides no explanation as to why human and other nontarget species are not exposed to pesticides when crops are treated. There is abundant medical and scientific evidence that confirms that pesticides cause significant numbers of cancers in the U.S. and throughout the world.[23] Lomborg also neglects to report that some herbicides stimulate the production of toxic chemicals in some plants, and that these toxicants can cause cancer.[24]

In keeping with Lomborg's view that agriculture and the food supply are improving, he states that "fewer people are starving."[25] Lomborg criticizes the validity of the two World Health Organization reports that confirm more than 3 billion people are malnourished.[26] This is the largest number and proportion of malnourished people ever in history! Apparently Lomborg rejects the WHO data because they do not support his basic thesis. Instead, Lomborg argues that only people who suffer from calorie shortages are malnourished, and ignores the fact that humans die from deficiencies of protein, iron, iodine, and vitamin A, B, C, and D.[27]

Further confirming a decline in food supply, the FAO reports that there has been a three-fold decline in the consumption of fish in the human diet during the past seven years.[28] This decline in fish per capita is caused by over-fishing, pollution, and the impact of a rapidly growing world population that must share the diminishing fish supply.

In discussing the status of water supply and sanitation services, Lomborg is correct in stating that these services were improved in the developed world during the 19th century, but he ignores the available scientific data when he suggests that these trends have been "replicated in the developing world" during the 20th century. Countless reports confirm that developing countries discharge most of their untreated urban sewage directly into surface waters.[29] For example, of India's 3,119 towns and cities, only eight have full waste water treatment facilities.[30] Furthermore, 114 Indian cities dump untreated sewage and partially cremated bodies directly into the sacred Ganges River. Downstream the untreated water is used for drinking, bathing, and washing.[31] In view of the poor sanitation, it is no wonder that water borne infectious diseases account for 80% of all infections worldwide and 90% of all infections in developing countries.[32]

Contrary to Lomborg's view, most infectious diseases are increasing world-wide.[33] The increase is due not only to population growth but also because of increasing environmental pollution.[34] Food-borne infections are increasing rapidly worldwide and in the United States. For example, during 2000 in the U.S. there were 76 million human food-borne infections with 5,000 associated deaths.[35] Many of these infections are associated with the increasing contamination of food and water by livestock wastes in the United States.[36]

In addition, a large number of malnourished people are highly suscepti-ble to infectious diseases, like tuberculosis (TB), malaria, schistosomiasis, and AIDS.[37] For example, the number of people infected with tuberculosis in the U.S. and the world is escalating, in part because medicine has not kept up with the new forms of TB. Currently, according to the World Health Organi-zation,[38] more than 2 billion people in the world are infected with TB,[39] with nearly 2 million people dying each year from it.[40]

Consistent with Lomborg's thesis that world natural resources are abundant, he reports that the U.S. Energy Information Agency for the period 2000 to 2020 projects an almost steady oil price over the next two decades at about $22 per barrel. This optimistic projection was crossed late in 2000 when oil rose to $30 or more per barrel in the United States and the world.[41] The best estimates today project that world oil reserves will last approximately 50 years, based on current production rates.[42]

Lomborg takes the World Wildlife Fund (WWF) to task for their estimates on the loss of world forests during the past decade and their emphasis on resulting ecological impacts and loss of biodiversity. Whether the loss of forests is slow, as Lomborg suggests, or rapid as WWF reports, there is no question that forests are disappearing worldwide. Forests not only provide valuable products but they harbor a vast diversity of species of plants, animals and microbes. Progress in medicine, agriculture, genetic engineering, and environmental quality depend on maintaining the species diversity in the world.[43]

This reviewer takes issue with Lomborg's underlying thesis that the size and growth of the human population is not a major problem. The difference between Lomborg's figure that 76 million humans were added to the world population in 2000, or the 80 million reported by the Population Reference Bureau,[44] is not the issue, though the magnitude of both projections is of serious concern. Lomborg neglects to explain that the major problem with world population growth is the young age structure that now exists. Even if the world adopted a policy of only two children per couple tomorrow, the world population would continue to increase for more than 70 years before stabilizing at more than 12 billion people.[45] As an agricultural scientist and ecologist, I wish I could share Lomborg's optimistic views, but my investigations and those of countless scientists lead me to a more conservative outlook. The supply of basic resources, like fertile cropland, water, energy, and an unpolluted atmosphere that support human life is declining rapidly, as nearly a quarter million people are daily added to the Earth. We all desire a high standard of living for each person on Earth, but with every person added, the supply of resources must be divided and shared. Current losses and degradation of natural resources suggest concern and a need for planning for future generations of humans. Based on our current understanding of the real state of the world and environment, there is need for conservation and protection of vital world resources.

References

1. Lomborg, B. 2001. *The Skeptical Environmentalist*. Cambridge University Press, 61.
2. FAO, 1961–1999. *Quarterly Bulletin of Statistics*. Food and Agriculture Organization of the United Nations.
3. Ibid.; PRB 2000. *World Population Data Sheet*. Washington, DC: Population Reference Bureau.
4. USBC, 2000. *Statistical Abstract of the United States 2000*. Washington, DC: U.S. Bureau of the Census, U.S. Government Printing Office.
5. PRB, 2000; WRI 1994. *World Resources 1994–95*. Washington, DC: World Resources Institute.
6. Helmlich, R. 2001. Economic Research Service, USDA, Washington, DC, personal communication.
7. UCBC, 2000, op. cit.
8. Lal, R., and B. A. Stewart, 1990. *Soil Degradation*. New York: Springer-Verlag: Troeh, F. R., Hobbs, J. A., & Donahue, R. L. 1991. *Soil and Water Conservation* (2nd ed.). Englewood Cliffs, NJ: Prentice Hall.
9. USDA, 1994. *Summary Report 1992 National Resources Inventory*. Washington, DC: Soil Conservation Service, U.S. Department of Agriculture.
10. Pimentel, D., and N. Kounang, 1998. "Ecology of Soil Erosion in Ecosystems," *Ecosystems*, 1, 416–426.
11. Lal and Stewart, 1990; Troeh et al., 1991, op. cit.

12. Khoshoo, T. N. & Tejwani, K. G. 1993. "Soil Erosion and Conservation in India (status and policies)." In Pimentel, D. (ed.) *World Soil Erosion and Conservation.* pp. 109–146. Cambridge: Cambridge University Press.

13. Tolba, M. K. 1989. "Our Biological Heritage Under Siege." *BioScience,* 39: 725–728.

14. Romanova, A. K., Kuznetsova, L. G., Golovina, E. V., Novichkova, N. S., Karpilova, I. F., & Ivanov, B. N. 1987. *Proceedings of the Indian National Science Academy, B (Biological Sciences),* 53(5–6): 505–512.

15. Troeh, F. R., & Thompson, L. M. 1993. *Soils and Soil Fertility* (5th ed.). New York: Oxford University Press.

16. Robertson, G. P. 2000. "Dinitrification." In *Handbook of Soil Science.* M. E. Summer (Ed). pp. C181–190. Boca Raton, FL: CRC Press.

17. Ibid.; Mapp, H. P. 1999. "Impact of Production Changes on Income and Environmental Risk in the Southern High Plains." *Journal of Agricultural and Applied Economics,* 31(2): 263–273; Gentry, L. E., David, M. B., Smith-Starks, K. M., and Kovacics, 2000. "Nitrogen Fertilizer and Herbicide Transport from Tile Drained Fields." *Journal of Environmental Quality,* 29(1): 232–240.

18. Smith, V. H., Tilman, G. D. and Nekola, J. C. 1999. "Eutrophication: Impacts of Excess Nutrient Inputs on Freshwater, Marine, and Terrestrial Ecosystems." *Environment and Pollution,* 100(1/3): 179–196.

19. Ibid.

20. Goolsby, D. A., Battaglin, W. A., Aulenbach, B. T. and Hooper, R. P. 2000. "Nitrogen Flux and Sources in the Mississippi River Basin." *Science and the Total Environment,* 248(2–3): 75–86.

21. NAS, 2000. *Clean Coastal Waters: Understanding and Reducing the Effects of Nutrient Pollution.* Washington, DC: National Academy of Sciences Press.

22. Lomborg, 2001, op. cit., 10.

23. WHO, 1992. *Our Planet, Our Health: Report of the WHO Commission on Health and Environment.* Geneva: World Health Organization: Ferguson, L. R. 1999. "Natural and Man-Made Mutagens and Carcinogens in the Human Diet." *Mutation Research, Genetic Toxicology and Environmental Mutagenesis,* 443(1/2): 1–10; NAS, 2000. *The Future Role of Pesticides in Agriculture.* Washington, DC: National Academy of Sciences Press.

24. Culliney, T. W., Pimentel, D., & Pimentel, M. H. 1992. "Pesticides and Natural Toxicants in Foods." *Agriculture, Eco-systems and Environment,* 41, 297–320.

25. Lomborg, 2001, op. cit., 328.

26. WHO, 1996. *Micronutrient Malnutrition—Half of the World's Population Affected* (Pages 1–4 No. Press Release WHO No. 78). World Health Organization; WHO, 2000a. *Malnutrition Worldwide.* http://www.who.int/nut/malnutrition_worldwide.htm, July 27, 2000.

27. Sommer, A. and K. P. West, 1996. *Vitamin A Deficiency: Health, Survival and Vision.* New York: Oxford University Press; Tomashek, K. M., Woodruff, B. A., Gotway, C. A., Bloand, P. & Mbaruku, G. 2001. "Randomized Intervention Study Comparing Several Regimens for the Treatment of Moderate Anemia Refugee Children in Kigoma Region, Tanzania." *American Journal of Tropical Medicine and Hygiene,* 64(3/4): 164–171.

28. FAO, 1991. *Food Balance Sheets.* Rome: Food and Agriculture Organization of the United Nations; FAO, 1998. *Food Balance Sheets.* http://armanncorn: 98ivysub@faostat.fao.org/lim...ap.pl?

29. WHO, 1993. "Global Health Situation." *Weekly Epidemiological Record,* World Health Organization 68 (12 February): 43–44; Wouters, A. V. 1993. "Health Care Utilization Patterns in Developing Countries: Role of the Technology Environment in 'Deriving' the Demand for Health Care." *Boletin de la Oficina*

Sanitaria Panamericana, 115(2): 128–139; Biswas; M. R. 1999. "Nutrition, Food, and Water Security." *Food and Nutrition Bulletin*, 20(4): 454–457.

30.   WHO, 1992, op. cit.

31.   NGS, 1995, *Water: A Story of Hope*. Washington, DC: National Geographic Society.

32.   WHO, 1992, op. cit.

33.   Ibid.

34.   Pimentel, D., Tort, M., D'Anna, L., Krawic, A., Berger, J., Rossman, J., Mugo, F., Doon, N., Shriberg, M., Howard, E. S., Lee, S., & Talbot, J. 1998. "Ecology of Increasing Disease: Population Growth and Environmental Degradation." *BioScience*, 48, 817–826.

35.   Taylor, M. R. & Hoffman, S. A. 2001. "Redesigning Food Safety: Using Risk Analysis to Build a Better Food Safety System." *Resources*. Summer, 144: 13–16.

36.   DeWaal, C. S., Alderton, L., and Jacobson, M. J. 2000. *Outbreak Alert! Closing the Gaps in Our Federal Food-Safety Net*. Washington, DC: Center for Science in the Public Interest.

37.   Chandra, R. K. 1979. "Nutritional Deficiency and Susceptibility to Infection." *Bulletin of the World Health Organization*, 57(2): 167–177; Stephenson, L. S., Latham, M. C. & Ottesen, E. A. 2000a. "Global Malnutrition." *Parasitology*. 121: S5–S22; Stephenson, L. S., Latham, M. C. & Ottesen, E. A. 2000b. "Malnutrition and Parasitic Helminth Infections." *Parasitology*. S23–S38.

38.   WHO, 2001. "World Health Organization. Global Tuberculosis Control." *WHO Report 2001*. Geneva, Switzerland, WHO/CDS/TB/2001. 287 (May 30, 2001).

39.   WHO, 2000b. "World Health Organization. Tuberculosis." *WHO Fact Sheet 2000 No104*. Geneva, Switzerland, www.who.int/gtb (May 30, 2001).

40.   WHO, 2001, op. cit.

41.   BP, 2000. *British Petroleum Statistical Review of World Energy*. London: British Petroleum Corporate Communications Services; Duncan, R. C. 2001. "World Energy Production, Population Growth, and the Road to the Olduvai Gorge." *Population and Environment*, 22(5), 503–522.

42.   Youngquist, W. 1997. *Geodestinies: The Inevitable Control of Earth Resources Over Nations and Individuals*. Portland, OR: National Book Company; Duncan, 2001, op. cit.

43.   Myers, N. 1996. "The World's Forests and Their Ecosystem Services." In G. C. Dailey (Ed.), *Ecosystem Services: Their Nature and Value* (pp. 1–19 in press). Washington, DC: Island Press.

44.   PRB, 2000, op. cit.

45.   Population Action International, 1993. *Challenging the Planet: Connections Between Population and the Environment*. Washington, DC: Population Action International.

POSTSCRIPT

Do Environmentalists Overstate Their Case?

The issue of whether science or ideology is at the heart of the environmental debate is a vexing one. The issue is framed by the juxtaposition of three groups. The first are those individuals, commonly called political or environmental activists, who emerged in the late 1960s and early 1970s following the success of the early civil rights movement. Taking its inspiration from the 1962 publication of Rachel Carson's *The Silent Spring*, which exposed the dangers of the pesticide DDT, many politically active individuals found a new cause. When the book received legitimacy because of President Kennedy's order that his Science Advisory Committee address the issues raised therein, the environmental movement was underway. The second group of government policy makers were then a part of the mix, and the third group, scientists, were soon to come on board. The first global environmental conference sponsored by the United Nations was held in Stockholm in 1972 to address atmospheric pollution on Scandinavian lakes. Emerging from the conference was a commitment of the international policy-making community to put environmental issues on the new global agenda. Environmentalism was now globalized.

Since the early 1970s through a variety of forums and arenas, the issue has been on the forefront of this global agenda. As with any issue where debates focus not only on how to address problems but whether, in fact, the problems really exist in the first place, many disparate formal and informal interest groups have become involved in all aspects of the debate—from trying to make the case that a problem exists and will ultimately have dire consequences if left unsolved, to specific prescriptions for solving the issue. The intersection of science, public policy, and political activism then becomes like the center ring at a boxing match, where contenders vie for success. Objectivity clashes with passion as well-intentioned and not so well intentioned individuals attempt to influence the debate and the ultimate outcome. In many cases, the doomsdayers gain the upper hand as their commitment to change seems greater than those who urge caution until all the scientific evidence is in.

The reaction to Lomborg illustrates this point perfectly. He has become the arch villain to environmentalists. One such Web site proclaims in its headline "Something Is Rotten in Denmark" and then proceeds to "fight fire with fire" in attacking him (www.gristmagazine.com). Another source provides a variety of links to the debate fueled by *The Skeptical Environmentalist*. Google.com shows 22,100 references to the young Danish political scientist at last count. Amazon.com provides an array of related books that fall into the same genre.

In sum, one is struck by both the forcefulness with which Lomborg makes his case and the even greater passion with which the scientific community responds. While the latter may be more accurate with respect to the true state of the world, to paraphrase the essence of the debate, Lomborg does provide a valuable service by reminding us that at the heart of any meaningful prescription for effective public policy is an accurate assessment of the nature of the problem. Science, not ideology, provides the instruments for such an assessment.

One principal source that consistently sounds the alarm on environment issues is the Worldwatch Institute. Its Web site, worldwatch@worldwatch.org, yields an extraordinary amount of resources on environmental issues.

In the first few decades after environment was placed on the global agenda, Julian Simon was one of the few, and certainly the most read, critics of environmentalists for their ideological approach to environmental problems, causing them, in Simon's view, to ignore science when science yielded an answer different from the one sought by the environmentalists. His *The Ultimate Resource* and *The Ultimate Resource 2* represented two harshly critical books that sought to show how science had taken a back seat to ideology. Since his death in 1998, his role as principal vocal critic of extremists in the environmental movement has been assumed by Ronald Bailey, science correspondent for the monthly magazine *Reason*. His *Global Warming and Other Eco-Myths* (Forum, 2002) charges the environmentalists with using "False Science to Scare Us to Death," which is part of the book's subtitle. The titles of earlier books also suggest his basic message: *Earth Report 2000: Revisiting the True State of the Planet* (McGraw-Hill, 1999); *ECOSCAM: The False Prophets of Ecological Apocalypse* (St. Martin's Press, 1993); and *The True State of the Planet* (The Free Press, 1995).

ISSUE 5

Is Environmental Degradation Worsening?

YES: United Nations Environmental Programme, from *Global Environment Outlook 2000* (Earthscan, 1999)

NO: Peter Huber, from "Wealth Is Not the Enemy of the Environment," *Vital Speeches of the Day* (April 1, 2000)

ISSUE SUMMARY

YES: The United Nations Environmental Programme presents a comprehensive and gloomy assessment of the global environment at the turn of the millennium.

NO: Peter Huber, senior fellow of the Manhattan Institute, contends that humankind is saving the Earth with the technologies that the "soft greens" most passionately oppose.

This issue focuses on a global arena that has occupied much newsprint and air-time in the last 30 years, the global environment. The production, distribution, and consumption of the resources that are vital to our daily lives take place in the context of a set of constraints placed on every one of us by the environment in which we live. Its influence crosses every kind of human endeavor, from our ability to grow enough food to our capacity to enjoy clean air and water.

It is not surprising, therefore, that early in the post–World War II period the international community came to appreciate that modern technology had given the human race the ability to "move mountains" literally and figuratively, and so humans did, for better and for worse. Problems that previously were confined to a small geographical area—soot from steam engines, pollution downstream and downwind from local factories, the clearing away of a small tract of land for a village's agricultural needs, the Dust Bowl of the 1930s—now manifested themselves in larger and more pervasive ways. Because factories started erecting larger smokestacks, the air pollution that rained down on the towns adjacent to the local factories was now carried much farther downwind, even across national boundaries. Moreover, this pollution was transformed during its journey into a much more insidious monster: acid rain.

It was the effects of this last pollutant that forced the international community to consider whether or not other human activities had a similar price associated with them. In the early 1970s, the United Nations answered a plea from the Scandinavian countries to investigate the mysterious death of millions of fish in their many lakes and streams. The formal gathering, termed the 1972 Conference on the Human Environment and held in Stockholm, ushered in a new global strategy for addressing what was to become a new set of global issues.

Since Stockholm, the international community has been much more sensitive to the potentially adverse effects of modern human behavior on the Earth's natural and ecological systems. The Stockholm model has been replicated well over 20 times since 1972 in the global community's efforts to ensure that the human race survives the various onslaughts on its planet.

No serious observer now disputes whether human activity leaves its mark on the Earth's various systems. The question is whether or not the impact is within an acceptable range of tolerance and whether or not the planet can recover within a reasonable period of time. The term *Spaceship Earth* has become part of our lexicon, and its visual image, the brilliant photograph of Earth from an early NASA mission, has been forever embedded on our minds. The implication of this new addition to our vocabulary is that everyone on this planet must collectively endure the consequences of human activity across the globe. As a result, it is easy for some people to be seduced by cries of "the sky is falling, the sky is falling," uttered by doomsayers who see the demise of the planet in every expanded human endeavor. Others, far fewer in number, respond as small voices in the wilderness, accusing the doomsayers of having "Chicken Little syndrome." Between these two extremes, though, are the majority of the international community's leaders. These officials have embarked on a series of steps in each environmental area to ascertain whether or not a problem really exists and, if so, the nature and degree of the problem, what a preferred future alternative should look like, and how policymakers might successfully address the problem and achieve the desired end state.

The two selections for this issue address the general question of the state of the environment. The first source is a major report of the United Nations Environment Programme (UNEP), the official UN agency created at Stockholm in 1972 to monitor the environment. This report was prepared with the participation of over 850 individuals and more than 30 environmental institutes worldwide, as well as with every other relevant UN organization. As such, it represents the current official scorecard for the state of the planet's national and ecological systems. The picture it paints is one of grave concern. The second selection is much different in both its message and method of delivery. Peter Huber offers an alternative view of the effects of much modern human activity in a speech before the Harvard Club. To Huber, science and technology have combined to allow the human race to make a rather positive footprint on much of this planet.

Synthesis

Two over-riding trends characterize the beginning of the third millennium. First, the global human ecosystem is threatened by grave imbalances in productivity and in the distribution of goods and services. A significant proportion of humanity still lives in dire poverty, and projected trends are for an increasing divergence between those that benefit from economic and technological development, and those that do not. This unsustainable progression of extremes of wealth and poverty threatens the stability of the whole human system, and with it the global environment.

Secondly, the world is undergoing accelerating change, with internationally-coordinated environmental stewardship lagging behind economic and social development. Environmental gains from new technology and policies are being overtaken by the pace and scale of population growth and economic development. The processes of globalization that are so strongly influencing social evolution need to be directed towards resolving rather than aggravating the serious imbalances that divide the world today. All the partners involved—governments, intergovernmental organizations, the private sector, the scientific community, NGOs [non-governmental organizations] and other major groups—need to work together to resolve this complex and interacting set of economic, social and environmental challenges in the interests of a more sustainable future for the planet and human society.

While each part of the Earth's surface is endowed with its own combination of environmental attributes, each area must also contend with a unique, but interlinked, set of current and emerging problems. *GEO-2000 [Global Environmental Outlook 2000]* provides an overview of this range of issues. This synthesis provides a summary of the main conclusions of *GEO-2000*.

The State of the Environment: A Global Overview

Climate Change

In the late 1990s, annual emissions of carbon dioxide were almost four times the 1950 level and atmospheric concentrations of carbon dioxide had reached their highest level in 160 000 years. According to the Intergovernmental Panel on Climate Change, 'the balance of evidence suggests that there is a discernible human influence on global climate'. Expected results include a shifting of climatic zones, changes in species composition and the productivity of ecosystems, an increase in extreme weather events and impacts on human health.

Through the United Nations Framework Convention on Climate Change and the Kyoto Protocol, efforts are under way to start controlling and reducing greenhouse gas emissions. During the Third Conference of the Parties in Buenos Aires in 1998, a plan of action was developed on how to use the new international policy instruments such as emission trading and the Clean Development Mechanism. However, the Kyoto Protocol alone will be insufficient to stabilize carbon dioxide levels in the atmosphere.

Stratospheric Ozone Depletion

Major reductions in the production, consumption and release of ozone-depleting substances (ODS) have been, and continue to be, achieved by the Montreal Protocol and its related amendments. The abundance of ODS in the lower atmosphere peaked in about 1994 and is now slowly declining. This is expected to bring about a recovery of the ozone layer to pre-1980 levels by around 2050.

Illegal trading, still a problem, is being addressed by national governments but substantial quantities of ODS are still being smuggled across national borders. The Multilateral Fund and the Global Environment Facility are helping developing countries and countries in transition to phase out ODS. Since 1 July 1999, these countries have, for the first time, had to start meeting obligations under the Montreal Protocol.

Nitrogen Loading

We are fertilizing the Earth on a global scale through intensive agriculture, fossil fuel combustion and widespread cultivation of leguminous crops. Evidence is growing that the huge additional quantities of nitrogen being used are exacerbating acidification, causing changes in the species composition of ecosystems, raising nitrate levels in freshwater supplies above acceptable limits for human consumption and causing eutrophication in many freshwater habitats. In addition, river discharges laden with nitrogen-rich sewage and fertilizer runoff tend to stimulate algal blooms in coastal waters, which can lead to oxygen starvation and subsequent fish kills at lower depths, and reduce marine biodiversity through competition. Nitrogen emissions to the atmosphere contribute to global warming. Consensus among researchers is growing that the scale of disruption to the nitrogen cycle may have global implications comparable to those caused by disruption of the carbon cycle.

Chemical Risks

With the massive expansion in the availability and use of chemicals throughout the world, exposure to pesticides, heavy metals, small particulates and other substances poses an increasing threat to the health of humans and their environment. Pesticide use causes 3.5 to 5 million acute poisonings a year. Worldwide, 400 million tonnes of hazardous waste are generated each year. About

75 percent of pesticide use and hazardous waste generation occurs in developed countries. Despite restrictions on toxic and persistent chemicals such as DDT, PCBs and dioxin in many developed countries, they are still manufactured for export and remain widely used in developing countries. Efforts are under way to promote cleaner production, to limit the emissions and phase out the use of some persistent organic pollutants, to control waste production and trade, and improve waste management.

Disasters

The frequency and effects of natural disasters such as earthquakes, volcanic eruptions, hurricanes, fires and floods are increasing. This not only affects the lives of millions of people directly, through death, injury and economic losses, but adds to environmental problems. As just one example, in 1996–98 uncontrolled wildfires swept through forests in Brazil, Canada, China's north-eastern Inner Mongolia Autonomous Region, France, Greece, Indonesia, Italy, Mexico, Turkey, the Russian Federation and the United States. The health impacts of forest fires can be serious. Experts consider a pollution index of 100 $\mu g/m^3$ unhealthy; in Malaysia, the index reached 800 $\mu g/m^3$. The estimated health cost of forest fires to the people of Southeast Asia was US$1 400 million. Fires are also a serious threat to biodiversity, especially when protected areas are burnt. Early warning and response systems are still weak, particularly in developing countries; there is an urgent need for improved information infrastructures and increased technical response capabilities.

El Niño

Unusual weather conditions over the past two years are also attributed to the *El Niño* Southern Oscillation (ENSO). The 1997/98 *El Niño* developed more quickly and resulted in higher temperatures in the Pacific Ocean than ever recorded before. The presence of this mass of warm water dominated world climate patterns up to mid-1998, causing substantial disruption and damage in many areas, including temperate zones. Extreme rainfall and flooding, droughts and forest fires were among the major impacts. Forecasting and early warning systems, together with human, agricultural and infrastructural protection, have been substantially improved as a result of the most recent *El Niño*.

Land, Forests and Biodiversity

Forests, woodlands and grasslands are still being degraded or destroyed, marginal lands turned into deserts, and natural ecosystems reduced or fragmented, further threatening biodiversity. New evidence confirms that climate change may further aggravate soil erosion in many regions in the coming decades, and threaten food production. Deforestation continues at high rates in developing countries, mainly driven by the demand for wood products and the need for land for agriculture and other purposes. Some 65 million hectares of forest were lost between 1990 and 1995, out of a total of 3500 million hectares. An increase of 9 million hectares in the developed world only slightly offset this loss. The quality of the remaining forest is threatened by a range of pressures including acidification, fuelwood and water abstraction, and fire. Reduced or degraded habitats threaten biodiversity at gene, species and ecosystems level, hampering the provision of key products and services. The widespread introduction of exotic species is a further major cause of biodiversity loss. Most of the threatened species are land-based, with more than half occurring in forests. Freshwater and marine habitats, especially coral reefs, are also very vulnerable.

Freshwater

Rapid population growth combined with industrialization, urbanization, agricultural intensification and water-intensive lifestyles is resulting in a global water crisis. About 20 percent of the population currently lacks access to safe drinking water, while 50 percent lacks access to a safe sanitation system. Falling water tables are widespread and cause serious problems, both because they lead to water shortages and, in coastal areas, to salt intrusion. Contamination of drinking water is mostly felt in megacities, while nitrate pollution and increasing loads of heavy metals affect water quality nearly everywhere. The world supply of freshwater cannot be increased; more and more people depend on this fixed supply; and more and more of it is polluted. Water security, like food security, will become a major national and regional priority in many areas of the world in the decades to come.

Marine and Coastal Areas

Urban and industrial development, tourism, aquaculture, waste dumping and discharges into marine areas are degrading coastal areas around the world and destroying ecosystems such as wetlands, mangroves and coral reefs. Climatic changes also affect the quality of ocean water as well as sea levels. Low-lying areas, including many small islands, risk inundation. The global marine fish catch almost doubled between 1975 and 1995, and the state of the world's fisheries has now reached crisis point. About 60 percent of the world's fisheries are at or near the point at which yields decline.

Atmosphere

There is a major difference between air pollution trends in developed and developing countries. Strenuous efforts have begun to abate atmospheric pollution in many industrialized countries but urban air pollution is reaching crisis dimensions in most large cities of the developing world. Road traffic, the burning of coal and high-sulphur fuels, and forest fires are the major causes of air pollution. People in developing countries are also exposed to high levels of indoor pollutants from open fires. Some 50 percent of chronic respiratory illness is now thought to be associated with air pollution. Large areas of forest and farmland are also being degraded by acid rain.

Urban Impacts

Many environmental problems reinforce one another in small, densely-populated areas. Air pollution, garbage, hazardous wastes, noise and water contamination turn these areas into environmental hot spots. Children are the most vulnerable to the inevitable health risks. Some 30–60 percent of the urban population in low-income countries still lack adequate housing with sanitary facilities, drainage systems and piping for clean water. Continuing urbanization and industrialization, combined with a lack of resources and expertise, are increasing the severity of the problem. However, many local authorities are now joining forces to promote the concept of the sustainable city.

Policy Responses: A Global Overview

As awareness of environmental issues and their causes develops, the focus of policy questions shifts towards the policy response itself: what is being done, is it adequate and what are the alternatives? *GEO-2000* includes a unique assessment of environmental policies worldwide.

Environmental laws and institutions have been strongly developed over the past few years in almost all countries. Command and control policy via direct regulation is the most prominent policy instrument but its effectiveness depends on the manpower available, methods of implementation and control, and level of institutional coordination and policy integration. In most regions, such policies are still organized by sector but environmental planning and environmental impact assessment are becoming increasingly common everywhere.

While most regions are now trying to strengthen their institutions and regulations, some are shifting towards deregulation, increased use of economic instruments and subsidy reform, reliance on voluntary action by the private sector, and more public and NGO participation. This development is fed by the increasing complexity of environmental regulation and high control costs as well as demands from the private sector for more flexibility, self-regulation and cost-effectiveness.

GEO-2000 confirms the overall assessment of *GEO-1:* the global system of environmental management is moving in the right direction but much too slowly. Yet effective and well tried policy instruments do exist that could lead much more quickly to sustainability. If the new millennium is not to be marred by major environmental disasters, alternative policies will have to be swiftly implemented.

One of the major conclusions of the policy review concerns the implementation and effectiveness of existing policy instruments. The assessment of implementation, compliance and effectiveness of policy initiatives is complicated and plagued by gaps in data, conceptual difficulties and methodological problems.

Multilateral environmental agreements (MEAs) have proven to be powerful tools for attacking environmental problems. Each region has its own regional and sub-regional agreements, mostly relating to the common management or protection of natural resources such as water supply in river basins and transboundary air pollution. There are also many global-level agreements, including those on climate change and biodiversity that resulted from the United Nations Conference on Environment and Development, held in Rio de Janeiro, Brazil, in 1992.

In addition to the binding MEAs, there are non-binding agreements (such as *Agenda 21*) and environmental clauses or principles in wider agreements (such as regional trade treaties). A major trend in MEAs over the years has been a widening focus from issue-specific approaches (such as provisions for shared rivers) to trans-sectoral approaches (such as the Basel Convention), to globalization and to the general recognition of the linkage between environment and development. Another trend is still unfolding: the step-by-step establishment of common principles (such as the Forest Principles) in different sectors.

The *GEO-2000* review of MEAs highlights two issues:

- the effectiveness of MEAs depends strongly on the institutional arrangements, the financial and compliance mechanisms, and the enforcement systems that have been set up for them;
- it is still difficult to assess accurately the effectiveness of MEAs and nonbinding instruments because of the lack of accepted indicators.

Regional Trends

Africa

Poverty is a major cause and consequence of the environmental degradation and resource depletion that threaten the region. Major environmental challenges include deforestation, soil degradation and desertification, declining biodiversity and marine resources, water scarcity, and deteriorating water and air quality. Urbanization is an emerging issue, bringing with it the range of human health and environmental problems well known in urban areas throughout the world. Growing 'environmental debts' in many countries are a major concern because the cost of remedial action will be far greater than preventive action.

Although many African countries are implementing new national and multilateral environmental policies, their effectiveness is often low due to lack of adequate staff, expertise, funds and equipment for implementation and enforcement. Current environmental policies are mainly based on regulatory instruments but some countries have begun to consider a broader range, including economic incentives implemented through different tax systems. Although cleaner production centres have been created in a few countries, most industries have made little effort to adopt cleaner production approaches. However, some multinational corporations, large-scale mining companies and even local enterprises have recently voluntarily adopted precautionary environmental standards.

There is growing recognition that national environmental policies are more likely to be effectively implemented if they are supported by an informed and involved public. Environmental awareness and education programmes are expanding almost everywhere, while indigenous knowledge receives greater recognition and is increasingly used. Environmental information systems are still weak.

There is fairly high interest in many of the global MEAs, and several regional MEAs have been developed to support the global ones. The compliance and implementation rate is, however, quite low, mainly due to lack of funds.

Asia and the Pacific

Asia and the Pacific is the largest region and it is facing serious environmental challenges. High population densities are putting enormous stress on the environment. Continued rapid economic growth and industrialization is likely to cause further environmental damage, with the region becoming more degraded, less forested, more polluted and less ecologically diverse in the future.

The region, which has only 30 percent of the world's land area, supports 60 percent of the world population. This is leading to land degradation, especially in marginal areas, and habitat fragmentation. Increasing habitat fragmentation has depleted the wide variety of forest products that used to be an important source of food, medicine and income for indigenous people. Forest fires caused extensive damage in 1997–98.

Water supply is a serious problem. Already at least one in three Asians has no access to safe drinking water and freshwater will be the major limiting factor to producing more food in the future, especially in populous and arid areas. Energy demand is rising faster than in any other part of the world. The proportion of people living in urban centres is rising rapidly, and is focused on a few urban centres. Asia's particular style of urbanization—towards megacities—is likely to increase environmental and social stresses.

Widespread concern over pollution and natural resources has led to legislation to curb emissions and conserve natural resources. Governments have been particularly active in promoting environmental compliance and enforcement although the latter is still a problem in parts of the region. Economic incentives and disincentives are beginning to be used for environmental protection and the promotion of resource efficiency. Pollution fines are common and deposit-refund schemes are being promoted to encourage reuse and recycling. Industry groups in both low- and high-income countries are becoming increasingly sensitive to environmental concerns over industrial production. There is keen interest in ISO14 000 standards for manufacturing and in eco-labelling.

In most countries, domestic investment in environmental issues is increasing. A major thrust, particularly among developing countries, is on water supply, waste reduction and waste recycling. Environment funds have also been established in many countries and have contributed to the prominent role that NGOs now play in environmental action. Many countries are in favour of public participation, and in some this is now required by law. However, education and awareness levels amongst the public are often low, and the environmental information base in the region is weak.

Whilst there is uneven commitment to global MEAs, regional MEAs are important. They include a number of important environmental policy initiatives developed by sub-regional cooperative mechanisms.

One of the greatest challenges is to promote liberal trade yet maintain and strengthen the protection of the environment and natural resources. Some governments are now taking action to reconcile trade and environmental interests through special policies, agreements on product standards, enforcement of the Polluter Pays Principle, and the enforcement of health and sanitary standards for food exports.

Europe and Central Asia

Environmental trends reflect the political and socio-economic legacy of the region. In Western Europe, overall consumption levels have remained high but measures to curb environmental degradation have led to considerable improvements in some, though not all, environmental parameters. Sulphur dioxide emissions, for example, were reduced by more than one-half between 1980 and 1995. In the other sub-regions, recent political change has resulted in sharp though probably temporary reductions in industrial activity, reducing many environmental pressures.

A number of environmental characteristics are common to much of the region. Large areas of forests are damaged by acidification, pollution, drought and forest fires. In many European countries, as much as half the known vertebrate species are under threat and most stocks of commercially-exploited fish in the North Sea have been seriously over-fished. More than half of the large cities in Europe are overexploiting their groundwater resources. Marine and coastal areas are susceptible to damage from a variety of sources. Road transport is now the main source of urban air pollution, and overall emissions are high—Western Europe produces nearly 15 percent of global CO_2 emissions and eight of the ten countries with the highest per capita SO_2 emissions are in Central and Eastern Europe.

Regional action plans have been effective in forging policies consistent with the principles of sustainable development and in catalysing national and local action. However, some targets have yet to be met and plans in Eastern Europe and Central Asia are less advanced than elsewhere because of weak institutional capacities and the slower pace of economic restructuring and political reform.

Public participation in environmental issues is considered satisfactory in Western Europe, and there are some positive trends in Central and Eastern Europe. Many countries, however, still lack a proper legislative framework for public participation although the Convention on Access to Environmental Information and Public Participation in Environmental Decision Making signed by most of the ECE [Economic Commission for Europe] countries in 1998 should improve the situation. Access to environmental information has significantly increased with the formation of the European Environment Agency and other information resource centres in Europe. The level of support for global and regional MEAs, in terms of both ratification and compliance, is high.

There has been significant success, particularly in Western Europe, in implementing cleaner production programmes and eco-labelling. Within the European Union, green taxation and mitigating the adverse effects of subsidies are important priorities. Legislation is being adopted on entirely new subjects. Examples include the Nitrates Directive, the Habitat Directive and the Natura 2000 plan for a European Ecological Network. Implementation is, however, proving difficult.

The transition countries need to strengthen their institutional capacities, improve the enforcement of fees and fines, and build up the capacity of enterprises to introduce environmental management systems. The major challenge for the region as a whole is to integrate environmental, economic and social policies.

Latin America and the Caribbean

Two major environmental issues stand out in the region. The first is to find solutions to the problems of the urban environment—nearly three-quarters of the population are already urbanized, many in mega-cities. The air quality in most major cities threatens human health and water shortages are common. The second major issue is the depletion and destruction of forest resources, especially in the Amazon basin. Natural forest cover continues to decrease in all countries. A total of 5.8 million hectares a year was lost during 1990–95, resulting in a 3 percent total loss for the period. This is a major threat to biodiversity. More than 1,000 vertebrate species are now threatened with extinction.

The region has the largest reserves of cultivable land in the world but soil degradation is threatening much cultivated land. In addition, the environmental costs of improved farm technologies have been high. During the 1980s, Central America increased production by 32 percent but doubled its consumption of pesticides. On the plus side, many countries have substantial potential for curbing their contributions to the build-up of greenhouse gases, given the region's renewable energy sources and the potential of forest conservation and reforestation programmes to provide valuable carbon sinks.

During the past decade, concern for environmental issues has greatly increased, and many new institutions and policies have been put in place. However, these changes have apparently not yet greatly improved environmental management which continues to concentrate on sectoral issues, without integration with economic and social strategies. The lack of financing, technology, personnel and training and, in some cases, large and complex legal frameworks are the most common problems.

Most Latin American economies still rely on the growth of the export sector and on foreign capital inflows, regardless of the consequences to the environment. One feature of such policies is their failure to include environmental costs. Economic development efforts and programmes aimed at fighting poverty continue to be unrelated to environmental policy, due to poor interagency coordination and the lack of focus on a broader picture. On the industrial side, some producers have adopted ISO 14 000 standards as a means of demonstrating compliance with international rules.

An encouraging aspect is the trend towards regional collaboration, particularly on transboundary issues. For example, a Regional Response Mechanism for natural disasters has been established with telecommunication networks that link key agencies so that they can make quick assessments of damage, establish needs and mobilize resources to provide initial relief to affected communities. There is considerable interest in global and regional MEAs, and a high level of ratification. However, the level of implementing new policies to comply with these MEAs is generally low.

North America

North Americans use more energy and resources per capita than people in any other region. This causes acute problems for the environment and human health. The region has succeeded, however, in reducing many environmental impacts through stricter legislation and improved management. Whilst emissions of many air pollutants have been markedly reduced over the past 20 years, the region is the largest per capita contributor to greenhouse gases, mainly due to high energy consumption. Fuel use is high—in 1995 the average North American used more than 1600 litres of fuel a year (compared to about 330 litres in Europe). There is continuing concern about the effects of exposure to pesticides, organic pollutants and other toxic compounds. Changes to ecosystems caused by the introduction of non-indigenous species are threatening biodiversity and, in the longer term, global warming could move the ideal range for many North American forest species some 300 km to the north, undermining the utility of forest reserves established to protect particular plant and animal species. Locally, coastal and marine resources are close to depletion or are being seriously threatened.

The environmental policy scene is changing in North America. In Canada, most emphasis is on regulatory reform, federal/provincial policy harmonization and voluntary initiatives. In the United States, the impetus for introducing new types of environmental policies has increased and the country is developing market-based policies such as the use of tradeable emissions permits and agricultural subsidy reform. Voluntary policies and private sector initiatives, often in combination with civil society, are also gaining in importance. These include voluntary pollution reduction initiatives and programmes to ensure responsible management of chemical products. The region is generally active in supporting and complying with regional and global MEAs.

Public participation has been at the heart of many local resource management initiatives. Environmental policy instruments are increasingly developed in consultation with the public and the business community. Participation by NGOs and community residents is increasingly viewed as a valuable part of any environmental protection programme.

Increasing accountability and capacity to measure the performance of environmental policies is an overarching trend. Target setting, monitoring, scientific analysis and the public reporting of environmental policy performance are used to keep stakeholders involved and policies under control. Access to information has been an important incentive for industries to improve their environmental performance.

Despite the many areas where policies have made a major difference, environmental problems have not been eliminated. Economic growth has negated many of the improvements made so far and new problems—such as climate change and biodiversity loss—have emerged.

West Asia

The region is facing a number of major environmental issues, of which degradation of water and land resources is the most pressing. Groundwater resources are in a critical condition because the volumes withdrawn far exceed natural recharge rates. Unless improved water management plans are put in place, major environmental problems are likely to occur in the future.

Land degradation is a serious problem, and the region's rangelands—important for food security—are deteriorating, mainly as a result of overstocking what are essentially fragile ecosystems. Drought, mismanagement of land resources, intensification of agriculture, poor irrigation practices and uncontrolled urbanization have also contributed. Marine and coastal environments have been degraded by overfishing, pollution and habitat destruction. Industrial pollution and management of hazardous wastes also threaten socio-economic development in the region with the oil-producing countries generating two to eight times more hazardous waste per capita than the United States. Over the next decade, urbanization, industrialization, population growth, abuse of agro-chemicals, and uncontrolled fishing and hunting are expected to increase pressures on the region's fragile ecosystems and their endemic species.

The command and control approach, through legislation, is still the main environmental management tool in almost all states. However, several new initiatives, such as public awareness campaigns, have been taken to protect environmental resources and control pollution. In addition, many enterprises such as refineries, petrochemical complexes and metal smelters have begun procedures for obtaining certification under the ISO 14 000 series. Another important approach to resource conservation has been a growing interest in recycling scarce resources, particularly water. In many states on the Arabian Peninsula, municipal wastewater is subjected at least to secondary treatment, and is widely used to irrigate trees planted to green the landscape.

Success in implementing global and regional MEAs in the region is mixed and commitment to such policy tools quite weak. At a national level there has, however, been a significant increase in commitment to sustainable development, and environmental institutions have been given a higher priority and status.

Polar Regions

The Arctic and Antarctic play a significant role in the dynamics of the global environment and act as barometers of global change. Both areas are mainly affected by events occurring outside the polar regions. Stratospheric ozone depletion has resulted in high levels of ultraviolet radiation, and polar ice caps, shelves and glaciers are melting as a result of global warming. Both areas act as sinks for persistent organic pollutants, heavy metals and radioactivity, mostly originating from other parts of the world. The contaminants accumulate in food chains and pose a health hazard to polar inhabitants. Wild flora and fauna are also affected by human activities. For example, capelin stocks have collapsed twice in the Arctic since the peak catch of 3 million tonnes in 1977. In the Southern Ocean, the Patagonian toothfish is being over-fished and there is a large accidental mortality of seabirds caught up in fishing equipment. On land, wild communities have been modified by introductions of exotic species and, particularly in northern Europe, by overgrazing of domestic reindeer.

In the Arctic, the end of Cold War tensions has led to new environmental cooperation. The eight Arctic countries have adopted the Arctic Environmental Protection Strategy which includes monitoring and assessment, environmental emergencies, conservation of flora and fauna, and protection of the marine environment. Cooperation amongst groups of indigenous peoples has also been organized. The Antarctic environment benefits from the continuing commitment of Parties to the Antarctic Treaty aimed at reducing the chance of the region becoming a source of discord between states. The Treaty originally focused on mineral and living resources but this focus has now shifted towards broader environmental issues. A similar shift is expected in the Arctic, within the broader context of European environmental policies. In both polar areas, limited financial resources and political attention still constrain the development and implementation of effective policies.

Peter Huber  **NO**

Wealth Is Not the Enemy of the Environment

I thought I'd focus my remarks . . . on trying to prove to you how very green New York really is. And I don't mean Central Park. I mean the skyscrapers.

But let me begin by stepping back from the city and saying one or two contrarian things about the continent and the planet.

Begin with a little publicized environmental fact. The prevailing winds blow west to east across North America. On the continent itself, we burn enough fossil fuel to release some 1.6 billion metric tons of carbon a year into the air. Yet— as best anyone can measure these things, carbon dioxide levels drop as you move across the continent. One estimate in *Science* magazine puts America's terrestrial uptake of carbon at about 1.7 billion metric tons. What's going on?

Besides burning fossil fuels, we're building suburbs and roads and cities. Suburban sprawl is much in the news these days. Estimates again vary widely, but without doubt, a lot of what used to be farmland around cities is being trans- formed into suburb. Our cities occupy over twice as much land today as they did in 1920.

Which, on its own, would certainly be making the carbon balance even worse. If we're leveling forest to make room for sprawling cities, it has to be getting worse, right? But we aren't leveling forests. We're reforesting this continent.

U.S. forest cover reached a low of about 600 million acres in 1920. It has been rising ever since. Precisely how fast is hard to pin down. But all analyses show that America's forest cover today is somewhere between 20 million and 140 million acres higher than it was in 1920. At least 10 million acres have been reforested in the last decade alone.

Now how can we square these facts?

We burn huge amounts of fossil fuel, but atmospheric carbon levels drop. Our cities sprawl, but we have more forest. Am I making all this up? No. The facts are what they are. The interesting question is: how come?

Understanding these apparent paradoxes sharply defines, the two types of green I want contrast, hard green and soft.

Now let's work our way back to these facts from where we now sit, which is right here, in the heart of [Manhattan].

Here's something the Sierra Club and I agree on almost perfectly: the skyscraper as America's great green gift to the planet. It packs more people on to less land, which leaves more wilderness, undisturbed, in other places, where the people aren't. The city gets Wall Street, Saks, The Met, and the Times Square crowds, which leaves more fly-over country for bison and cougars. It's Saul Steinberg's celebrated *New Yorker* cover, painted green.

Conserve land and you conserve environment. Save "earth"—land itself—and you save the wilderness. There are other things to consider too, and I'll get to them. But if you're honestly interested in wilderness and environmental quality, land is the most important. When I say land, of course, I mean surface—including stream, river, lake and coastal waters, as well.

This is just basic ecology. Life on earth lives at the surface. The less humans disturb the surface—the land—the better for all other species. It's as simple as that.

That's why the Sierra Club and I basically agree about the city: though profligate in its consumption of most everything else, is very frugal with land. The one thing your average New Yorker does not occupy is 40 acres and a mule.

The original conservationists and Theodore Roosevelt [T. R.] had the focus just right. Recognize: to conserve wilderness, you have to conserve wilderness. Not trash. Not aluminum. Not zinc. Not energy. But rather forest, river, stream, coastal waters.

Today's soft say the two go hand in hand—conserve trash—stop using plastics—or conserve energy—ride a bike, not a car—and you conserve wilderness further up the line. But do you?

So how do humans destroy wilderness?

Lots of ways. We build cities and highways, of course. But in fact, they're quite secondary. Our cities, towns, suburbs, roads, and all the interstate highways currently cover under 3 percent of the U.S. land mass—under 60 million acres.

But for every acre of land we occupy ourselves, we use 6 acres for crops. Another 8 acres are designated as range—larders for our livestock. Most of the wilderness loss worldwide is to third-world agriculture.

However bucolic they may appear, farms aren't green. Endless miles of wheat are not bio-diverse prairie. Rangeland isn't pristine wilderness. Cattle denude it of native perennial grasses, to be replaced by sagebrush, mesquite, and juniper, or simply reduce it to desert. Acres of cassava are not acres of rain forest.

Shrinking agriculture is best, and we've done it.

Shrink the footprint of agriculture, and you can really save or restore a lot of wilderness. A 7 percent shrinkage in our agricultural footprint would return as much land to wilderness as wiping all our cities, suburbs, highways and roads off the map.

And shrinking agriculture is exactly what we've been doing, since about 1920. The result has been truly remarkable. Roughly 80 million more acres of North American cropland were harvested sixty years ago than are harvested today.

That explains one of the paradoxes I started with. Yes, our cities are sprawling. People are leaving the country and moving to the city. The cities themselves have negative population growth. They sprawl because they draw people off the rest of the land. We've flipped from a population that was 80 percent rural two centuries ago, to one that is 80 percent urban today. Sprawling cities absorb a bit of farmland nearby, but return a lot more land to the wilderness farther away.

Most "soft" greens—like the Sierra Club—have the right instincts about cities. They recognize that cities pack a lot of people into a small amount of space. That leaves more wilderness undisturbed.

But what they get right about cities, soft greens get wrong about almost everything else. Most of the policies soft greens prescribe are environmental disasters, because most force us to use more land.

Let me put this as bluntly as I can. From Al Gore to the Sierra Club, the green establishment strives to make the human footprint on the wilderness . . . bigger. If we do what they urge us to do with agriculture, our footprint on the land will at least double. If we do what they urge us to do with energy, it will more than double again.

We are saving the earth on this continent with the technologies that the soft greens most passionately oppose.

Until quite recently, humanity's advance meant retreat for the wilderness. The surface—land, river, and shallow coastal waters—supplied all our food, building materials, and fuel. The more we grew, the more land we seized.

In 1850, with one-tenth today's population, the United States cultivated as much cropland as we do today. Almost all of it was managed just as the modern soft would prefer: organic, pesticide-free, and genetically pure. For all that, the plows and the farmer wiped out the perennial grasses, crowded out the last ranges of the bison, eradicated the cougars and wolves, and created the dust bowl.

For much of this century, however, that process has been reversed. With genetic seed improvements, fertilizers, pesticides, and the relentless efficiencies of corporate farming, we almost trebled the land-to-food productivity of our agriculture.

Pesticides, packaging, preservatives, and refrigeration dramatically reduce the agricultural footprint, too.

One of the things that most concerned T.R. when he was pressing for forest conservation in 1905 was the future supply of lumber for railroad ties. Railroads were essential to the economy, and ties consumed enormous amounts of wood.

But that same year—1905—the railroads began coating their ties with creosote. That stopped the termites. The chemical preservative effectively slashed demand by two-thirds, by tripling their lifespan. Since then, wood preservatives and termite eradication have done far more to save forests in America than the recycling of newspapers.

Pesticides, preservatives, food irradiation, and plastic packaging all have comparable effects: they sharply reduce losses along the food chain, from farmer's field to dining room table, with commensurate reductions in agricultural sprawl.

Far more is still possible. The sun delivers an average energy density of roughly 180 watts per square meter of the United States. Wild plants currently convert about 0.35 w/m2 of that into stored energy, a dreadful 1:500 energy conversion efficiency.

Biotechnology could now double and redouble the agricultural productivity of land, in much the same way as silicon technology has doubled and redoubled the informational productivity of sand.

We've done even better with construction. A British frigate in 1790 required some 3,000 trees to build; the Exxon Valdez required none. Concrete, steel, and plastics substitute for hardwoods in our dwellings and furniture, which leaves the wood itself to the forest.

Far more substitution of this kind could be achieved. The total current U.S. wood harvest—90 percent of which is for construction, not energy—exceeds by several-fold the tonnage of all of the steel, copper, lead, nickel, zinc and other metals extracted in the United States. The raw materials for concrete are not at all scarce, and limitless amounts can be extracted from comparatively minuscule areas of land.

Soft greens endorse a few of these opportunities, but oppose most of them. They urge us to eat less meat, because letting cows eat our food before we do creates agricultural sprawl, but they oppose pesticides and preservatives, without which fungi, insects, and rodents do the cow's job instead. They urge us to improve the efficiency of motors and engines, but they oppose bio-technology, which improves the solar conversion efficiency of corn and cow. Sorts prefer natural, organic, free-range, wild—all of which require more land for less food.

The one solar technology that perennially fascinates soft greens is Pholto-voltaics (PV). Indeed, I owe my hard/soft taxonomy of green to Amory Lovins, of the Rocky Mountain Institute, who gained fame in the 1970s promoting what he called "soft" sources of energy like wind and solar over "hard" ones like uranium and coal.

Compared to a green plant, PV is certainly impressive technology. Selenium-doped silicon wafers mounted in glass or plastic can convert sunlight to useful energy better than chlorophyll in a typical leaf. About 60 times better: current PV technology can capture about 20 w/m2 in the United States.

New York, unfortunately, consumes 55 w/m2 of energy. So to power New York with PV, you have to cover every square inch of the city's horizontal surface with wafer—and then extend the PV sprawl over at least twice that area again, somewhere or other upstate.

PV technology could improve, of course. But the laws of nature and physics never allow anything approaching 100 percent capture; a reliable and economical 20 percent would be extraordinary, and would still require enormous amounts of ancillary material and space for energy storage, to keep the lights up when the sun is down.

The numbers for liquid fuels are even worse. The soft alternative here is "biomass." Shifting the prefix from "agri" to "bio" does make things sound greener from the get-go. "Tilling, not drilling. Biology, not geology. Living carbon, not dead carbon. Vegetables, not minerals." I am quoting here from the institute for local self reliance, which exhorts us to return to what it calls the "carbohydrate economy."

Which is to say: agriculture. We are to burn wood, garbage, bacterial mats, sunflower oil, buffalo gourd, peanut shells, chicken droppings, and tallow from lambs.

Farmers love the idea, which makes it doubly attractive to politicians. Green acres in your tank: natural, organic, free range, and wild. "Our goal," declare carbohydrate-economy pundits, "should be to get back to the ratio of 1920 when plant matter constituted one-third of all materials."

We know what the carbohydrate energy economy would look like: we've already lived it. In 1790, agriculture employed over 80 percent of the people; the founding fathers obtained 80 percent of their energy from wood. They had a "renewable" source of oil, too: the sperm whale. Their children and grandchildren almost fished it to extinction, until colonel Edwin Drake finally thought to mount his harpoon on a derrick in Titusville, Pennsylvania. Well into this century, vast areas of woods and forest were still being cut down, just to warm homes and cook food. The forests retreated steadily.

Any return to that way of life would mean resuming the retreat. Your car engine consumes far more calories than your muscles do.

No conceivable mix of solar, biomass, or wind technology could meet even half our current energy demand without (at the very least) doubling the human footprint on the surface of the continent. Sunlight is a thin, diffuse form of energy, and it's found nowhere but on the surface. Humanity burns the energy equivalent of about 1/7th of all the energy captured by all the green plants every year through photosynthesis. And that energy is already being used—to power the wilderness.

How did we escape from the carbohydrate economy in the first place? We began to dig. All U.S. oil wells and refineries together cover about 160,000 acres of land, or about one-fifth the area of King Ranch in southeastern Texas. All coal mines, including all strip mines, cover some two million acres, or well under 1 percent of the area occupied by U.S. crop land.

How did our energy footprint ever get so small?

Coal is dead trees—fossilized biomass. A coal mine yields something like 5000 w/m2 of land, an oil field more like 10,000. This makes them 5 to 10 thousand times more land frugal than existing PV technology, and 100 times more frugal than the highest PV or biomass efficiency even theoretically attainable.

There's no great mystery to why the numbers shake out that way: oil and coal are three-dimensional, sub-surface sources of energy; sunlight is all two-dimensional surface.

The environmental implications are enormous. For every acre occupied by dwellings and offices, we need about 14 acres of farm and range to produce the quite modest amount of energy—edible food—required by the people on that original acre. But we need a mere 0.3 acres of land to deliver the much greater quantities of energy people use as fuel.

In sum, it takes far less land to dig up energy than to grow it. That's what's so dreadfully wrong with all "renewables." By definition, renewables tap energy flow: solar and its immediate byproducts, plant growth or wind. But plant growth is either wilderness we should aim to conserve, or it's a farmer's field already under the plow, that the wilderness might otherwise reclaim.

Real estate on rooftops does offer some ecologically dead surface for solar capture, but there's nowhere near enough of it, and it isn't economical.

Technology won't ever close the gap. Biotechnology is young, which suggests great potential for improving the biomass energy economy. But conventional hard fuels are so much more concentrated forms of energy, and so much more profitable that they attract far more investment, which drives continuous innovation. Soft does improve, but hard improves even faster.

It just isn't a fair race, and it never can be. Sunlight is too thin, and it starts out spread across the surface—exactly where the wilderness is. Going after the sunlight means going after the energy that fuels the wilderness. It means over-building the wilderness itself.

Coal, oil, and gas, by contrast, represent ancient biomass already concentrated by nature, and buried in vast, three-dimensional reservoirs. From which it can now be extracted quite easily, through comparatively tiny incisions in the skin of earth. Given a choice, the wilderness will prefer the keyhole surgery every time.

A first soft green response is to suggest that renewables can be harvested interstitially, from within the sprawl that has already occurred. Mount PV arrays on existing rooftops. Process agricultural "waste" into usable fuels. Erect 400 foot-tall wind turbines amongst the rolling vineyards.

These are good rejoinders, so far as they go, but they don't go very far. Despite decades of subsidy and government promotion, "renewables" (other than conventional hydro) now generate barely 0.7 percent of our electricity. Add in the forest industry's on-site burning of waste wood, and renewables contribute perhaps 3 percent of all the energy we consume.

The economics just don't wash. Again, this isn't very surprising—it is, in fact, exactly the same reason that makes hard fuels kinder to the environment. It's just a lot cheaper to gather highly concentrated forms of energy, than dispersed ones.

A second soft rejoinder has nothing to do with the environment, really—it's a purely economic argument. We're going to run out of the stuff under the ground.

They've been saying that about energy since the '70s, just as they said it about food since the time of Thomas Malthus. But the fact is, we don't run out, and all serious economists understand why. Human ingenuity stretches resources and finds substitutes.

What terrifies the honest softs isn't that we're going to run out of fossils, but that we aren't. If we were going to run out any time soon, we wouldn't have to worry so much about global warming, would we? Running out would stop our emissions better than Al Gore possibly could.

In fact, we still have enormous reserves of fossil fuels, coal most especially. And beyond coal lie vast reservoirs of heavy oils in tar sands, which bioengineered bacteria may make economically accessible. And uranium, and more.

Overall, the price of hard energy keeps dropping, which tells us all we can or need to know about the depths. Meanwhile, the price of land on the environmentally critical surface, and the value we attach to wilderness, keep rising.

The softs' last and most emphatic rejoinder is that hard fuels just devour surface in other, more insidious ways. Uranium becomes Chernobyl. Oil destroys Prince William Sound or—by way of global warming—the whole planet. Hard agriculture's real footprint includes all the land and water contaminated by fertilizers, pesticides, or wayward genes that kill monarch butterflies.

These are valid arguments. But to win the debate they have to be pushed further than they reasonably can be. Hard agriculture and hard fuels start out with footprints at least ten times smaller than the soft alternatives, and more often hundreds or thousands of times smaller. Their secondary sprawl, from pollution and such, has to be very bad indeed to make it as bad as soft alternatives.

And soft alternatives entail secondary sprawl of their own. Technologies that use more material and more surface are generally going to end up polluting more, as well.

The bicyclist emits greenhouse gas, as he puffs along the road—a lot of it, per pound of useful payload moved. Soft agriculture has run-offs, and relies on all-natural pesticides, bred into the pest-resistant crops themselves, and designed to kill predatory insects in much the same way as the chemicals Dow or Dupont would have us spray on less hardy plants. PV's are manufactured from toxic metals. More eagles have been killed by wind turbines than were lost in the Exxon Valdez oil spill. The Audubon Society labeled Enron's proposed wind farm in the Tehachapi mountains north of L.A. the "condor Cuisinart." The carbohydrate economy is the deforestation economy: third world practices make that quite clear.

All the while, hard technology keeps getting cleaner and more efficient. And in one very important sense, it's the hard fuels that are "renewable," not the soft ones.

Much of the land used when we extract fuels from beneath the surface is restored to wilderness after the fuels have been mined. Not so with soft "renewables": the surface-intensive technologies they depend on have to stay in place permanently—that's essential to the "renewing." Only the subsurface fuels are "renewable" so far as renewing the surface of the land is concerned.

Finally, when we have to allow for offsetting environmental benefits, too, soft alternatives certainly have their share, many of them purely aesthetic.

But hard technologies have their secondary benefits as well. Returning land to wilderness, as hard technologies do, can take care of a lot of pollution. Whatever impact pesticides have, freeing up 100 million acres to be reclaimed by forest will likely protect more birds than trying to bankrupt Dow Chemical through Superfund. The most beautiful way to purify water is probably the most effective too: maintain unfarmed, unlogged watersheds.

Carbon vividly illustrates this. As I mentioned at the outset, America is apparently sinking more carbon out of the air than it is emitting into it. What's doing the sinking? In large part, regrowth of forests on land no longer farmed or logged, and faster growth of existing plants and forests, which are fertilized by nitrogen oxides and carbon dioxide "pollutants."

The total forest ecosystem in the United States holds about 52 billion metric tons of carbon. Net growth rate of 3 percent a year is enough to consume all carbon emissions of the U.S. economy, and either in forests themselves or on surrounding grasslands and farms, that is about the net growth rate we seem to have.

Our carbon-energy cycle, begins with carbon sequestered in fossil fuels, and ends with carbon sequestered in trees. The sun supplied the energy that first converted atmospheric carbon to become fossil fuel, and today again, the sun supplies the energy that takes the carbon back out of the air and into the trees. Give it geologic time, and today's new biomass will eventually end up back in the depths once again, as new coal and oil. Well sealed landfills will help advance that process.

The soft green establishment would have us believe that wealth is the enemy of the wilderness. The facts are the opposite.

Wealth limits our residential sprawl: it urbanizes us.

Wealth makes possible hard technologies, which drastically reduce our agricultural and energy footprints on the land. Finally, and perhaps most importantly, wealth is the only means yet discovered to limit human fecundity. The richer people grow, the fewer children they bear.

The city itself is a center of negative population growth.

Western industrial nations have stabilized their population growth overall. And, of course, it is the wealthy nations, not the poor, that pour their land into wilderness conservation.

Wealth is green. Poverty isn't.

I started off by saying you're sitting in the greenest spot on earth, but I misspoke. The greenest spot is just south of here. It's Wall Street.

POSTSCRIPT

Is Environmental Degradation Worsening?

Since the human race developed the capacity to move mountains, both literally and figuratively, it has done so, for good and for ill. Thus, environmental degradation, once confined to the local community, has moved into the global arena, beginning with the well-known 1972 Stockholm conference. Since then, various environmental problems have been the subject of much debate. Is there a problem? If so, what is its origin? How should it be addressed? For each environmental issue, the global community is at different stages in its fight to prevent, minimize, or eliminate some environmental condition. For example, global warming remains a controversy despite an international agreement, the Kyoto Protocol of 1997. Greater progress has been made with respect to ozone layer depletion, in that 126 countries agreed to action in 1992. Acid rain has been the focus of regional action because of its regional character. And biodiversity issues have begun to be addressed, as have desertification, deforestation, and many others.

The point is that the international community must first reach some satisfactory level of consensus on the existence and nature of a problem before it can begin to find agreeable solutions. Thus, at any given moment in time, global policymakers might be at the problem identification stage (does it exist and, if so, what are its parameters?), the solution identification stage (choosing from a range of policy options), or the policy implementation stage (action undertaken to solve the problem).

The latter circumstance makes it difficult for the thoughtful reader to make some summary judgments about where we are in addressing questions of environmental degradation. Nonetheless, the *Global Environmental Outlook 2000* report was an impressive undertaking. Admittedly, the individuals and groups who took part in the process brought to the table a concern about the environment and, presumably, a belief that it is in trouble. But given the complexity of the terrain and the distinctiveness of the issue areas, one is hard pressed to pass off the report as the work of a small group of individuals who brought their own agendas to the enterprise while ignoring paradigms that were inconsistent with their own beliefs. Accordingly, rejecting the essence of their findings, not to mention specific conclusions, requires an effort of the same magnitude.

But one need not reject the basic premise of the Huber article either. Much smaller in scope and ambition, it suggests that hard agriculture (substantial technological inputs) and hard fuels (nonrenewable sources) make an imprint "at least 10 times smaller than soft alternatives." For Huber, a major way of leaving a much more environmentally friendly planet is to continue to utilize science in the pursuit of fulfilling food and energy needs, and environmental degradation might just take care of itself.

Perhaps the basic difference between those who see changes occurring in the global environment but reject the doomsayers' assessment of their dire consequences and those who believe that "the sky is falling" can really be captured in the following question. Are the changes that all of us must face in the light of the environmental transformation of the planet really of such an extreme magnitude and of such a supreme personal sacrifice so as to constitute labeling such planetary change as degradation?

This is essentially the argument made in a recent book that has captured the attention of observers on both sides of the issue. Bjørn Lomborg, a professor of statistics at the University of Aarhus in Denmark, has published a major critique of the environmentalists, *The Skeptical Environmentalist: Measuring the Real State of the World* (Cambridge University Press, 2001). Taking a cue from the late Julian L. Simon, Lomborg uses his professional knowledge of statistics to argue that much of the doomsday rhetoric about environmental degradation is based on flawed analysis or inappropriate assumptions, and assumes a "sky is falling" mentality. He further asks whether or not the costs of solving the environmental problems are higher than those of the pollution itself. Lomborg's work has sparked a hostile reaction in print by many who counter that his set of analyses are flawed.

Marvin S. Soroos' *The Endangered Atmosphere: Preserving a Global Commons* (University of South Carolina Press, 1997) addresses the entire range of threats to the atmosphere posed by various external factors. *Global Environmental Politics*, 2d ed., by Gareth Porter and Janet Welsh Brown (Westview Press, 1996) examines environmental policy making across a wide range of issues. Lynton K. Caldwell's *International Environmental Policy: From the Twentieth to the Twenty-First Century*, 3rd ed., in collaboration with Paul S. Weiland (Duke University Press, 1996) represents the contribution of the senior-most scholar in the field of environmental policy making.

The Worldwatch Institute publishes periodic studies of the environment, including an annual *State of the World* volume, which examines the contemporary environmental scene of that year. In the Institute's September/October issue of its *World Watch* magazine, 14 authors examine the environmental impact of population growth.

The cover story in *Environment* described the link between poverty and degradation ("Poverty and Degradation," vol. 44, issue 1, January/February 2002).

The annual *World Resources* volume is another good source. Finally, Stephen Collett's "Environmental Protection and the Earth Summit: Paving the Path to Sustainable Development," in Michael T. Snarr and D. Neil Snarr, eds., 2nd ed., *Introducing Global Issues* (Lynne Rienner, 2002), presents a succinct and easily readable description of current global environmental policy-making efforts.

In addition to Web sites mentioned earlier in this volume, there are two worthy of mention. Lycos—Environmental News Service provides up-to-date accounts of late-breaking environmental news (ens.lycos.com). The Environment: A Global Challenge provided numerous articles online (library: thinkquest.org).

ISSUE 6

Should the World Continue to Rely on Oil as a Major Source of Energy?

YES: Hisham Khatib et al., from *World Energy Assessment: Energy and the Challenge of Sustainability* (United Nations Development Programme, 2002)

NO: Lester R. Brown, from "Turning on Renewable Energy," *Mother Earth News* (April/May 2004)

ISSUE SUMMARY

YES: Hisham Khatib, honorary vice chairman of the World Energy Council and a former Jordanian minister of energy and minister of planning, and his coauthors conclude that reserves of traditional commercial fuels, including oil, "will suffice for decades to come."

NO: Lester Brown, former head of the Worldwatch Institute, suggests in this adaptation from his *Plan B: Rescuing a Planet Under Stress and a Civilization in Trouble*, that there is good news about the global shift to renewable energy.

The new millennium witnessed an oil crisis almost immediately, the third major crisis in the last 30 years. (1972–72 and 1979 were the dates of earlier problems.) The crisis of 2000 manifested itself in the United States via much higher gasoline prices and in Europe via both rising prices and shortages at the pump. Both were caused by the inability of national distribution systems to adjust to the Organization of Petroleum Exporting Countries' (OPEC's) changing production levels. The 2000 panic eventually subsided, but the issue is far from resolved.

These three major fuel crises are discrete episodes in a much larger problem facing the human race, particularly the industrial world. That is, oil, the earth's current principal source of energy, is a finite resource that ultimately will be totally exhausted. And unlike earlier energy transitions, where a more attractive source invited a change (such as from wood to coal and from coal to oil), the next energy transition is being forced upon the human race in the absence of an attractive alternative. In short, we are being pushed out of our almost total reliance on oil toward a new system with a host of unknowns. What will the new system be? Will it be a single source or some combination? Will it be a more attractive source? Will the source be readily available at a reasonable price, or will a new cartel emerge that controls much of the supply? Will its production and consumption lead to major new environmental consequences? Will it require major changes to our lifestyles and standards of living? When will we be forced to jump?

Before considering new sources of fuel, other questions need to be asked. Are the calls for a viable alternative to oil premature? Are we simply running scared without cause? Did we learn the wrong lessons from the three earlier energy crises? More specifically, were these crises artificially created or a consequence of the actual physical unavailability of the energy source? Have these crises really been about running out of oil globally, or were they due to other phenomena at work, such as poor distribution planning by oil companies or the use of oil as a political weapon by oil-exporting countries?

For well over half a century now, Western oil-consuming countries have been predicting the end of oil. Using a model known as Hubbert's Curve (named after a U.S. geologist who designed it in the 1930s), policymakers have predicted that the world would run out of oil at various times; the most recent prediction is that oil will run out a couple of decades from now. Simply put, the model visualizes all known available resources and the patterns of consumption on a time line until the wells run dry. Despite such prognostication, it was not until the crisis of the early 1970s that national governments began to consider ways of both prolonging the oil system and finding a suitable replacement. Prior to that time, governments as well as the private sector encouraged energy consumption. "The more, the merrier" was an oft-heard refrain. Increases in energy consumption were associated with economic growth. After Europe recovered from the devastation of World War II, for example, every 1 percent increase in energy consumption brought a similar growth in economic output. To the extent that governments engaged in energy policymaking, it was designed solely to encourage increased production and consumption. Prices were kept low, and the energy was readily available. Policies making energy distribution systems more efficient and consumption patterns both more efficient and lowered were almost non-existent.

Today the search for an alternative to oil still continues. Nuclear energy, once thought to be the answer, may play a future role but at a reduced level. Both water power and wind power remain possibilities, as do biomass, geothermal, and solar energy. Many also believe that the developed world is about to enter the hydrogen age in order to meet future energy needs. The question before us, therefore, is whether the international community has the luxury of some time before all deposits of oil are exhausted.

The two selections for this issue suggest different answers to this last question. Lester Brown argues that renewable energy sources should and will define the energy future. He spells out the various options, suggesting the position that wind should be "the centerpiece in the new energy economy." Hisham Khatib and his colleagues do not see the world running out of a reliance on fossil fuel, most principally oil, for a long time to come.

Energy Security

The world has generally seen considerable development and progress in the past 50 years. Living standards have improved, people have become healthier and longer-lived, and science and technology have considerably enhanced human welfare. No doubt the availability of abundant and cheap sources of energy, mainly in the form of crude oil from the Middle East, contributed to these achievements. Adequate global energy supplies, for the world as a whole as well as for individual countries, are essential for sustainable development, proper functioning of the economy, and human well-being. Thus the continuous availability of energy—in the quantities and forms required by the economy and society—must be ensured and secured.

Energy security—the continuous availability of energy in varied forms, in sufficient quantities, and at reasonable prices—has several aspects. It means limited vulnerability to transient or longer disruptions of imported supplies. It also means the availability of local and imported resources to meet growing demand over time and at reasonable prices.

Beginning in the early 1970s energy security was narrowly viewed as reduced dependence on oil consumption and imports, particularly in OECD [Organization for Economic Cooperation and Development] and other major oil-importing countries. Since that time considerable changes in oil and other energy markets have altered the picture. Suppliers have increased, as have proven reserves and stocks, and prices have become flexible and transparent, dictated by market forces rather than by cartel arrangements. Global tensions and regional conflicts are lessening, and trade is flourishing and becoming freer. Suppliers have not imposed any oil sanctions since the early 1980s, nor have there been any real shortages anywhere in the world. Instead, the United Nations and other actors have applied sanctions to some oil suppliers, but without affecting world oil trade or creating shortages.

All this points to the present abundance of oil supplies. Moreover, in today's market environment energy security is a shared issue for importing and exporting countries. As much as importing countries are anxious to ensure security by having sustainable sources, exporting countries are anxious to export to ensure sustainable income (Mitchell, 1997).

However, although all these developments are very encouraging, they are no cause for complacency. New threats to energy security have emerged in recent years. Regional shortages are becoming more acute, and the possibility of insecurity of supplies—due to disruption of trade and reduction in strategic reserves, as a result of conflicts or sabotage—persists, although it is decreasing. These situations point to a need to strengthen global as well as regional and national energy security. . . . There is also a need for a strong plea, under the auspices of the World Trade Organization (WTO), to refrain from restrictions on trade in energy products on grounds of competition or differences in environmental or labour standards. . . .

Energy Adequacy

. . . Most of the world's future energy requirements, at least until the middle of the 21st century, will have to be met by fossil fuels (figure 1). Many attempts have been made to assess the global fossil fuel resource base. Table 1 shows the results of two.

In 1998 world consumption of primary energy totalled almost 355 exajoules, or 8,460 million tonnes of oil equivalent (Mtoe)—7,630 Mtoe of fossil fuels, 620 Mtoe of nuclear energy, and 210 Mtoe of hydropowcr. To this should be added around 47 exajoules (1,120 Mtoe) of biomass and other renewables, for a total of 402 exajoules (9,580 Mtoe). The huge resource base of fossil and nuclear fuels will be adequate to meet such global requirements for decades to come.

Figure 1

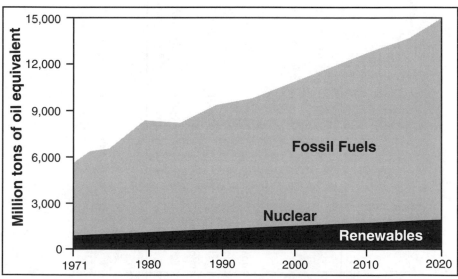

Share of Fuels in Global Energy Supply, 1971–2020

Source: IEA, 1998.

Table 1

Global Energy Resource Base (Exajoules Except Where Otherwise Indicated)

Term	World Energy Council estimates		Institute for Applied Systems Analysis estimates			Consumption
	Proven reserves	Ultimately recoverable	Reserves	Resources	Resource base	1998
Conventional oil	6,300 (150)	8,400 (200)	6,300 (150)	6,090 (145)	12,390 (295)	142.8 (3.4)
Unconventional oil	—	23,100 (550)	8,190 (195)	13,944 (332)	22,050 (525)	n.a.
Conventional gas	5,586 (133)	9,240 (220)	5,922 (141)	11,718 (279)	17,640 (420)	85 (2.0)
Unconventional gas	—	—	8,064 (192)	10,836 (258)	18,900 (450)	n.a.
Coal and lignite	18,060 (430)	142,800 (3,400)	25,452 (606)	117,348 (2,794)	142,800 (3,400)	93 (2.2)
Uranium	3.4×10^9 tonnes	17×10^9 tonnes	(57)	(203)	(260)	64,000 tonnes

— Not available; n.a. Not applicable.

Note: Numbers in parentheses are in gigatonnes of oil equivalent.

a. Because of uncertainties about the method of conversion, quantities of uranium have been left in the units reported by the sources.

Source: WEC, 1998; IIASA, 1998.

VALUING THE COST OF ELECTRICITY SUPPLY SECURITY

The cost of electricity to a consumer—the consumer's valuation of the electricity supply (ignoring consumer surplus)—equals payments for electricity consumed plus the economic (social) cost of interruptions.

Supply insecurity causes disutility and inconvenience, in varying degrees and in different ways, to different classes of consumers—domestic, commercial, and industrial. The costs and losses (L) for the average consumer from supply interruptions are a function of the following:

- Dependence of the consumer on the supply (C).
- Duration of the interruptions (D).
- Frequency of their occurrence during the year (F).
- Time of day in which they occur (T).

That is, $L = C (D^d \times F^f, T^t)$, where d, f, and t are constants that vary from one consumer category to another.

The table shows estimates of the annual cost of electricity supply interruptions for the U.S. economy.

Economic cost of electricity supply interruptions for non-deferrable economic activities the United States, 1997

Consumer class and average duration of interruption	Cost to consumer per outage (U.S. dollars)	Cost to consumer per lengthy outage (U.S. dollar)	Estimated total annual losses (billions of U.S. dollars)
Residential (20 minutes)	0–20	50–250	0.9–2.7
Commercial (10 minutes)	25–500	5–20 (per minute)	2.9–11.7
Industrial (less than 30 seconds)	200–500 (small plant) 1,000–10,000 (large plant)	5,000–50,000 (per 8-hour day)	1.1–13.5

Note: Assumes nine outages a year for each class of consumer.

Source: Newton-Evans Research Company, 1998.

Crude Oil

Proven oil reserves have increased steadily over the past 20 years, mainly because oil companies have expanded their estimates of the reserves in already discovered fields. This optimism stems from better knowledge of the fields, increased productivity, and advances in technology. New technologies have led to more accurate estimates of reserves through better seismic (three- and four-dimensional) exploration, have improved drilling techniques (such as horizontal and offshore drilling), and have increased recovery factors—the share of oil that can be recovered—from 30 percent to 40–50 percent (Campbell and Laherrere, 1998).

Huge amounts of untapped unconventional oil also exist, augmenting conventional oil reserves.* Some 1.2 trillion barrels of heavy oil are found in the Orinoco oil belt in Venezuela. And the tar sands of Canada and oil shale deposits of the Russian Federation may contain 300 billion barrels of oil.

The U.S. Geological Survey assessed ultimate oil and gas reserves at the beginning of 1993 (IEA, 1998; WEC, 1998). The results, which tally with the World Energy Council (WEC) and International Energy Agency (IEA) figures (see table 1), point to ultimate conventional oil reserves of 2,300 billion barrels, with cumulative production until 1993 amounting to 700 billion barrels and unidentified reserves to 470 billion. No shortage of conventional liquid fuels is foreseen before 2020. Any deficiencies after that can be met by the ample reserves of unconventional oil.

Natural Gas

The U.S. Geological Survey also assessed ultimate natural gas reserves in 1993 (Masters, 1994). It estimated ultimate reserves at 11,448 trillion cubic feet (11,214 exajoules, or 267 gigatonnes of oil equivalent [Gtoe]), with cumulative production until 1993 amounting to 1,750 trillion cubic feet (1,722 exajoules, or 41 Gtoe). Cumulative world gas production through the end of 1995 was only 17.1 percent of the U.S. Geological Survey's estimate of conventional gas reserves.

Natural gas consumption is projected to grow 2.6 percent a year mostly as a result of growth in electricity generation in non-OECD countries. Despite this growth, cumulative production is expected to be no more than 41 percent of the U.S. Geological Survey's estimate of conventional gas reserves by 2020. This points to a resource base large enough to serve global requirements for natural gas well into the second half of the 21st century.

Coal

Coal is the world's most abundant fossil fuel, with reserves estimated at almost 1,000 billion tonnes, equivalent to 27,300 exajoules, or 650,000 Mtoe (WEC, 1998). At the present rate of production, these reserves should last for more than 220 years. Thus the resource base of coal is much larger than that of oil and gas. In addition, coal reserves are more evenly distributed across the world. And coal is cheap. Efforts are being made to reduce production costs and to apply clean coal technologies to reduce the environmental impact.

Coal demand is forecast to grow at a rate slightly higher than global energy growth. Most of this growth will be for power generation in non-OECD countries, mostly in Asia. Although trade in coal is still low, it is likely to increase slowly over time. Long-term trends in direct coal utilisation are difficult

---

*Conventional oil resources refer to those sources that are known and economically recoverable with present technologies. Unconventional oil resources refer to those sources that are less certain and/or are not economically recoverable with present technologies.—Eds.

to predict because of the potential impact of climate change policies. Coal gasification and liquefaction will augment global oil and gas resources in the future.

Nuclear Energy

Although nuclear energy is sometimes grouped with fossil fuels, it relies on a different resource base. In 1998 nuclear energy production amounted to 2,350 terawatt-hours of electricity, replacing 620 Mtoe of other fuels. Uranium requirements amounted to 63,700 tonnes in 1997, against reasonably assured resources (reserves) of 3.4 million tonnes. Ultimately recoverable reserves amount to almost 17 million tonnes. Considering the relative stagnation in the growth of nuclear power, the enormous occurrences of low-grade uranium, and the prospects for recycling nuclear fuels, such reserves will suffice for many decades.

Renewables

Renewable energy sources—especially hydroelectric power, biomass, wind power, and geothermal energy—account for a growing share of world energy consumption. Today hydropower and biomass together contribute around 15 percent.

Hydroelectric power contributes around 2,500 terawatt-hours of electricity a year, slightly more than nuclear power does. It replaces almost 675 Mtoe of fuels a year, although its direct contribution to primary energy consumption is only a third of this. But it has still more potential. Technically exploitable hydro resources could potentially produce more than 14,000 terawatt-hours of electricity a year, equivalent to the world's total electricity requirements in 1998 (WEC, 1998). For environmental and economic reasons, however, most of these resources will not be exploited.

Still, hydropower will continue to develop. Hydropower is the most important among renewable energy sources. It is a clean, cheap source of energy, requiring only minimal running costs and with a conversion efficiency of almost 100 percent. Thus its annual growth could exceed the growth of global energy demand, slightly improving hydropower's modest contribution towards meeting world requirements.

Renewable energy sources other than hydro are substantial. These take the form mainly of biomass. Traditional biomass includes fuelwood—the main source of biomass energy—dung, and crop and forest residues. Lack of statistics makes it difficult to accurately estimate the contribution of renewables to the world's primary energy consumption. But it is estimated that the world consumed around 1.20 Gtoe in 1998. About two-thirds of this was from fuelwood, and the remainder from crop residues and dung. Much of this contribution is sustainable from a supply standpoint. But the resulting energy services could be substantially increased by improving conversion efficiencies, which are typically very low.

The contribution of biomass to world energy consumption is expected to increase slightly. It is mainly used as an energy source in developing countries.

While energy demand in these countries is steadily increasing, some of the demand is being met by switching from traditional to commercial energy sources.

Biomass energy technology is rapidly advancing. Besides direct combustion, techniques for gasification, fermentation, and anaerobic digestion are all increasing the potential of biomass as a sustainable energy source. The viability of wind energy is increasing as well. Some 2,100 megawatts of new capacity was commissioned in 1998, pushing global wind generating capacity to 9,600 megawatts. Wind power accounted for an estimated 21 terawatt-hours of electricity production in 1999. While that still amounts to only 0.15 percent of global electricity production, the competitiveness of wind power is improving and its growth potential is substantial. Use of geothermal energy for electricity generation is also increasing, with a present generating capacity of more than 8,300 megawatts.

The Resource Outlook

To summarise, no serious global shortage of energy resources is likely during at least the first half of the 21st century. Reserves of traditional commercial fuels—oil, gas, and coal—will suffice for decades to come. When conventional oil resources are depleted, the huge unconventional oil and gas reserves will be tapped as new extraction and clean generating technologies mature. Coal reserves are also huge: the resource base is more than twice that of conventional and unconventional oil and gas. Clean technologies for coal will allow greater exploitation of this huge resource base, mainly in electricity production, but also through conversion into oil and gas, minimising environmentally harmful emissions.

The uranium resource base is also immense, and it is unlikely, at least in the short term, to be tapped in increasing amounts. The ultimately recoverable uranium reserves will easily meet any nuclear power requirements during this century.

The renewable resource base is also promising. Only part of the global hydro potential has been tapped. Hydropower plants will continue to be built as demand for electricity grows and the economics of long-distance, extra-high-voltage transmission improve. Biomass has substantial potential and will continue to be used not only as a traditional fuel but also in increasingly sophisticated ways, through thermochemical and biochemical applications. New renewable sources, particularly wind power, will gradually increase the contribution of renewables to global energy supplies as the economies and technologies of these environmentally attractive sources continue to improve.

In short, the world's energy supplies offer good prospects for energy security in the 21st century. The fossil fuel reserves amount to 1,300 Gtoe and the fossil fuel resource base to around 5,000 Gtoe (see table 1), amounts sufficient to cover global requirements throughout this century, even with a high-growth scenario. That does not mean there will be no temporary or structural energy shortages, but as long as the energy resources are being explored and exploited, these shortages will not be due to resource inadequacy.

References

Campbell, C. J., and J. H. Laherrere. 1998. "The End of Cheap Oil." *Scientific American* 278: 60–65.

IEA (International Energy Agency). 1998. *World Energy Outlook*. Paris.

Masters, C. D. 1994. *World Petroleum Assessment and Analysis*. Proceedings of the 14th World Petroleum Congress. New York: John Wiley & Sons.

Mitchell, J. V. 1997. *Will Western Europe Face an Energy Shortage?* Strasbourg: Energy Council of France.

WEC (World Energy Council). 1998. *Survey of Energy Resources*. London.

World Bank. 1999. *World Development Report 1999/2000: Entering the 21st Century*. New York: Oxford University Press.

Lester R. Brown

← **NO**

Turning on Renewable Energy

As world population has doubled and the global economy has expanded sevenfold over the last half-century, our claims on the environment have become excessive. We are asking more of the Earth than it can give on an ongoing basis and creating a "bubble" economy—one in which economic output is artificially inflated by overconsumption of the Earth's natural resources.

We are cutting trees faster than they can regenerate, overgrazing rangelands and converting them into deserts, overpumping aquifers and draining rivers dry. On our cropland, erosion exceeds new soil formation, slowly depriving the soil of its inherent fertility. We are taking fish from the ocean faster than they can reproduce. We are depleting our nonrenewable fossil fuels and releasing carbon dioxide (CO_2) faster than nature can absorb it, creating a greenhouse effect. As atmospheric CO_2 levels rise, so does the Earth's temperature.

The resulting mega-threat—climate change—is not getting the attention it deserves, particularly from the United States, the nation responsible for one-fourth of all carbon emissions. Washington wants to wait until all the evidence on climate change is in, by which time it will be too late to prevent a wholesale warming of the planet. As the Earth's temperature rises, it affects all life on the planet. Climate change will cause intense heat waves, more destructive storms, lower crop yields, glacier melting and rising seas.

To head off disaster, we must design more efficient transportation systems; raise efficiency standards for buildings, appliances and automobiles; and develop and promote renewable energy technology.

The good news is that although this is a staggering challenge, it is entirely doable, and many countries are now taking action. Detailed studies by governments and environmental groups reveal the potential for reducing carbon emissions while saving money in the process. Cutting global carbon emissions in haft by 2015 is entirely within range. Ambitious though this seems, it is commensurate with the threat climate change poses.

National and local governments, corporations and environmental groups are coming up with ambitious plans to cut carbon emissions. Prominent among these is a plan by British Prime Minister Tony Blair to reduce carbon emissions 60 percent in the United Kingdom by 2050. Blair and Sweden's Prime Minister, Göran Persson, are jointly urging the European Union to adopt the 60-percent goal, the amount scientists deem necessary to stabilize global atmospheric CO_2 levels.

A plan developed for Canada by the David Suzuki Foundation and the Climate Action Network—Canadian nongovernmental organizations promoting environmental sustainability—would halve carbon emissions by 2030 and would do so only with profitable investments in energy efficiency. This plan was inspired by U.S.-based Interface, the world's largest manufacturer of industrial carpeting. During the 1990s, the company's Canadian affiliate cut its carbon emissions by two-thirds through examining every facet of its business—from electricity consumption to trucking procedures. The company has since saved more than $400,000 a year in energy expenditures. CEO Ray Anderson says, "Interface Canada has reduced greenhouse gas emissions by 64 percent from the peak, and made money in the process, in no small measure because our customers support environmental responsibility."

Another push for renewable energy in Canada comes from the Ontario Clean Air Alliance, an environmental group that has devised a four-pronged strategy to phase out the province's five coal-fired power plants by 2010. Jack Gibbons, director of the Alliance, says of coal burning, "It's a 19th-century fuel that has no place in 21st-century Ontario."

And Germany, which has set the pace for reducing carbon emissions among industrial countries, is now talking about lowering its emissions 40 percent by 2020. This country already is far more energy-efficient than the United States, whose carbon emissions are projected to continue to *increase*. A lack of leadership, not a lack of technology, is why the United States' goal for cutting carbon emissions contrasts with Germany's.

In April 2003, the World Wildlife Fund released a peer-reviewed analysis that proposed reducing carbon emissions from U.S. electric-power generation 60 percent by 2020. This proposal focuses on more energy-efficient power-generation equipment; more efficient household appliances, industrial motors and other equipment; and a shift from coal to natural gas. If implemented, it could result in national savings averaging $20 billion a year until 2020.

The accelerating rise in the Earth's temperature calls for simultaneously raising efficiency standards and shifting to renewables in order to cut carbon emissions in half. The initial large gains are likely to come with efficiency improvements from mandating efficiency standards for household appliances, automobiles and the construction of new buildings.

Setting New Standards

Of course, each nation will have to fashion its own plan for raising energy productivity. Nevertheless, a number of potential common components exist. These include banning nonrefillable beverage containers, eliminating incandescent light bulbs, doubling the fuel efficiency of automobiles and redesigning urban transport systems.

Canada's Prince Edward Island has banned the use of nonrefillable beverage containers, and Finland's stiff tax on nonrefillables has led to 98-percent container reuse for soft drinks. These actions reduce energy use, water use and garbage generation. A refillable glass bottle used over and over again requires 90 percent less energy per use than an aluminum can, even if the can is recycled. Banning inefficient nonrefillables is a win-win policy because it will cut both energy use and garbage flow.

Another simple step is to replace incandescent light bulbs with compact fluorescent bulbs (CFLs), which only use one-third the electricity and last 10 times longer. In the United States, 20 percent of electricity is used for lighting. If each household replaces incandescents with CFLs, electricity needed for lighting could be cut in half. CFLs yield a risk-free return of 25 percent to 40 percent a year. Their cost is falling, and they typically pay for themselves in electricity savings in a few years. Worldwide, replacing incandescent light bulbs with CFLs could save enough electricity to close hundreds of coal-fired power plants, and it could be accomplished within three years, if we decided to do it.

A third way to raise energy efficiency is to produce more efficient automobiles. In the United States, if all motorists shifted to cars with hybrid engines, such as the Toyota Prius or the Honda Insight, gasoline use could be cut in half. Sales of hybrid cars, introduced into the U.S. market in 1999, reached an estimated 46,000 in 2003, and the Prius was named "2004 Car of the Year" by *Motor Trend* magazine. (For more on hybrid vehicles, see Page 23.) Higher gasoline prices and a tax deduction for purchasing these vehicles are boosting sales and making the cars more cost-competitive. With U.S. auto manufacturers coming onto the market, hybrid-vehicle sales are projected to reach 1 million by 2007.

Cutting carbon emissions also means restructuring our transportation systems. Most automobile-centered systems are highly inefficient because most cars carry only the driver. Constructing well-developed light-rail systems, hydrogen-fueled buses as needed, and bicycle- and pedestrian-friendly lanes could increase mobility, reduce air pollution and provide exercise. These improvements are much needed in a world where 3 million people die each year from urban air pollution, and where half or more of the adults in exercise-deprived, affluent societies are overweight. Fewer automobiles also means some parking lots could be converted into parks.

Harnessing the Wind

Wind energy offers a powerful alternative to fossil fuels—it is abundant, inexhaustible, widely distributed and clean, which is why it has been the world's fastest-growing energy source over the last decade. Wind energy doesn't produce sulfur dioxide or nitrous oxides that cause acid rain, and it does not disrupt the Earth's climate. It also doesn't generate health-threatening mercury or pollute streams like coal-fired power plants.

Harnessing the wind also is cheap: Advances in wind-turbine design have reduced the cost of wind power to less than 4 cents per kilowatt-hour at prime wind sites—well below the price of nuclear power or coal. On prime sites, wind power can now even compete with gas, currently the cheapest source of electricity generation.

Even more exciting, with each doubling of world wind-generating capacity, costs fall by 15 percent. The recent growth rate of 31 percent a year means costs are dropping by 15 percent about every 30 months. While natural-gas prices are highly volatile, the cost of wind power is declining. And, there is no OPEC (Organization of the Petroleum Exporting Countries) for wind.

By the end of 2002, world wind-generating capacity had increased sixfold to 31,100 megawatts—enough to meet the residential needs of Norway, Sweden, Finland, Denmark and Belgium combined. Germany, with more than 12,000 megawatts of wind power at the end of 2002, leads the world in generating capacity. Spain and the United States, at 4,800 and 4,700 megawatts, are second and third, respectively. Tiny Denmark is fourth with 2,900 megawatts, and India is fifth with 1,700 megawatts. A second wave of major players is coming onto the field, including the United Kingdom, France, Italy, Brazil and China. Europe has enough easily accessible offshore wind energy to meet all of its electricity needs, and China can easily double its current electricity generation from wind alone.

Globally, ambitious efforts to develop wind power are beginning to take shape. Germany is proposing a 30-percent cut in greenhouse-gas emissions throughout Europe by 2020—developing the continent's wind-energy resources is at the heart of this carbon-reduction effort. And the United States is following Europe's lead. A 3,000-megawatt wind farm in South Dakota, designed to partly power the industrial Midwest surrounding Chicago, is one of the largest energy projects of any kind. Cape Wind is planning a 420-megawatt wind farm off the coast of Cape Cod, Massachusetts, and a newly formed energy company, called Winergy, has plans for some 9,000 megawatts in a network of wind farms stretching along the Atlantic coast.

In the United States, a national wind-resource inventory published in 1991 indicates **enough harnessable wind energy exists in just three states— North Dakota, Kansas and Texas—to satisfy national electricity needs.** Today, this greatly understates U.S. potential: Recent advances in wind-turbine design and size have dramatically expanded the wind-power industry.

It is time to consider an all-out effort to develop wind resources, given the enormous wind-generating potential and the associated benefits of climate stabilization. Instead of doubling wind-power generation every 30 months or so, perhaps we should aim to double wind-electric generation each year for the next several years. Costs would drop precipitously, giving wind-generated electricity an even greater advantage over fossil fuels.

Cheap electricity from wind is likely to become the principle source for electrolyzing water to produce hydrogen. Hydrogen can be transported through pipelines to power residential and industrial buildings; it also can be stored in power plants and used when the wind ebbs. The hydrogen storage and distribution system—most likely an adaptation of existing natural-gas systems— provides a way of both storing and transporting wind energy.

The incentives for switching to a wind/hydrogen system could come partly from restructuring global energy subsidies—shifting the $210 billion in annual fossil-fuel subsidies to the development of wind energy, hydrogen generators and kits to convert engines from gasoline to hydrogen. The investment capital could come from private capital markets and from companies already in the energy business: Energy giants Shell and BP have begun investing in wind power, and major corporations such as General Electric and ABB, a company that produces technology systems, are now in the wind-power business.

Solar Energy Rises

In recent years, a vast new market for solar power has opened in developing nations that are not yet linked to an electrical grid. About 1.7 billion people in developing nations do not have electricity, but as the cost of solar cells declines, it often is cheaper to provide electricity from solar cells than from a centralized source.

In Andean villages, solar-power systems are replacing candles. For villagers paying installation costs over 30 months, the monthly payment is roughly the same as the cost of a month's supply of candles. Once the solar cells are paid for, the villagers essentially have a free source of power that can supply electricity for decades.

At the end of 2002, more than 1 million homes in villages in the developing world were getting their electricity from solar cells. But this is less than 1 percent of the estimated 1.7 billion people who do not have electricity. The principal obstacle slowing the spread of solar-cell installations is not the cost, but the lack of small-scale credit programs to finance them. As this credit shortfall is overcome, purchases of solar cells could climb far above the rate of recent years.

The residential use of solar cells also is expanding in industrial countries. In Japan, where companies have commercialized a solar roofing material, some 70,000 homes now have solar installations. Consumers in Germany receive low-interest loans and a favorable guaranteed price when feeding excess electricity into the grid. In industrial nations, most installations reduce the consumer's dependence on grid-supplied electricity, much of it originating from coal.

The governments with the strongest incentives for solar cells have the largest solar-cell manufacturing industries. Japan leads in solar-cell manufacturing and controls 43 percent of the global solar-cell market; residential installations produced roughly 100 megawatts in 2001. Germany produced 75 megawatts that year, and the European Union, led by Germany's vigorous program, is in second place behind Japan with 25 percent of the world's total output. The United States is third—with 32 megawatts of installations and 24 percent of the market. India is fourth with 18 megawatts.

Solar-generated electricity still is much more costly than power from wind or coal-fired plants, but industry experts estimate that with each doubling of cumulative production, the price drops roughly 20 percent. Over the last seven years, solar-cell sales expanded an average of 31 percent annually, doubling every 2.6 years. Only modest government incentives are needed to accelerate the growth of solar power and make it a major player in the world energy economy.

Building the Hydrogen Economy

The evolution of the fuel cell—a device that uses an electrochemical process to convert hydrogen into electricity—is setting the stage for the evolution of a hydrogen-based economy. The fuel cell is twice as efficient as the internal combustion engine and emits only water vapor. The fuel cell facilitates the shift to a single fuel—hydrogen—that neither pollutes nor disrupts the Earth's climate. Stationary fuel cells can be installed in the basements of buildings to heat, cool and generate electricity for lights and appliances. Mobile fuel cells can power cars and portable electronic devices, such as cell phones and laptop computers. Hydrogen can come from many sources, including water, natural gas or gasoline.

Iceland already has a plan to convert from fossil fuels to hydrogen. In 2003, the government, working with a consortium of companies led by Shell and DaimlerChrysler, took the first step by beginning to convert the capital city of Reykjavik's fleet of 80 buses from internal combustion to fuel cell engines. Shell built a hydrogen station to service the buses, using inexpensive hydroelectricity to produce clean hydrogen. In the next stage, Iceland's automobiles will be converted to fuel cell engines. And in the final stage, the Icelandic fishing fleet—the centerpiece of its economy—also will convert to fuel cells. Iceland already heats most of its homes and buildings with geothermal energy and gets most of its electricity from hydropower and geothermal power, and by 2050, plans to be the first modern economy to declare independence from fossil fuels.

On Yakushima Island, an 875-square-kilometer island off the southern tip of Japan, retired corporate executive Masatsugu Taniguchi is creating a hydrogen economy. The island receives more than 300 inches of rainfall a year, so Taniguchi plans to build small dams to convert the abundant hydropower into electricity to power hydrogen generators. The first goal will be to meet the needs of the island's 14,000 residents. Then, Taniguchi plans to ship hydrogen to mainland Japan. He says the island can export enough to run 500,000 automobiles.

More than 50 fueling stations equipped with hydrogen technology have opened around the world. In the Munich, Germany, airport, a hydrogen station fuels 15 airport buses with hydrogen-burning internal combustion engines. The United States now has more than a dozen hydrogen stations, mostly in California, although they are open only for demonstrations and research, not for public use.

Hydrogen is the fuel of choice for the new fuel cell engines every major automobile manufacturer is developing. In 2002, the Honda FCX and the Toyota FCHV-4 became the first fuel-cell-powered automobiles to appear on the market. DaimlerChrysler also manufacturers a fuel cell car called the F-cell, and Ford is following close behind. The evolution of fuel cells and advances in wind-turbine design create the hope that U.S. farmers and ranchers, who own most of the country's wind rights, could one day supply much of the electricity and fuel for cars in the United States.

In the end, the central question with hydrogen is whether it is made using renewable energy to electrolyze water, or with climate-disrupting fossil fuels. Natural gas likely will be the main source of hydrogen in the near future, but, given its abundance, wind has the potential to become the principal source in the new energy economy. The hydrogen storage and distribution system provides ways of storing and transporting wind energy efficiently—it is a natural marriage. Thus, countries that are rich in wind and rather sparsely populated, such as Canada, Argentina and Russia, could export hydrogen. Eastern Siberia, for example, could supply vast amounts of hydrogen to China, South Korea and Japan.

In the United States, energy consultant Harry Braun made a proposal at an April 2003 Renewable Hydrogen Roundtable to quickly shift to a wind/hydrogen economy. He noted that if wind turbines were mass-produced like automobiles, the cost of wind-generated electricity would drop to 1 or 2 cents per kilowatt-hour.

Rather than wait for fuel cell engines, Braun suggests using hydrogen in internal combustion engines of the sort developed by German auto manufacturer BMW. He calculates that the production of hydrogen and high-efficiency, hydrogen-fueled engines would bring the cost of hydrogen down to $1.40 per equivalent-gallon of gasoline. If we make this conversion a priority, it can happen in two to three years.

Building a Better Future

In looking at new energy sources, wind seems certain to be the centerpiece in the new energy economy. Its wide distribution offers a practical alternative to oil, and the industry has evolved to the point where it can expand dramatically over the next decade and become the world's leading electricity source. If you think change can't happen that fast, just look at the recent adoption of other popular technologies, such as cell phones. In 1990, 11 million cell phones were in use globally. By 2002, the number had reached 1.2 billion, outnumbering the 1.1 billion fixed-line phones. In 12 years, cell phones went from being a novelty to dominating the market—their sales growth illustrates how market forces can drive the adoption of an appealing technology. The cell-phone market grew by 50 percent a year during the 1990s; wind power has grown 31 percent per year since 1995.

If we decided for climate-stabilization reasons that we wanted to double wind-electric generation each year, wind could become the dominant source of electricity. The United States, for example, now has nearly 5,000 megawatts of wind-generating capacity. Doubling that each year would take it to 640,000 megawatts in seven years and make it the leading source of electricity. And this is not beyond the capacity of the industry: In 2001, wind-electric generating capacity grew by 67 percent. The total investment needed to reach this level of generation, using the rule of thumb of $1 million per megawatt (which is now on the high side), would be $640 billion over a seven-year span, or roughly $90 billion a year. For perspective, Americans currently spend $190 billion a year on gasoline.

While subsidies are being shifted from fossil fuels to renewables and the hydrogen economy infrastructure, it would make eminent sense to reduce income taxes and raise taxes on climate-disrupting energy sources at the same time. This tax shifting, already under way in several nations in Europe, helps consumers of energy—both individuals and corporations—understand the full costs of burning fossil fuels.

Although shifting subsidies and taxes are at the heart of the energy transformation that is needed, other policy tools can either increase efficiency or accelerate the shift to renewables and the hydrogen-based economy. For instance, national and local governments, corporations, universities and individual homeowners can buy green power. In the United States, even if green power is not offered locally, a national Green Power Partnership electricity market operated by the Environmental Protection Agency enables anyone to buy green power. As more users sign up, the incentive for energy companies to produce green power increases.

As wind energy expands, the next step would be to close coal-fired power plants or use them to back up wind. Coal-fired plants are the most climate-disruptive energy source because coal is almost pure carbon. Coal burning also is the principal source of the mercury deposits that contaminate freshwater lakes and streams. The prevalence of mercury-contaminated fish has led 44 state governments in the United States to issue warnings to consumers to limit or avoid eating fish from their locales. In 2001, the Centers for Disease Control and Prevention warned that 375,000 babies born each year in the United States are at risk of impaired mental development because of mercury exposure.

Although some industry groups and governmental bodies complain that reducing carbon emissions is costly and a burden on the economy, study after study concludes it is possible to reduce carbon emissions while making money in the process. The experience of individual companies confirms this. DuPont, one of the world's largest chemical manufacturers, already has cut its greenhouse-gas emissions from its 1990 level by 65 percent. In an annual report, CEO Charles Holliday Jr. proudly reports savings of $1.5 billion in energy-efficiency gains from 1990 to 2002.

It has become clear that incorporating renewable energy is one of the most profitable investments many companies can make, and as the true costs of climate change—withering crops, rising sea levels and wildlife extinction—become apparent, companies that ignore the need to phase out fossil fuels will ultimately disappear. The companies that prosper will be the ones that adapt to a modern economy fueled by clean, renewable energy.

POSTSCRIPT

Should the World Continue to Rely on Oil as a Major Source of Energy?

The twenty-first century ushered in another in a series of energy crises that have plagued the developed world since 1972. Gas prices jumped to record heights, reaching between $2 and $3 a gallon in mid-2004, and the prospects of a return to $1.00-a-gallon levels seem increasingly remote. Once again, cries for eliminating dependence on foreign oil and for developing alternatives to the twentieth century's principal energy source were heard.

Yet the message of Khatib et al.'s assessment of foreseeable world energy supplies is "Don't panic just yet." The study reveals no serious energy shortage during the first half of the twentieth-first century. In fact, the report suggests that oil supply conditions have actually improved since the crises of the 1970s and early 1980s. Khatib et al. go further in their assessment, concluding that fossil fuel reserves are "sufficient to cover global requirements throughout this century, even with a high-growth scenario."

Francis R. Stabler, in "The Pump Will Never Run Dry," *The Futurist* (November 1998), argues that technology and free enterprise will combine to allow the human race to continue its reliance on oil far into the future. To be sure, Stabler's view of the future availability of gas is in the minority. One supporter is Julian L. Simon who argues in his *The Ultimate Resource 2* (1996) that even God may not know exactly how much oil and gas are "out there." Chapter 11 of Simon's book is entitled "When Will We Run Out of Oil? Never!" Simon takes the reader through a twelve-step process to demonstrate that the doomsayers are wrong. Another Stabler supporter is Bjørn Lomborg in *The Skeptical Environment: Measuring the Real State of the World* (Cambridge University Press, 2001). Lomborg argues hat the world seems to find more fossil energy than it consumes, and he concludes that "we have oil for at least 40 years at present consumption, at least 60 years' worth of gas, and 230 years' worth of coal."

Simon and Lomborg are joined by Michael C. Lynch in a published article on the Web under global oil supply entitled "Crying Wolf: Warnings about Oil Supply," http://sepwww.stanford.edu/sep/jon/world-oil.dir/lynch.html. John Jennrich argues that earlier predictions that the world would run out of oil have been way off the mark and will continue to be so in the future ("Fueling the Future," in Ronald Bailey, *Global Warming and Other Eco-Myths*, Forum, 2002). He points to a quote by David Nemtzow, president of the Alliance to Save Energy, for his counter-argument. Said Nemtzow, "It's not the scarcity of oil that endangers pristine and public lands; it's the profusion."

Seth Dunn, on the other hand, follows conventional wisdom in his article, "Decarbonizing the Energy Economy" (in Lester R. Brown et al., *State of the World 2001*, W. W. Norton, 2001). That is, because oil is a finite resource, its supply will end some day, and that day will be sooner rather than later. In fact, he suggests that new renewable energy sources are in the same position as oil a century ago, that is, "gaining footholds" in the energy market. Dunn had argued elsewhere (Christopher Flavin and Seth Dunn, "Reinventing the Energy System," *State of the World 1999,* Worldwatch Institute, 1999) that the global economy has been built on the rapid depletion of non-renewable resources, and such consumption levels cannot possibly be maintained throughout the twenty-first century, as they were the previous century. Although Flavin and Dunn's arguments have probably received a receptive audience among most scholars who are concerned with the increasing scarcity of nonrenewable resources, they require the reader to accept a set of assumptions about the acceleration of future energy consumption. But one can easily be seduced by the logic of their argument, because it "just seems to make sense."

Lester R. Brown, et al. in *Beyond Malthus* (1999) suggest that most writers point to between the years 2010 and 2025 as the time when world oil production will peak. The consequence, if that is accurate, is a need for alternative sources. The student of energy politics, however, must be careful not to ignore how advances in energy source exploration and extraction have tended to expand known reserves. Is the future lesson that the tide has finally turned and no significant reserves remain to be discovered? Or is the lesson that history will repeat itself and modern science will yield more oil and gas deposits, as well as make their extraction cost effective? In the selection for this issue, Brown argues that the search is on for renewable energy sources.

Finally, James J. MacKensie has provided a comprehensive yet succinct article on the peaking of oil in "Oil as a Finite Resource: When Is Global Production Likely to Peak?" (World Resources Institute). Seth Dunn of the Worldwatch Institute in *State of the World 2001* suggests that a new energy system is fast approaching because of a series of revolutionary new technologies and approaches.

The msn.com Web site provides numerous citations of articles on both sides of the issue.

119

ISSUE 7

Will the World Be Able to Feed Itself in the Foreseeable Future?

YES: Sylvie Brunel, from "Increasing Productive Capacity: A Global Imperative," in Action Against Hunger, *The Geopolitics of Hunger, 2000–2001: Hunger and Power* (Lynne Rienner Publishers, 2001)

NO: Lester R. Brown, from "Eradicating Hunger: A Growing Challenge," in Lester R. Brown et al., *State of the World 2001* (W.W. Norton, 2001)

ISSUE SUMMARY

YES: Sylvie Brunel, former head of Action Against Hunger, argues that "there is no doubt that world food production . . . is enough to meet the needs of" the world's peoples.

NO: Lester R. Brown, former president of the Worldwatch Institute, suggests that little if any progress is being made to eradicate pervasive global hunger, despite increases in food productivity.

Visualize two pictures. The first snapshot, typical of photographs that have graced the covers of the world's magazines, reveals a group of people in Africa, including a significant number of small children, who show dramatic signs of advanced malnutrition and even starvation. The second picture (really several in sequence) shows an apparently wealthy couple finishing a meal at a rather expensive restaurant. The waiter removes their plates still half-full of food, an untouched loaf of French bread, and assorted other morsels from the table, and deposits them in the kitchen garbage can. These scenarios once highlighted a popular film about world hunger. The implication was quite clear. If only the wealthy would share their food with the poor, no one would go hungry. Today the simplicity of this image is obvious.

This issue addresses the question of whether or not the world will be able to feed itself by the middle of the twenty-first century. A prior question, of course, is whether or not enough food is grown throughout the world today to handle current nutritional and caloric needs of all the planet's citizens. News accounts of chronic food shortages somewhere in the world seem to have been appearing with regularly consistency for about 25 years. This time has witnessed graphic accounts in news specials about the consequences of insufficient food, usually somewhere in sub-Saharan Africa. Also, several national and international studies have been commissioned to address world hunger. An American study organized by President Carter, for example, concluded that the root cause of hunger was poverty.

One might deduce from all of this activity that population growth had outpaced food production and that the planet's agricultural capabilities are no longer sufficient. Yet, the ability of most countries to grow enough food has not yet been challenged. During the 1970–2000 period, only one region of the globe, sub-Saharan Africa, was unable to have its own food production keep pace with population growth. All other regions of the world experienced food increases greater than human growth.

This is instructive because, beginning in the early 1970s, a number of factors conspired to lessen the likelihood that all humans would go to bed each night adequately nourished. Weather in major food-producing countries turned bad; a number of countries, most notably Japan and the Soviet Union, entered the world grain importing business with a vengeance; the cost of energy used to enhance agricultural output rose dramatically; and less capital was available to poorer countries as loans or grants for purchasing agricultural inputs or the finished product (food) itself. Yet the world has had little difficulty growing sufficient food, enough to provide every person with two loaves of bread per day as well as other commodities. Major food-producing countries even cut back the amount of acreage devoted to agriculture.

Why then did famine and other food-related maladies appear with increasing frequency? The simple answer is that food is treated as a commodity, not a nutrient. Those who can afford to buy food or grow their own do not go hungry. However, the world's poor became increasingly unable to afford either to create their own successful agricultural ventures or to buy enough food.

The problem for the next half-century, then, has several facets to it. First, can the planet physically sustain increases in food production equal to or greater than the ability of the human race to reproduce itself? This question can only be answered by examining both factors in the comparison—likely future food production levels and future fertility scenarios. A second question relates to the economic dimension associated with an efficient global food distribution system. Will those poorer countries of the globe that are unable to grow their own food have sufficient assets to purchase it, or will the international community create a global distribution network that ignores a country's ability to pay? And third, will countries that want to grow their own food be given the opportunity to do so?

The selections in this issue address the question of the planet's continuing ability to grow sufficient food to feed its growing population. Syvie Brunel contends that if world food production were distributed among all the world's peoples, there would be plenty of food. Lester R. Brown is pessimistic about the likelihood that hunger will be eradicated. His assessment of the international food situation in the quarter century since then U.S. Secretary of State Henry Kissinger observed that no one would go to bed hungry by 1984 prompt him to conclude that simply growing enough food is not enough, a fact that world leaders have not quite grasped yet.

Sylvie Brunel

▶ **YES**

Increasing Productive Capacity: A Global Imperative

Can the earth feed its inhabitants? Despite the most alarmist predictions—those of Lester Brown in his State of the World published each year by the World Watch Institute of Washington; those of Paul Ehrlich, author of *The Population Bomb;* or those of the Club of Rome—there is no doubt that world food production, if equally distributed among all the world's peoples, is enough to meet the needs of them all. It is true that the increase in world agricultural production has slowed in recent years, a situation that has led to an immediate flood of alarmist predictions. The world, however, is not heading toward famine. And this is for a number of reasons.

The Increase in World Agricultural Production Continues to Outpace Population Growth

Only persons who are ill informed or of bad faith can argue that the trend is toward a decline in food production. They may even succeed in proving that claim. It is enough for them to select as the base year one in which harvests were particularly good and a second year in which they declined steeply in order to show a "disturbing" trend. Hervé Kempf demonstrated, for example, how a comparison of 1984 and 1991 would show an increase in cereal production of only 0.7 percent per year, which would be "disturbing," since it is far below the 1.7% annual rate of population increase. By selecting the preceding years, one can show, on the contrary, that world agriculture has never been more productive: A comparison between 1983 and 1990 shows an increase in cereal production of 2.7 percent per year. These two statistics are clearly equally deceptive, and Joseph Klatzmann, in a refreshing little book, repeatedly warned against the danger of blindly trusting statistical data taken out of context.

If we examine world agricultural production over a long period, it becomes clear that the production curve exceeds the population growth curve. While world population did indeed double in one generation, grain production increased more than threefold, from 600 million to approximately 1,900 million

From Sylvie Brunel, "Increasing Productive Capacity: A Global Imperative," in Action Against Hunger, *The Geopolitics of Hunger, 2000–2001: Hunger and Power* (Lynne Rienner, 2001). Copyright © 2001 by Action Against Hunger—USA. Reprinted by permission of Lynne Rienner Publishers, Inc. Notes omitted.

tons per year. Each human being has available in theory 20 percent more food than in the early 1970s, or 2,700 calories per person per day, which is far more than a person's estimated need of between 2,000 and 2,200 calories, depending on the sources. However, half of the current grain production does not directly benefit people: Approximately 20 percent is used to feed cattle, 5 percent is kept for seeds, and the remaining 25 percent is quite simply lost as a result of poor storage or destruction by rodents, insects, and so on, especially in developing countries. It is therefore not the impossibility of increasing agricultural production that threatens mankind, but rather the way in which this increase is achieved and for the benefit of whom.

Indeed, it is in fact not in the countries of the so-called Third World but rather in the developed countries that agricultural production has slowed, in other words precisely where the problems of hunger have been overcome. (At least they have been overcome in quantitative terms; in qualitative terms, obesity, on the one hand, and malnutrition caused by the economic and social marginalization of certain categories of persons, on the other, have become real societal problems.) The developed countries have chosen to voluntarily limit their agricultural production in order to adapt it to the level of demand at which production would be profitable, in other words, to the consumer market. The fact that there are some 800 million people suffering from malnutrition in the world in no way changes this calculation, since those persons are too poor to buy food.

The reduction or slowdown in the rate of increase in world food production is thus attributable mainly to the developed countries, for reasons that have nothing to do with ecological limitations. Pierre Le Roy estimates at 20 million tons the reduction in supply that results from Europe's policy of limitation of production (land left fallow), an amount that represents twice the total of all food imports by sub-Saharan Africa.

Food imports by the Third World are indeed increasing, rising from 20 million tons in 1960, or 2 percent of consumption, to 120 million tons in the mid 1990s, or 20 percent of consumption. Economic forecasts suggest that this dependency is likely to increase even further in the decades ahead and to rise to 160 million tons within two decades. The reasons for this growing dependence, which will create problems without precedent for the economies of poor countries that will face increasingly onerous food import bills, are both negative and positive.

The negative factor of continuing population growth and spreading urbanization in the countries of the South, where nearly half the population now lives in cities, explains why more and more people are consuming food that their farmers are incapable of providing. The positive factor of the increase in average living standards in the developing countries and the emergence of a middle class that consumes more meat and dairy products places increasing pressure on the demand for cereals, in particular secondary cereals for stock feed.

Two-Speed Agricultural Policies

Why cannot the Third World feed itself, even though self-sufficiency in food was the grand slogan of the 1970s and 1980s?

The "technical" impossibility of increasing agricultural production in the South is not the problem: The earth is far from reaching its maximum agricultural potential, and the Food and Agriculture Organization (FAO) has pointed out that the useful agricultural surface in developing countries (700 million hectares) could be doubled without encroaching on protected areas such as forests or areas in which people live. Latin America and Africa hold the greatest potential in this regard. In addition, the potential for increased production through more intensive farming methods remains considerable. Only 11 kilograms of fertilizer are used per hectare in Africa, compared with 66 kilograms in Latin America and 139 kilograms in Asia, and only 5 percent of land is irrigated in Africa (most of this in countries that are unable to take advantage of it, such as Sudan and Madagascar), compared to 37 percent in Asia and 14 percent in Latin America. This situation offers tremendous potential for growth.

But the political and economic choices made by the countries of the Third World have thus far been detrimental to agriculture, and in particular to small peasant farming. Investments in agriculture have been concentrated in regions in which purchasing power is greatest and are characterized by a concern to protect the income of farmers, which has been steadily declining. As a result, these investments are moving in the direction of a two-tiered world that is becoming increasingly unequal in terms of access to food.

On one hand, the developed countries enjoy rapid growth, and despite the fact that farmers represent on average no more than 3 percent of the active population, their food supply is abundant and diversified, prices are low, and import levels are low as a result of the massive support given to the agricultural sector (in the mid-1990s, Organization for Economic Cooperation and Development (OECD) countries each year spent more than two hundred billion dollars to support their agricultural sectors). In that part of the world, the concern is no longer the fear of shortage, but rather the quality of the food consumed. The agrofood industry, now powerful after being forced to steadily increase its output over the past decades in order to keep up with the steadily rising demand, is today facing another challenge, namely, shifting to production methods that focus less on quantity and more on the quality of the inputs used and on the quality of the final product. Producers are also concerned about the methods used to satisfy the demand of consumers, who now want food that is not only abundant but also varied and, above all, healthy. The successive scandals of mad-cow disease, salmonella poisoning in chickens, hormone-treated beef cattle; the rejection by consumer groups of genetically modified plants; animal feed that includes mud from cleaning stations; and the questions raised about the production of eggs by battery hens are indicative of a new era in which insistence on quality is now a greater challenge than the demand for quantity.

On the other side are the poor and vulnerable countries, where malnutrition is endemic and where a majority of the population still depends on agriculture. The food supply remains insufficient, however, because of the poor yields that result from the low level of technology used, the absence of incentives to produce because of economic policies that discourage agriculture, and unfavorable exchange rates that make the importation of agricultural imports expensive. It is therefore precisely in those countries that agricultural production needs to be increased. First, increased production would reduce the cost of food, particularly in the large urban centers, and thereby make it accessible to this large sector of the population that is too poor to eat properly. Second, increased production would reduce the food import bills of countries that are increasingly dependent on imports, mainly from the rich countries.

The Food Supply Is a Regional, Not a Global Problem

At the global level, the food supply is increasing for reasons that are both positive and negative. On the positive side, the agricultural sector in Eastern Europe, which needed to be restructured following the collapse of the Iron Curtain, is now on the road to recovery. On the negative side, the economic difficulties of East Asia have led to a decline in food imports by that region.

At the regional level, the structural overproduction that the world has been experiencing for the past twenty years hardly prevents sub-Saharan Africa and South Asia from experiencing hunger. Of the approximately 800 million malnourished people in the world, more than 200 million live in Africa (nearly 40 percent of the population) and 530 million in South Asia (or one person out of five).

What is therefore responsible for this disastrous and paradoxical situation at a time when, in order to reduce the supply of food, rich countries are destroying mountains of surplus food each year and forcing their farmers to leave a portion of their land uncultivated, through subsidies for land that is left fallow?

One answer is wars and conflicts, particularly in Africa, that disrupt agricultural production. A second, even more important answer is mass poverty, which prevents an entire sector of the world's population (one out of every five inhabitants of the Third World) from obtaining adequate food. That sector is incapable of producing enough food to meet its needs and lacks the means to purchase it, even when the food is available and can be bought. Worldwide, some 1.5 billion people live below the poverty level. Mass poverty is all the more serious, as it is always combined with ignorance: It is always the poorest classes that commit the most harmful errors of nutrition, since they lack the advantage of basic education. The errors of nutrition committed by pregnant women and children, who make up the primary groups at risk of hunger, should be the focus of particular attention, since these errors have disastrous consequences for the future of the entire society.

According to the United Nations Children's Fund (UNICEF), half of the world's malnourished children live in Asia, which has 100 million of the 200 million total, including 70 million in India. That country alone has two and a half times more malnourished children than all of sub-Saharan Africa.

Writing about South Asia . . . , Gilbert Etienne remarked on the extent to which the problem of hunger remains unresolved because of mass poverty and the slowdown in investments in agriculture. This is despite the notable progress achieved on the Asian continent.

The case of Africa gives cause for even greater concern:

1. Unlike the situation in other continents, the rate of malnutrition is not declining. Quite the opposite, in fact, because chronic malnutrition still affects nearly 40 percent of the region's population.
2. The high proportion of young people in the population, a sign of a vigorous population still characterized by high birthrates, has led to a high proportion of unemployed in relation to the number of those who are in a position to contribute to production. The burden on the economies of African countries is therefore particularly heavy, especially since most of these countries suffer from an acute lack of financial resources.
3. The continent's dependency on food from foreign countries is due to the low productivity of its own agriculture and the growing number of Africans now living in urban areas (more than one in three compared with one in ten a generation ago). This dependency is effectively addressed neither by imports (9 to 10 million tons per year), because of the lack of adequate financial resources, nor by food aid (approximately 2 million tons), which has been falling drastically for some years now. Consequently, the food needs of Africans are not being satisfied, since the widespread poverty of a sector of the urban population does not permit that sector to obtain food at market prices. At the same time many rural dwellers are unable to provide for themselves during the period between harvests, on account of the low productivity levels and inadequate access to food.
4. A large number of people in Africa are affected by war or internal conflict. Even in countries in which the populations could, in theory, be properly fed, there is an adverse impact on the food situation of the population because of the insecurity of the economic actors, the weakness of the State, and the destruction or confiscation of crops. In this regard, Africa is by far the continent most affected by conflicts, which also result in massive populations of refugees and displaced persons who depend on international aid for their survival.
5. Poverty and the pressure on land and resources of the high population growth rate are not matched by corresponding investments in agriculture that would bring about increases in yields. Africa is thus the continent in which the problems of deforestation, desertification, and soil erosion are most acute. It is also the continent in which access to drinking water and irrigation is still very limited.

The situation is not desperate, however: Despite the lack of investment in small peasant farming, agricultural production in Africa has risen by 2 percent per year since the early 1960s. Grain production has more than doubled, from 30 million to 66 million tons. This rate of increase is insufficient to meet the needs of a growing population (3 percent per year) because of the way in which it has been achieved (mainly by increasing the area of land under cultivation).

Nevertheless it shows that more intensive farming methods are needed in Africa and that this approach has the potential to significantly increase agricultural output. When the FAO states that the "load capacity of the land" in many countries has now been exceeded, it is basing its conclusions on the use of traditional production methods, such as use of the hoe more often than not, lack of fertilizers, lack of irrigation, and use of a diverse range of traditional varieties of grain to compensate for climatic and pedological constraints, with but modest results. The average yield for Africa remains 1,000 kilograms per hectare of millet and corn, which shows how much room for improvement there would be if African governments were to decide to treat their farmers a little better and to invest in their agricultural potential.

What are some of the ways in which agricultural production can be increased? It is interesting to note that food problems do not occur in countries that are at peace, that enjoy democracy, and in which farmers operate under conditions of relative legal and administrative security, even when these countries are densely populated and located in unfavorable climatic zones. It is better to live in Burkina Faso than in the Democratic Republic of Congo, even though the Congo is infinitely better endowed than Burkina Faso in terms of rainfall and available land. Similarly, the "white revolution" in Mali is revitalizing those regions that produce rice, millet, and cotton and enriching their farmers, even as hunger still plagues Madagascar, the former breadbasket of southern Africa, which has been making error after economic error over the last quarter of a century.

In order to bring about peace and security in Africa, a resumption of cooperation is necessary. However, the level of official development assistance has never been lower. Will the renegotiation of the Lomé Convention relaunch the partnership between Europe and Africa for the concerted development of agriculture?

Eradicating Hunger: A Growing Challenge

In 1974, U.S. Secretary of State Henry Kissinger made a pledge at the World Food Conference in Rome: "By 1984, no man, woman, or child would go to bed hungry." Those attending the conference, including many political leaders and ministers of agriculture, came away inspired by this commitment to end hunger.

More than 26 years later, hunger is still very much part of the social landscape. Today, 1.1 billion of the world's 6 billion people are undernourished and underweight. Hunger and the fear of starvation quite literally shape their lives. A report from the U.N. Food and Agriculture Organization (FAO) describes hunger: "It is not a transitory condition. It is chronic. It is debilitating. Sometimes, it is deadly. It blights the lives of all who are affected and undermines national economies and development processes . . . across much of the developing world."

Kissinger's boldly stated goal gave the impression that there was a plan to eradicate hunger. In fact, there was none. Kissinger himself had little understanding of the difficult steps needed to realize his goal. Unfortunately, this is still true of most political leaders today.

In 1996, governments again met in Rome at the World Food Summit to review the food prospect. This time delegates from 186 countries adopted a new goal of reducing the number who were hungry by half by 2015. But as in 1974, there was no plan for how to do this, nor little evidence that delegates understood the scale of the effort needed. FAO projections released in late 1999, just three years after the new, modest goal was set, acknowledge that the objective for 2015 is not likely to be reached because "the momentum is too slow and the progress too uneven."

Assertions such as Kissinger's and those of other political leaders may make people feel good, but if they are not grounded in a carefully thought out plan of action and supported by the relevant governments, they ultimately undermine confidence in the public process. This in turn can itself undermine progress.

In its most basic form, hunger is a productivity problem. Typically people are hungry because they do not produce enough food to meet their needs or because they do not earn enough money to buy it. The only lasting solution is to raise the productivity of the hungry—a task complicated by the ongoing shrinkage in cropland per person in developing countries.

From Lester R. Brown, "Eradicating Hunger: A Growing Challenge," in Lester R. Brown et al., *State of the World 2001: A Worldwatch Institute Report on Progress Toward a Sustainable Society* (W. W. Norton, 2001). Copyright © 2001 by The Worldwatch Institute. Reprinted by permission of W. W. Norton & Company, Inc. Notes omitted.

The war against hunger cannot be won with business as usual. Given the forces at work, it will no longer be possible to stand still. If societies do not take decisive steps, they face the possibility of being forced into involuntary retreat by continuing population growth, spreading land hunger, deepening hydrological poverty, increasing climate instability, and a shrinking backlog of unused agricultural technology. Eradicating hunger—never easy—will now take a superhuman effort.

A Hunger Report: Status and Prospects

As noted, 1.1 billion people are undernourished and underweight as the new century begins. The meshing of this number with a World Bank estimate of 1.3 billion living in poverty, defined as those living on $1 a day or less, comes as no surprise. Poverty and hunger go hand in hand.

The alarming extent of hunger in the world today comes after a half-century during which world food output nearly tripled. The good news is that the share of population that is hungry is diminishing in all regions except Africa. Since 1980, both East Asia and Latin America have substantially reduced the number and the share of their populations that are hungry. In the Indian subcontinent, results have been mixed, with the number of hungry continuing to increase but the share declining slightly. In Africa, however, both the number and the share of hungry people have increased since 1980.

The decline in East Asia was led by China, which brought the share of its people who are hungry down from 30 percent in 1980 to 11 percent in 1997. China's economic reforms initiated in 1978 led to a remarkable surge in agricultural output, one that boosted the grain harvest from roughly 200 kilograms per person a year to nearly 300 kilograms. This record jump in production in less than a decade led to the largest reduction in hunger on record. For most countries, the best nutrition data available are for children, who are also the segment of society most vulnerable to food scarcity. In Latin America, the share of children who are undernourished dropped from 14 percent in 1980 to 6 percent in 2000.

These gains in eradicating hunger in East Asia and Latin America leave most of the world's hungry concentrated in two regions: the Indian subcontinent and sub-Saharan Africa. In India, with more than a billion people, 53 percent of all children are undernourished. In Bangladesh, the share is 56 percent. And in Pakistan, it is 38 percent. In Africa, the share of children who are undernourished has increased from 26 percent in 1980 to 28 percent today. In Ethiopia, 48 percent of all children are underweight. In Nigeria, the most populous country in Africa, the figure is 39 percent.

Within the Indian subcontinent and sub-Saharan Africa, most of the hungry live in the countryside. The World Bank reports that 72 percent of the world's 1.3 billion poor live in rural areas. Most of them are undernourished; many are sentenced to a short life. These rural poor usually do not live on the productive irrigated plains, but on the semiarid/arid fringes of agriculture or in the upper reaches of watersheds on steeply sloping land that is highly erodible. Eradicating hunger depends on stabilizing these fragile ecosystems.

Recognizing that malnutrition is largely the result of rural poverty, the World Bank is replacing its long-standing agricultural development strategies, which were centered around crop production, with rural development strategies that use a much broader approach. The Bank planners believe that a more systemic approach to eradicating poverty in rural areas—one that embraces agriculture but that also integrates human capital development, the development of infrastructure, and social development into a strategy for rural development—is needed to shrink the number living in poverty. One advantage of encouraging investment in the countryside in both agribusiness and other industries is that it encourages breadwinners to stay in the countryside, keeping families and communities intact. In the absence of such a strategy, rural poverty simply feeds urban poverty.

Demographically, most of the world's poor live in countries where populations continue to grow rapidly, countries where poverty and population growth are reinforcing each other. The Indian subcontinent, for example, is adding 21 million people a year, the equivalent of another Australia. India's population has nearly tripled over the last half-century, growing from 350 million in 1950 to 1 billion in 2000. According to the U.N. projections, India will add 515 million more people by 2050, in effect adding roughly twice the current U.S. population. Pakistan's numbers, which tripled over the last half-century, are now expected to more than double over the next 50 years, going from 156 million to 345 million in 2050. And Bangladesh is projected to add 83 million people during this time, going from 129 million to 212 million. The subcontinent, already the hungriest region on Earth, is thus expected to add another 787 million people by mid-century.

No single factor bears so directly on the prospect of eradicating hunger in this region as population growth. When a farm passes from one generation to the next, it is typically subdivided among the children. With the second generation of rapid population growth and associated land fragmentation now unfolding, farms are shrinking to the point where they can no longer adequately support the people living on them.

Between 1970 and 1990, the number of farms in India with less than 2 hectares (5 acres) of land increased from 49 million to 82 million. Assuming that this trend has continued since then, India may now have 90 million or more families with farms of less than 2 hectares. If each family has six members, then 540 million people—over half India's population—are trapped in a precarious balance with their land.

Whether measuring changes in farm size or in grainland per person, the results of continuing rapid population growth are the same. Pakistan's projected growth from 156 million today to 345 million by 2050 will shrink its grainland per person from 0.08 hectares at present to 0.03 hectares, an area scarcely the size of a tennis court. African countries, with the world's fastest population growth, are facing a similar reduction. For example, as Nigeria's population increases from 111 million today to a projected 244 million in 2050, its per capita grainland, most of it semiarid and unirrigated, will shrink from 0.15 hectares to 0.07 hectares. Nigeria's food prospect, if it stays on this population trajectory, is not promising.

In Bangladesh, average farm size has already fallen below 1 hectare. According to one study, Bangladesh's "strong tradition of bequeathing land in fixed proportions to all male and female heirs has led to increasing landlessness and extreme fragmentation of agricultural holdings." In addition to the millions who are now landless, millions more have plots so small that they are effectively landless.

Further complicating efforts to expand food production are water shortages. Of the nearly 3 billion people to be added to world population in the next 50 years, almost all will be added in countries already facing water shortages, such as India, Pakistan, and many countries in Africa. In India, water tables are falling in large areas as the demands of the current population exceed the sustainable yield of aquifers. For many countries facing water scarcity, trying to eradicate hunger while population continues to grow rapidly is like trying to walk up a down escalator.

Even as the world faces the prospect of adding 80 million people a year over the next two decades, expanding food production is becoming more difficult. In each of the three food systems—croplands, rangelands, and oceanic fisheries—output expanded dramatically during the last half of the twentieth century. Now all that has changed.

Between 1950 and 2000, the world production of grain, the principal product of croplands, expanded from 631 million tons to 1,860 million tons, nearly tripling. In per capita terms, production went from 247 kilograms per person in 1950 to an all-time high of 342 kilograms in 1984, a gain of nearly 40 percent as growth in the grain harvest outstripped that of population. After 1984, production slowed, falling behind population. Production per person declined to 308 kilograms in 2000, a drop of 10 percent from the peak in 1984. (See Figure 1.) This decline is concentrated in the former Soviet Union, where the economy has shrunk by half since 1990, and in Africa where rapid population growth has simply outrun grain production.

Roughly 1.2 billion tons of the world grain harvest are consumed directly as food, with most of the remaining 660 million tons being consumed indirectly in livestock, poultry, and aquacultural products. The share of total grain used for feed varies widely among the "big three" food producers—ranging from a low of 4 percent in India to 27 percent in China and 68 percent in the United States.

Over the last half-century, the world's demand for animal protein has soared. Expanded output of meat from rangelands and of seafood from oceanic fisheries has satisfied most of this demand. World production of beef and mutton increased from 24 million tons in 1950 to 67 million tons in 1999, a near tripling. Most of the growth, however, occurred between 1950 and 1990, when output went up 2.5 percent a year. Since then, beef and mutton production has expanded by only 0.6 percent a year. (See Figure 2.)

An estimated four fifths of the 67 million tons of beef and mutton produced worldwide in 1999, roughly 54 million tons, comes from animals that forage on rangelands. If the grain equivalent of the forage-based output is set at seven kilograms of grain per kilogram of beef or mutton, which is the conversion rate in feedlots, the beef and mutton produced on rangeland are the equivalent of 378 million tons of grain.

Figure 1

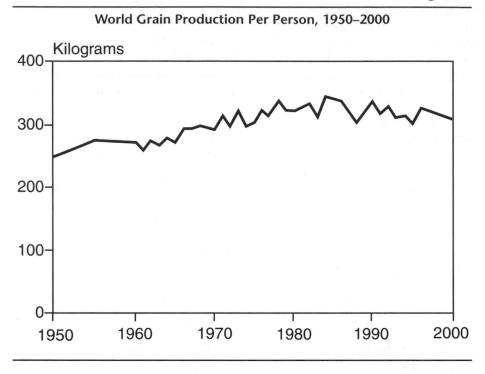

World Grain Production Per Person, 1950–2000

Source: USDA

The growth in the oceanic fish catch exceeded even that of beef and mutton, increasing from 19 million tons in 1950 to 86 million tons in 1998, the last year for which data are available. This fourfold growth, too, was concentrated in the period from 1950 to 1990, a time during which the annual growth in the oceanic catch—at 3.8 percent—was easily double that of world population. As a result, seafood consumption per person worldwide roughly doubled, climbing from 8 kilograms in 1950 to 16 kilograms in 1990. Since then, it has fallen by some 10 percent. Assuming the fish farm conversion of less than two kilograms of grain for each kilogram of live weight added, then the grain equivalent of the 86-million-ton fish catch in 1998 was 172 million tons of grain.

The new reality is that fishers and ranchers can no longer contribute much to the growth of the world's food supply. For the first time since civilization began, farmers must carry the burden alone.

For a sense of the relative importance of rangelands and oceanic fisheries in the world food economy, compare the grain equivalent of their output with the world grain harvest. With rangelands accounting for the equivalent of 378 million tons of grain and with fisheries at 172 million tons, rangelands contributed 16 percent of the world grain supply and oceanic fisheries 7 percent. (See Table 1.)

Figure 2

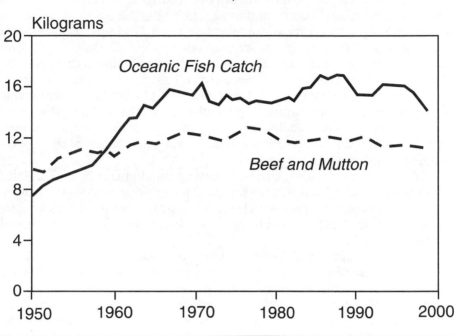

**Oceanic Fish Catch and Beef and Mutton Production
Per Person, 1950–99**

Source: FAO

Thus rangelands and oceanic fisheries provide the equivalent of nearly one fourth of the world grain supply. With their output no longer expected to expand, all future growth in the food supply must come from the 77 percent of total grain equivalent that is represented by croplands. With little new land to plow, the world's ability to eradicate hunger in the years ahead will depend

Table 1

**Cropland, Rangeland, and Oceanic Fishery Contributions to World Food
Supply, Measured in Grain Equivalent, 1999**

Source	Quantity of Grain (million tons)	Share of Total (percent)
Grain production from cropland	1,855	77
Grain equivalent of rangeland beef and mutton	378	16
Grain equivalent of oceanic fish catch[1]	172	7
Total	2,405	100

[1]Fish production data from 1998.

Source: USDA, *Production, Supply, and Distribution,* electronic database, Washington, DC, updated September 2000; FAO, *Yearbook of Fishery Statistics: Capture Production* (Rome: various years).

heavily on how much cropland productivity can be raised. This is also one of the keys to eliminating poverty.

Eradicating hunger in the Indian subcontinent and Africa will not be easy. It is difficult to eradicate for the same reasons it exists in the first place—rapid population growth, land hunger, and water scarcity. But there are also new forces that could complicate efforts to eliminate hunger. For example, Bangladesh, a country of 129 million people, has less than one tenth of a hectare of grainland per person—one of the smallest allotments in the world—and is threatened by rising sea level. The World Bank projects that a 1-meter rise in sea level during this century, the upper range of the recent projections by the Intergovernmental Panel on Climate Change, would cost Bangladesh half its riceland. This, combined with the prospect of adding another 83 million people over the next half-century, shows just how difficult it will be for Bangladesh, one of the world's hungriest countries, to feed its people.

Water shortages are forcing grain imports upward in many countries. North Africa and the Middle East is now the world's fastest growing grain import market. In 1999, Iran eclipsed Japan, which until recently was the world's leading importer of wheat. And Egypt, another water-short country, has also edged ahead of Japan.

Many developing countries that are facing acute land and water scarcity will rely on industrialization and labor-intensive industrial exports to finance needed food imports. This brings a need to expand production in exporting countries so they can cover the import needs of the growing number of grain-deficit countries. Over the last-half-century, grain-importing countries, now the overwhelming majority, have become dangerously dependent on the United States for nearly half of their grain imports.

This concentration of dependence applies to each of the big three grains—wheat, rice, and corn. Just five countries—the United States, Canada, France, Australia, and Argentina—account for 88 percent of the world's wheat exports. Thailand, Viet Nam, the United States, and China account for 68 percent of all rice exports. For corn, the concentration is even greater, with the United States alone accounting for 78 percent and Argentina for 12 percent.

With more extreme climate events in prospect if temperatures continue rising, this dependence on a few exporting countries leaves importers vulnerable to the vagaries of weather. If the United States were to experience a summer of severe heat and drought in its agricultural heartland like the summer of 1988, when grain production dropped below domestic consumption for the first time in history, chaos would reign in world grain markets simply because the near-record reserves that cushioned the huge U.S. crop shortfall that year no longer exist.

The risk for the scores of low-income, grain-importing countries is that prices could rise dramatically, impoverishing more people in a shorter period of time than any event in history. The resulting rise in hunger would be concentrated in the cities of the Third World.

POSTSCRIPT

Will the World Be Able to Feed Itself in the Foreseeable Future?

Presumably, economist Thomas Robert Malthus was not the first to address the question of the planet's ability to feed its population. But his 1789 *Essay on the Principle of Population* is the most quoted of early writings on the subject. Malthus' basic proposition was that population, if left unchecked, would grow geometrically, while subsistence resources could grow only arithmetically. Malthus, who wrote his essay in response to an argument with his father about the ability of the human race to produce sufficient resources vital for life, created a stir back in the late eighteenth century. The same debate holds the public's attention today as population grows at a rate unparalleled at any other time in human history.

Brunel argues that an examination of world agricultural production over a long period of time reveals that, in fact, it outpaced population growth. In the last 30 years of the twentieth century, for example, each individual's amount of food, in theory, grew by 20 percent. This equated to 2,700 calories per person per day, more than a person's estimated need. There is a catch, however. Half of the current grain production does not benefit humans. Approximately 20 percent is used to feed cattle, 5 percent is kept for seeds, and the remaining 25 percent is lost. In short, ecological factors are not at fault; humans, particularly from the developed world, are the culprits. Therein lies the solution for Brunel.

Brown sees some but not all of the problem rooted in human behavior. His pessimism arises from the belief that world leaders have not come forward with a comprehensive master plan to address the problem and are extremely unlikely to do so in the foreseeable future.

A balanced look at the planet's capacity to feed the UN's projected 2050 population is L.T. Evans' *Feeding the Ten Billion* (Cambridge University Press, 1998). In it, the author takes the reader through the ages, showing how the human race has addressed the agricultural needs of each succeeding billion people. The biggest challenge during the next half century, according to Evans, is to solve two problems: producing enough food for a 67 percent increase in the population, and eliminating the chronic undernutrition afflicting so many people. Solving the first problem requires a focus on the main components of increased food supply. For Evans, these include: "(1) increase in land under cultivation; (2) increase in yield per hectare per crop; (3) increase in the number of crops per hectare per year; (4) displacement of lower yielding crops by higher yielding ones; (5) reduction of post-harvest losses; [and] (6) reduced use as feed for animals."

The second problem brings into play many socioeconomic factors beyond those that are typically associated with agricultural production. Growing enough food worldwide is a necessary but not sufficient condition for eliminating problems associated with hunger. Many studies have observed that the root cause of hunger is poverty. Addressing poverty, therefore, is a prerequisite for ensuring that the world's food supply is distributed such that the challenge of global hunger is met.

Three reports by the UN's Food and Agricultural Organization (FAO) are worth reading: "World Agriculture: Towards 2000" (1988); "The State of Food and Agriculture 2002" (2002); and "World Agriculture: Towards 2015/2030" (2002). The central message of these studies is that the planet will be able to feed a growing population in the foreseeable future, if certain conditions are met. Another FAO report, "The State of Food Insecurity in the World 2000" (2002), analyzes the latest data on hunger and discusses approaches to fulfilling the commitments of the World Food Summit of 1996. FAO's 2003 report, "The State of Food Insecurity in the World 2003," suggests that world hunger is rising again. The report states that only 19 countries managed to reduce the number of undernourished during the 1990s.

An optimistic viewpoint about future food prospects is *The World Food Outlook* by Donald Mitchell, Merlinda D. Ingco, and Ronald C. Duncan (Cambridge University Press, 1997). Their basic conclusion is that the world food situation has improved dramatically for most of the regions of the globe and will continue to do so. The only exception is sub-Saharan Africa, but there the problems go far beyond agriculture.

ISSUE 8

Is the Threat of Global Warming Real?

YES: Intergovernmental Panel on Climate Change, from "Climate Change 2001: The Scientific Basis," A Report of Working Group I of the Intergovernmental Panel on Climate Change (2001)

NO: Christopher Essex and Ross McKitrick, from *Taken By Storm: The Troubled Science, Policy and Politics of Global Warming* (Key Porter Books, 2002)

ISSUE SUMMARY

YES: The summary of the most recent assessment of climatic change by a UN-sponsored group of scientists concludes that an increasing set of observations reveals that the world is warming and much of it is due to human intervention.

NO: Christopher Essex and Ross McKitrick, Canadian university professors of applied mathematics and economics, respectively, attempt to prove wrong the popularly held assumption that scientists know what is happening with respect to climate and weather, and thus understand the phenomenon of global warming.

At the UN-sponsored Earth Summit in Rio de Janeiro in 1992, a Global Climate Treaty was signed. According to S. Fred Singer, in *Hot Talks, Cold Science: Global Warming's Unfinished Debate* (Independent Institute, 1998), the treaty rested on three basic assumptions. First, global warming has been detected in the records of climate of the last 100 years. Second, a substantial warming in the future will produce catastrophic consequences—droughts, floods, storms, a rapid and significant rise in sea level, agricultural collapse, and the spread of tropical disease. And third, the scientific and policy-making communities know: (1) which atmospheric concentrations of greenhouse gases are dangerous and which ones are not, (2) that drastic reductions of carbon dioxide (CO_2) emissions as well as energy use in general by industrialized countries will stabilize CO_2 concentrations at close to current levels, and (3) that such economically damaging measures can be justified politically despite no significant scientific support for global warming as a threat.

Since the Earth Summit, it appears that scientists have opted for placement into one of three camps. The first camp buys into the three assumptions outlined above. In late 1995, 2,500 leading climate scientists announced in the first Intergovernmental Panel on Climatic Change (IPCC) report that the planet was warming due to coal and gas emissions. Scientists in a second camp suggest that while global warming has occurred and continues at the present, the source of such temperature rise cannot be ascertained yet. The conclusions of the Earth Summit were misunderstood by many in the scientific community, the second camp would suggest. For these scientists, computer models, the basis of much evidence for the first group, have not yet linked global warming to human activities.

A third group of scientists, representing a minority, argues that we cannot be certain that global warming is taking place, yet alone determine its cause. They present a number of arguments in support of their position. Among them is the contention that pre-satellite data (pre-1979) showing a century-long pattern of warming is an illusion because satellite data (post-1979) reveal no such warming. Furthermore, when warming was present, it did not occur at the same time as a rise in greenhouse gases. Scientists in the third camp are also skeptical of studying global warming in the laboratory. They suggest, moreover, that most of the scientists who have opted for one of the first two camps have done so as a consequence of laboratory experiments, rather than of evidence from the real world.

Despite what appear to be wide differences in scientific thinking about the existence of global warming and its origins, the global community has moved forward with attempts to achieve consensus among the nations of the world for taking appropriate action to curtail human activities thought to affect warming. A 1997 international meeting in Kyoto, Japan, concluded with an agreement for reaching goals established at the earlier Earth Summit. Thirty-eight industrialized countries, including the United States, agreed to reduction levels outlined in the treaty. However, the U.S. Senate never ratified the treaty, and the Bush administration decided not to support it. As this book went to press, 122 countries had signed the Kyoto Agreement, and Russian President Vladimir Putin has set in motion his country's joining this group. Once the latter signs, 55 percent of the developed world's carbon dioxide's emissions will be from signees, thus allowing the treaty to go into effect.

The first selection is the recent report by the IPCC, an international body of scientists created by the United Nations to address the issue of global warming. It provides the most recent analysis by a broad scientific community on the twin issues of global warming's existence and causes. It predicts that global temperature will likely rise by 1.4 to 5.8°C by the year 2100. The second selection by Christopher Essex and Ross McKitrick suggests a different conclusion to the question posed in this issue. To them, important components of what they term the "doctrine" of global warming—the conventional view described above—are either unproven or simply wrong. In their judgment, the science used by advocates of global warming may be good enough for popular journals read by interested citizens but it has not yet made a strong enough case for global policymakers.

Summary for Policymakers

This Summary for Policymakers (SPM), which was approved by IPCC [Intergovernmental Panel on Climate Change] member governments in Shanghai in January 2001,[1] describes the current state of understanding of the climate system and provides estimates of its projected future evolution and their uncertainties. Further details can be found in the underlying report, and the appended Source Information provides cross references to the report's chapters.

An increasing body of observations gives a collective picture of a warming world and other changes in the climate system.

Since the release of the Second Assessment Report (SAR[2]), additional data from new studies of current and palaeoclimates, improved analysis of data sets, more rigorous evaluation of their quality, and comparisons among data from different sources have led to greater understanding of climate change.

The global average surface temperature has increased over the 20th century by about 0.6°C.

- The global average surface temperature (the average of near surface air temperature over land, and sea surface temperature) has increased since 1861. Over the 20th century the increase has been 0.6 ± 0.2°C[3,4]. This value is about 0.15°C larger than that estimated by the SAR for the period up to 1994, owing to the relatively high temperatures of the additional years (1995 to 2000) and improved methods of processing the data. These numbers take into account various adjustments, including urban heat island effects. The record shows a great deal of variability; for example, most of the warming occurred during the 20th century, during two periods, 1910 to 1945 and 1976 to 2000.
- Globally, it is very likely[5] that the 1990s was the warmest decade and 1998 the warmest year in the instrumental record, since 1861.
- New analyses of proxy data for the Northern Hemisphere indicate that the increase in temperature in the 20th century is likely[5] to have been the largest of any century during the past 1,000 years. It is also likely[5] that, in the Northern Hemisphere, the 1990s was the warmest decade

and 1998 the warmest year. Because less data are available, less is known about annual averages prior to 1,000 years before present and for conditions prevailing in most of the Southern Hemisphere prior to 1861.

- On average, between 1950 and 1993, night-time daily minimum air temperatures over land increased by about 0.2°C per decade. This is about twice the rate of increase in daytime daily maximum air temperatures (0.1°C per decade). This has lengthened the freeze-free season in many mid- and high latitude regions. The increase in sea surface temperature over this period is about half that of the mean land surface air temperature.

Temperatures have risen during the past four decades in the lowest 8 kilometres of the atmosphere.

- Since the late 1950s (the period of adequate observations from weather balloons), the overall global temperature increases in the lowest 8 kilometres of the atmosphere and in surface temperature have been similar at 0.1°C per decade.
- Since the start of the satellite record in 1979, both satellite and weather balloon measurements show that the global average temperature of the lowest 8 kilometres of the atmosphere has changed by +0.05 ± 0.10°C per decade, but the global average surface temperature has increased significantly by +0.15 ± 0.05°C per decade. The difference in the warming rates is statistically significant. This difference occurs primarily over the tropical and sub-tropical regions.
- The lowest 8 kilometres of the atmosphere and the surface are influenced differently by factors such as stratospheric ozone depletion, atmospheric aerosols, and the El Niño phenomenon. Hence, it is physically plausible to expect that over a short time period (e.g., 20 years) there may be differences in temperature trends. In addition, spatial sampling techniques can also explain some of the differences in trends, but these differences are not fully resolved.

Snow cover and ice extent have decreased.

- Satellite data show that there are very likely[5] to have been decreases of about 10% in the extent of snow cover since the late 1960s, and ground-based observations show that there is very likely[5] to have been a reduction of about two weeks in the annual duration of lake and river ice cover in the mid- and high latitudes of the Northern Hemisphere, over the 20th century.
- There has been a widespread retreat of mountain glaciers in non-polar regions during the 20th century.
- Northern Hemisphere spring and summer sea-ice extent has decreased by about 10 to 15% since the 1950s. It is likely[5] that there has been about a 40% decline in Arctic sea-ice thickness during late summer to early autumn in recent decades and a considerably slower decline in winter sea-ice thickness.

Global average sea level has risen and ocean heat content has increased.

- Tide gauge data show that global average sea level rose between 0.1 and 0.2 metres during the 20th century.
- Global ocean heat content has increased since the late 1950s, the period for which adequate observations of sub-surface ocean temperatures have been available.

Changes have also occurred in other important aspects of climate.

- It is very likely[5] that precipitation has increased by 0.5 to 1% per decade in the 20th century over most mid- and high latitudes of the Northern Hemisphere continents, and it is likely[5] that rainfall has increased by 0.2 to 0.3% per decade over the tropical (10°N to 10°S) land areas. Increases in the tropics are not evident over the past few decades. It is also likely[5] that rainfall has decreased over much of the Northern Hemisphere sub-tropical (10°N to 30°N) land areas during the 20th century by about 0.3% per decade. In contrast to the Northern Hemisphere, no comparable systematic changes have been detected in broad latitudinal averages over the Southern Hemisphere. There are insufficient data to establish trends in precipitation over the oceans.
- In the mid- and high latitudes of the Northern Hemisphere over the latter half of the 20th century, it is likely[5] that there has been a 2 to 4% increase in the frequency of heavy precipitation events. Increases in heavy precipitation events can arise from a number of causes, e.g., changes in atmospheric moisture, thunderstorm activity and largescale storm activity.
- It is likely[5] that there has been a 2% increase in cloud cover over mid- to high latitude land areas during the 20th century. In most areas the trends relate well to the observed decrease in daily temperature range.
- Since 1950 it is very likely[5] that there has been a reduction in the frequency of extreme low temperatures, with a smaller increase in the frequency of extreme high temperatures.
- Warm episodes of the El Niño-Southern Oscillation (ENSO) phenomenon (which consistently affects regional variations of precipitation and temperature over much of the tropics, sub-tropics and some mid-latitude areas) have been more frequent, persistent and intense since the mid-1970s, compared with the previous 100 years.
- Over the 20th century (1900 to 1995), there were relatively small increases in global land areas experiencing severe drought or severe wetness. In many regions, these changes are dominated by inter-decadal and multi-decadal climate variability, such as the shift in ENSO towards more warm events.
- In some regions, such as parts of Asia and Africa, the frequency and intensity of droughts have been observed to increase in recent decades.

Some important aspects of climate appear not to have changed.

- A few areas of the globe have not warmed in recent decades, mainly over some parts of the Southern Hemisphere oceans and parts of Antarctica.
- No significant trends of Antarctic sea-ice extent are apparent since 1978, the period of reliable satellite measurements.

- Changes globally in tropical and extra-tropical storm intensity and frequency are dominated by inter-decadal to multi-decadal variations, with no significant trends evident over the 20th century. Conflicting analyses make it difficult to draw definitive conclusions about changes in storm activity, especially in the extra-tropics.
- No systematic changes in the frequency of tornadoes, thunder days, or hail events are evident in the limited areas analysed.

Emissions of greenhouse gases and aerosols due to human activities continue to alter the atmosphere in ways that are expected to affect the climate.

Changes in climate occur as a result of both internal variability within the climate system and external factors (both natural and anthropogenic). The influence of external factors on climate can be broadly compared using the concept of radiative forcing.[6] A positive radiative forcing, such as that produced by increasing concentrations of greenhouse gases, tends to warm the surface. A negative radiative forcing, which can arise from an increase in some types of aerosols (microscopic airborne particles) tends to cool the surface. Natural factors, such as changes in solar output or explosive volcanic activity, can also cause radiative forcing. Characterisation of these climate forcing agents and their changes over time is required to understand past climate changes in the context of natural variations and to project what climate changes could lie ahead. . . .

Concentrations of atmospheric greenhouse gases and their radiative forcing have continued to increase as a result of human activities.

- The atmospheric concentration of carbon dioxide (CO_2) has increased by 31% since 1750. The present CO_2 concentration has not been exceeded during the past 420,000 years and likely[5] not during the past 20 million years. The current rate of increase is unprecedented during at least the past 20,000 years.
- About three-quarters of the anthropogenic emissions of CO_2 to the atmosphere during the past 20 years is due to fossil fuel burning. The rest is predominantly due to land-use change, especially deforestation.
- Currently the ocean and the land together are taking up about half of the anthropogenic CO_2 emissions. On land, the uptake of anthropogenic CO_2 very likely[5] exceeded the release of CO_2 by deforestation during the 1990s.
- The rate of increase of atmospheric CO_2 concentration has been about 1.5 ppm[7] (0.4%) per year over the past two decades. During the 1990s the year to year increase varied from 0.9 ppm (0.2%) to 2.8 ppm (0.8%). A large part of this variability is due to the effect of climate variability (e.g., El Niño events) on CO_2 uptake and release by land and oceans.
- The atmospheric concentration of methane (CH_4) has increased by 1060 ppb[7] (151%) since 1750 and continues to increase. The present CH_4 concentration has not been exceeded during the past 420,000 years. The annual growth in CH_4 concentration slowed and became more variable in the 1990s, compared with the 1980s. Slightly more than half of current CH_4 emissions are anthropogenic (e.g., use of fossil fuels, cattle, rice agriculture and landfills). In addition, carbon monoxide

(CO) emissions have recently been identified as a cause of increasing CH_4 concentration.

- The atmospheric concentration of nitrous oxide (N_2O) has increased by 46 ppb (17%) since 1750 and continues to increase. The present N_2O concentration has not been exceeded during at least the past thousand years. About a third of current N_2O emissions are anthropogenic (e.g., agricultural soils, cattle feed lots and chemical industry).

- Since 1995, the atmospheric concentrations of many of those halocarbon gases that are both ozone-depleting and greenhouse gases (e.g., $CFCl_3$ and CF_2Cl_2), are either increasing more slowly or decreasing, both in response to reduced emissions under the regulations of the Montreal Protocol and its Amendments. Their substitute compounds (e.g., CHF_2Cl and CF_3CH_2F) and some other synthetic compounds (e.g., perfluorocarbons (PFCs) and sulphur hexafluoride (SF_6)) are also greenhouse gases, and their concentrations are currently increasing.

- The radiative forcing due to increases of the well-mixed greenhouse gases from 1750 to 2000 is estimated to be 2.43 Wm^{-2}: 1.46 Wm^{-2} from CO_2; 0.48 Wm^{-2} from CH_4; 0.34 Wm^{-2} from the halocarbons; and 0.15 Wm^{-2} from N_2O. . . .

- The observed depletion of the stratospheric ozone (O_3) layer from 1979 to 2000 is estimated to have caused a negative radiative forcing (−0.15 Wm^{-2}). Assuming full compliance with current halocarbon regulations, the positive forcing of the halocarbons will be reduced as will the magnitude of the negative forcing from stratospheric ozone depletion as the ozone layer recovers over the 21st century.

- The total amount of O_3 in the troposphere is estimated to have increased by 36% since 1750, due primarily to anthropogenic emissions of several O_3-forming gases. This corresponds to a positive radiative forcing of 0.35 Wm^{-2}. O_3 forcing varies considerably by region and responds much more quickly to changes in emissions than the long-lived greenhouse gases, such as CO_2.

Anthropogenic aerosols are short-lived and mostly produce negative radiative forcing.

- The major sources of anthropogenic aerosols are fossil fuel and biomass burning. These sources are also linked to degradation of air quality and acid deposition.

- Since the SAR, significant progress has been achieved in better characterising the direct radiative roles of different types of aerosols. Direct radiative forcing is estimated to be −0.4 Wm^{-2} for sulphate, −0.2 Wm^{-2} for biomass burning aerosols, −0.1 Wm^{-2} for fossil fuel organic carbon and +0.2 Wm^{-2} for fossil fuel black carbon aerosols. There is much less confidence in the ability to quantify the total aerosol direct effect, and its evolution over time, than that for the gases listed above. Aerosols also vary considerably by region and respond quickly to changes in emissions.

- In addition to their direct radiative forcing, aerosols have an indirect radiative forcing through their effects on clouds. There is now more evidence for this indirect effect, which is negative, although of very uncertain magnitude.

Natural factors have made small contributions to radiative forcing over the past century.

- The radiative forcing due to changes in solar irradiance for the period since 1750 is estimated to be about +0.3 Wm^{-2}, most of which occurred during the first half of the 20th century. Since the late 1970s, satellite instruments have observed small oscillations due to the 11-year solar cycle. Mechanisms for the amplification of solar effects on climate have been proposed, but currently lack a rigorous theoretical or observational basis.
- Stratospheric aerosols from explosive volcanic eruptions lead to negative forcing, which lasts a few years. Several major eruptions occurred in the periods 1880 to 1920 and 1960 to 1991.
- The combined change in radiative forcing of the two major natural factors (solar variation and volcanic aerosols) is estimated to be negative for the past two, and possibly the past four, decades. . . .

There is new and stronger evidence that most of the warming observed over the last 50 years is attributable to human activities.

The SAR concluded: "The balance of evidence suggests a discernible human influence on global climate." That report also noted that the anthropogenic signal was still emerging from the background of natural climate variability. Since the SAR, progress has been made in reducing uncertainty, particularly with respect to distinguishing and quantifying the magnitude of responses to different external influences. Although many of the sources of uncertainty identified in the SAR still remain to some degree, new evidence and improved understanding support an updated conclusion.

- There is a longer and more closely scrutinised temperature record and new model estimates of variability. The warming over the past 100 years is very unlikely[5] to be due to internal variability alone, as estimated by current models. Reconstructions of climate data for the past 1,000 years also indicate that this warming was unusual and is unlikely[5] to be entirely natural in origin.
- There are new estimates of the climate response to natural and anthropogenic forcing, and new detection techniques have been applied. Detection and attribution studies consistently find evidence for an anthropogenic signal in the climate record of the last 35 to 50 years.
- Simulations of the response to natural forcings alone (i.e., the response to variability in solar irradiance and volcanic eruptions) do not explain the warming in the second half of the 20th century. However, they indicate that natural forcings may have contributed to the observed warming in the first half of the 20th century.
- The warming over the last 50 years due to anthropogenic greenhouse gases can be identified despite uncertainties in forcing due to anthropogenic sulphate aerosol and natural factors (volcanoes and solar irradiance). The anthropogenic sulphate aerosol forcing, while uncertain, is negative over this period and therefore cannot explain the warming. Changes in natural forcing during most of this period are also estimated to be negative and are unlikely[5] to explain the warming.

- Detection and attribution studies comparing model simulated changes with the observed record can now take into account uncertainty in the magnitude of modelled response to external forcing, in particular that due to uncertainty in climate sensitivity.
- Most of these studies find that, over the last 50 years, the estimated rate and magnitude of warming due to increasing concentrations of greenhouse gases alone are comparable with, or larger than, the observed warming. Furthermore, most model estimates that take into account both greenhouse gases and sulphate aerosols are consistent with observations over this period.
- The best agreement between model simulations and observations over the last 140 years has been found when all the above anthropogenic and natural forcing factors are combined. These results show that the forcings included are sufficient to explain the observed changes, but do not exclude the possibility that other forcings may also have contributed.

In the light of new evidence and taking into account the remaining uncertainties, most of the observed warming over the last 50 years is likely[5] to have been due to the increase in greenhouse gas concentrations.

Furthermore, it is very likely[5] that the 20th century warming has contributed significantly to the observed sea level rise, through thermal expansion of sea water and widespread loss of land ice. Within present uncertainties, observations and models are both consistent with a lack of significant acceleration of sea level rise during the 20th century.

> Human influences will continue to change atmospheric composition throughout the 21st century.

Models have been used to make projections of atmospheric concentrations of greenhouse gases and aerosols, and hence of future climate, based upon emissions scenarios from the IPCC Special Report on Emission Scenarios (SRES). These scenarios were developed to update the IS92 series, which were used in the SAR and are shown for comparison here in some cases.

Greenhouse gases

- Emissions of CO_2 due to fossil fuel burning are virtually certain[5] to be the dominant influence on the trends in atmospheric CO_2 concentration during the 21st century.
- As the CO_2 concentration of the atmosphere increases, ocean and land will take up a decreasing fraction of anthropogenic CO_2 emissions. The net effect of land and ocean climate feedbacks as indicated by models is to further increase projected atmospheric CO_2 concentrations, by reducing both the ocean and land uptake of CO_2.
- By 2100, carbon cycle models project atmospheric CO_2 concentrations of 540 to 970 ppm for the illustrative SRES scenarios (90 to 250% above the concentration of 280 ppm in the year 1750). These projections include the land and ocean climate feedbacks. Uncertainties, especially about the magnitude of the climate feedback from the terrestrial biosphere, cause a variation of about −10 to +30% around each scenario. The total range is 490 to 1260 ppm (75 to 350% above the 1750 concentration).

- Changing land use could influence atmospheric CO_2 concentration. Hypothetically, if all of the carbon released by historical land-use changes could be restored to the terrestrial biosphere over the course of the century (e.g., by reforestation), CO_2 concentration would be reduced by 40 to 70 ppm.
- Model calculations of the concentrations of the non-CO_2 greenhouse gases by 2100 vary considerably across the SRES illustrative scenarios, with CH_4 changing by –190 to +1,970 ppb (present concentration 1,760 ppb), N_2O changing by +38 to +144 ppb (present concentration 316 ppb), total tropospheric O_3 changing by –12 to +62%, and a wide range of changes in concentrations of HFCs, PFCs and SF_6, all relative to the year 2000. In some scenarios, total tropospheric O_3 would become as important a radiative forcing agent as CH_4 and, over much of the Northern Hemisphere, would threaten the attainment of current air quality targets.
- Reductions in greenhouse gas emissions and the gases that control their concentration would be necessary to stabilise radiative forcing. For example, for the most important anthropogenic greenhouse gas, carbon cycle models indicate that stabilisation of atmospheric CO_2 concentrations at 450, 650 or 1,000 ppm would require global anthropogenic CO_2 emissions to drop below 1990 levels, within a few decades, about a century, or about two centuries, respectively, and continue to decrease steadily thereafter. Eventually CO_2 emissions would need to decline to a very small fraction of current emissions.

Aerosols

- The SRES scenarios include the possibility of either increases or decreases in anthropogenic aerosols (e.g., sulphate aerosols, biomass aerosols, black and organic carbon aerosols) depending on the extent of fossil fuel use and policies to abate polluting emissions. In addition, natural aerosols (e.g., sea salt, dust and emissions leading to the production of sulphate and carbon aerosols) are projected to increase as a result of changes in climate.

Radiative forcing over the 21st century

- For the SRES illustrative scenarios, relative to the year 2000, the global mean radiative forcing due to greenhouse gases continues to increase through the 21st century, with the fraction due to CO_2 projected to increase from slightly more than half to about three quarters. The change in the direct plus indirect aerosol radiative forcing is projected to be smaller in magnitude than that of CO_2.

Global average temperature and sea level are projected to rise under all IPCC SRES scenarios.

In order to make projections of future climate, models incorporate past, as well as future emissions of greenhouse gases and aerosols. Hence, they include estimates of warming to date and the commitment to future warming from past emissions.

Temperature

- The globally averaged surface temperature is projected to increase by 1.4 to 5.8°C over the period 1990 to 2100. These results are for the full range of 35 SRES scenarios, based on a number of climate models.[8,9]
- Temperature increases are projected to be greater than those in the SAR, which were about 1.0 to 3.5°C based on the six IS92 scenarios. The higher projected temperatures and the wider range are due primarily to the lower projected sulphur dioxide emissions in the SRES scenarios relative to the IS92 scenarios.
- The projected rate of warming is much larger than the observed changes during the 20th century and is very likely[5] to be without precedent during at least the last 10,000 years, based on palaeoclimate data.
- By 2100, the range in the surface temperature response across the group of climate models run with a given scenario is comparable to the range obtained from a single model run with the different SRES scenarios.
- On timescales of a few decades, the current observed rate of warming can be used to constrain the projected response to a given emissions scenario despite uncertainty in climate sensitivity. This approach suggests that anthropogenic warming is likely[5] to lie in the range of 0.1 to 0.2°C per decade over the next few decades under the IS92a scenario. . . .
- Based on recent global model simulations, it is very likely[5] that nearly all land areas will warm more rapidly than the global average, particularly those at northern high latitudes in the cold season. Most notable of these is the warming in the northern regions of North America, and northern and central Asia, which exceeds global mean warming in each model by more than 40%. In contrast, the warming is less than the global mean change in south and southeast Asia in summer and in southern South America in winter.
- Recent trends for surface temperature to become more El Niño-like in the tropical Pacific, with the eastern tropical Pacific warming more than the western tropical Pacific, with a corresponding eastward shift of precipitation, are projected to continue in many models.

Precipitation

- Based on global model simulations and for a wide range of scenarios, global average water vapour concentration and precipitation are projected to increase during the 21st century. By the second half of the 21st century, it is likely[5] that precipitation will have increased over northern mid- to high latitudes and Antarctica in winter. At low latitudes there are both regional increases and decreases over land areas. Larger year to year variations in precipitation are very likely[5] over most areas where an increase in mean precipitation is projected.

Notes

1. Delegations of 99 IPCC member countries participated in the Eighth Session of Working Group I in Shanghai on 17 to 20 January 2001.
2. The IPCC Second Assessment Report is referred to in this Summary for Policymakers as the SAR.
3. Generally temperature trends are rounded to the nearest 0.05°C per unit time, the periods often being limited by data availability.

4. In general, a 5% statistical significance level is used, and a 95% confidence level.

5. In this Summary for Policymakers and in the Technical Summary, the following words have been used where appropriate to indicate judgmental estimates of confidence: *virtually certain* (greater than 99% chance that a result is true); *very likely* (90–99% chance); *likely* (66–90% chance); *medium likelihood* (33–66% chance); *unlikely* (10–33% chance); *very unlikely* (1–10% chance); *exceptionally unlikely* (less than 1% chance). . . .

6. *Radiative forcing* is a measure of the influence a factor has in altering the balance of incoming and outgoing energy in the Earth-atmosphere system, and is an index of the importance of the factor as a potential climate change mechanism. It is expressed in Watts per square metre (Wm^{-2}).

7. ppm (parts per million) or ppb (parts per billion, 1 billion = 1,000 million) is the ratio of the number of greenhouse gas molecules to the total number of molecules of dry air. For example: 300 ppm means 300 molecules of a greenhouse gas per million molecules of dry air.

8. Complex physically based climate models are the main tool for projecting future climate change. In order to explore the full range of scenarios, these are complemented by simple climate models calibrated to yield an equivalent response in temperature and sea level to complex climate models. These projections are obtained using a simple climate model whose climate sensitivity and ocean heat uptake are calibrated to each of seven complex climate models. The climate sensitivity used in the simple model ranges from 1.7 to 4.2°C, which is comparable to the commonly accepted range of 1.5 to 4.5°C.

9. This range does not include uncertainties in the modelling of radiative forcing, e.g., aerosol forcing uncertainties. A small carbon-cycle climate feedback is included.

NO ←

Christopher Essex and
Ross McKitrick

Taken By Storm: The Troubled Science, Policy and Politics of Global Warming

After Doctrine: Making Policy Amid Uncertainty and Nescience

After Doctrine

. . . [The Doctrine] it consists of the following familiar ideas.

1. The earth is warming.
2. Warming has already been observed.
3. Humans are causing it.
4. All but a handful of scientists on the fringe believe it.
5. Warming is bad.
6. Action is required immediately.
7. Any action is better than none.
8. Uncertainty only covers the ulterior motives of individuals aiming to stop needed action.
9. Those who defend uncertainty are bad people.

Let's go over these points one by one:

Is the Earth Warming?

. . . This cannot seem like a very sensible question. It sets up a simplistic context in which to view climate change. No direct answer to the question is true; yes and no are both wrong. Someone who skipped straight to this chapter might be puzzled, but it is really a simple concept in the end. There is no single physical variable that describes warming and cooling, for the whole Earth; there is no single global temperature. This is just a basic truth of thermodynamics.

If there were some climate changes in the category of our sun going nova, or even something more moderate like a major ice age, then all of the infinity of local temperatures would be saying the same thing; namely, that it is heating or cooling everywhere and everywhen. But even the most strident and extreme

From TAKEN BY STORM by Christopher Essex and Ross McKitrick, pp. 289–297. Copyright © 2002 by Key Porter Books. Reprinted by permission.

scenarios put forward by the Panel of the United Nations (PUN) are Lilliputian in comparison. Contrary movements in the local temperature field everywhere would swamp any potential "signal."

The only way to force a simple warming or cooling picture onto a field of local temperatures is to say that, in some way, the places where and when it is cooling outweigh the places where and when it is warming. There is a problem doing this with temperature because there is no way to weight temperature at one point against temperature at another. Mathematicians would say it has no integration measure in space. It is an intensive thermodynamic variable, one of many quantities in thermodynamics that represents a condition and not an amount of something.

So if you want to take some average of the temperature field, it has very little physical meaning, unlike an average over energy, or the height of people in a classroom. Lots of—infinitely many—different averages over temperature are mathematically possible, but none of them has direct physical significance for the Earth. Any resulting average does not represent the temperature of anything at all. This is complicated by the huge variations in local temperatures up and down in comparison to what PUN claims to be looking for. You might get away with calling some average an index, but this huge variation virtually guarantees that different averages will behave quite differently from each other.

We have already seen that different published averages conducted over different pieces of the Earth's temperature field in different ways behave very differently from each other. This is to be expected. It would be remarkable if it were otherwise. Indeed, . . . a simple example of a temperature field in which two equally plausible mathematical averages over the same data from a temperature field showed opposite trends. It would not be hard to construct families of averages over the Earth's data that decline instead of the more popular trend in the opposite direction. This is not physics; it is mathematics.

Plain-vanilla warming or cooling is not how climate change actually works. There is a lot more going on than temperature in climate change. And that is why the big climate models are not just thermometers, but must also treat a huge spectrum of fiendishly complex dynamics as best they can.

Has Warming Been Observed Already?
Sure. You can find lots of places in the world where temperatures have gone up recently. You can find lots of places where temperatures have gone down, too. But what people have in mind here is that an "unnatural" warming has been observed here or there. To conclude that would require some idea of what the natural temperature in a location is, but there is no such single thing. Nor do we have any theory of the climate that would define natural levels or rates of change in temperatures in any location.

Are Humans Causing It?
Causing what? Causing someone's favourite global average temperature statistic (i.e., T-Rex) to go up, or causing climate change? The former doesn't make much sense, while the latter is more complicated. Commonly used methods that purport to show humans causing climate change rely on climate models,

which, . . . are not suitable for this kind of purpose. And the absence of an underlying theory means that any statistical method, . . . must fall short because they compare changes in some arbitrarily defined statistic to an arbitrarily defined alternative.

Moreover, as the signal is generated from a turbulent and chaotic deterministic dynamical process, we ultimately must be looking for a signal on an unknown chaotic carrier. In chaos cryptography, classic statistical methods cannot crack an encrypted signal carried on a chaotic carrier. The only way to do so is to have the equations of the chaotic carrier to work with. No one knows how to get hold of them for climate.

Do All But a Handful of Scientists on the Fringe Believe It?
This is the sort of claim that governments have found very useful to spread around. It certainly makes their job easier, since they can avoid the tough business of having to think through a complicated issue. . . . The interaction of government and Official Science has led to great declarations of certainty, as well as political marginalization of those who try to voice their doubts.

So it is understandable that people think only a handful of "marginal" critics have doubts. The critics are often referred to as "skeptics." A skeptic is someone that true believers do not want to invite to a seance. Neither are they generally invited to Big Panel meetings, though the two events are not quite the same.

One of the ways people have tried to argue that there is more than a handful of critics is to circulate petitions and get scientists to publicly own up to their doubts. We have not argued along this line. Instead we have simply shown, page after page, that certainty on the subject of the future direction of climate is impossible. There is no theoretical basis for predicting climate; models do not provide a substitute for theory, and the profound complexity of such things as turbulence and chaotic systems means that anyone who thinks we can predict the climate only courts the laughter of the gods. So if this part of the Doctrine is correct, then the scientific community is in a sorry state.

The truth of a scientific fact does not depend upon a democratic vote among "qualified" people. But that is what governments, Official Science and advocacy groups have pushed. Moreover, there is no consensus in the scientific community anyway. The existence of the PUN report does not establish that at all. It is like a collection of short stories on many topics, with a small summary attached in which a small group of people offers its interpretation. What each scientist wrote was neither conclusive nor comprehensive in terms of the overall position set out in the SPAMs. There was no simple proposition that the authors of the scientific report endorsed, and many would not agree with many aspects of the SPAMs.

There was no consensus among scientists, who are always debating debatable issues. And climate change is debatable, to say the least! What genuine scientists all agree on is that science is a personal journey. It is not an exercise in authoritarianism in which edicts on the truth sever the world into patriots and dissidents. Such a picture invokes aspects of politics offensive to free and thinking people. It is not science.

Is It Bad? Should We Act Immediately?

Items 5 and 6 founder on the basic problem that if we cannot forecast something, we do not know whether it will be bad. Moreover, we cannot say it is bad or not if we aren't even sure what "it" is. Clearly we cannot say that "it" is all bad, and getting even more bad, as some seem to want to do. We cannot function rationally this way. It degenerates into nothing more than modern-day soothsaying without proper skepticism about the connections of bad events to climate change.

We may never even know if "it" is happening! As we discussed in Chapter 8, the genre of "impacts" studies is marked by an inordinate focus on temperature (often naively understood), a false belief that we can predict climate on a local scale far into the future, and a persistent inability to recognize the ingenuity with which people adapt to, and prosper under, changes of all kinds.

Is Any Action Better Than None?

Item 7 gets to the heart of thinking about policy. [We earlier] talked about factors like costs and benefits and the question of whether Kyoto is worth doing. We think not. And there are many other related policy ideas that have been floated over the past few years that make equally little sense. Consequently item 7 cannot be true. Not every action is worth the expense.

Are Critics of the Doctrine Bad People With Ulterior Motives?

As for items 8 and 9, you are on your own with those. If you think the present book has been offered to the public as part of a plot by bad people who are just covering ulterior and malicious motives, well, so be it. Nothing we say now will convince you otherwise. And as we talked about in Chapter 7, uncertainty is not really the problem. Nescience is the problem. The difference between uncertainty and nescience is like the difference between looking for your lost puppy in your neighbourhood and looking for a purple dinosaur in an undefined parallel dimension.

Once that distinction between nescience and uncertainty is understood, we are in a position to have a realistic discussion about policy.

What Should We Do About Global Warming?

If you have skipped directly to this page, go back to the beginning of the book. This section will make no sense to you unless you have read what came before. It is certainly not a "Summary for Policymakers." Start at page 1 and take the time to read what follows. After all, a big part of the answer to the question above is that we ought to stop relying on quick summaries and actually take the time to think through this issue carefully.

The question as posed assumes that we are not already "doing" something when we are studying the issue. But at this point, study is precisely what we ought to be doing. There is a lot of science on which we can draw when trying to understand the climate change issue. But bear in mind a distinction: There is science that is good enough for, say, *Science* magazine, and there is science that is good enough for making policy. They are not the same things, even though they have been muddled together in this problem and many others involving big issues in science.

Articles that are good enough to publish do not have to be right or rock-solid or magisterial. They just have to be interesting to other scientists and free of any obvious errors, in the opinion of the editor and reviewers. Their contribution toward knowledge can be no more than that. An article like the one in which the hockey stick diagram (Chapter 5) was drawn fits these criteria. But that does not make it a sensible basis on which to change society.

The fact that an article was published does not relieve readers of the obligation to look critically upon its assertions and ask if they are really true or not, or to understand what the limitations are on the conclusions. Very few conclusions in such an article are free of a great many limitations and most authors go to great trouble to lay them out. However, there are also limitations implicit to a whole field that are not laid out. A reader is just supposed to know. It can happen that even some within a field lose track of these things.

There is no shortcut for decisionmakers that avoids the need to think carefully about the decisions they make. People who propose costly policy changes like the Kyoto Protocol cannot shield themselves from the responsibility attached to making such proposals simply by citing a published article or a scholarly authority. The peer review process provides a service to the scientific community for the purposes internal to science. This is not the same as providing a service to society for the purpose of ensuring policy decisions are sound. While we hope the two processes are not fundamentally contradictory, they are not the same. Peer review does not absolve policymakers of the responsibility for making bad decisions.

Nor should we suppose that we have climate change basically figured out, and all that remains is to shrink some bounds of uncertainty. The problem is much deeper. On the core issues we are confronted by an absence of knowledge. Under the circumstances, it is impossible to recommend a detailed plan of action when we do not know if action is even needed, nor what its effects would be. Therefore the only sensible recommendation is not to take action except for further study.

But study alone will not amount to progress if it continues along the same lines as those organized by the Big Panel. The fact that the Doctrine emerged on their watch, and that it has generated such deep public confusion and professional malaise over the global warming issue, is surely sufficient cause for us to consider a different way of studying the topic.

We will shortly propose a new method for public study of global warming, based on a different way of thinking about how science and public policy should be related. It requires that we shed some illusions about the easy presumption of certainty on complex topics, and the danger of applying political models of authority to scientific endeavours.

But first we must finish with the question in the heading. While familiar, it is badly posed. An improvement would be: *What, if anything, should we do about the fact that some scientists think adding carbon dioxide to the atmosphere will cause a deleterious change in the climate system?* The answer is that we should take the concern seriously enough to try and clarify the issue by posing some additional questions. How would we find out if this is true? Would it matter? Can we do anything about it? And would we want to, even if we could?

As to the first, the problem is that in the absence of a theory of climate, we would have no way of knowing if human-caused climate change were taking place, even if it were going on right before our eyes. Nor could we look back in time and determine that it had happened and that it was responsible for particular changes at any particular location, as opposed to natural long-term changes. We cannot expect to identify human-caused climate change even after the fact!

What we do know is that prosperous and free societies are best able to deal with changes, including adverse ones, regardless of cause. Therefore we ought to encourage freedom and prosperity around the world. All but a handful of scientists on the fringe believe this.

On the second question, we do not know how this or that particular average of temperatures ought to behave in the absence of fossil fuel use, so whether T-Rex is rising or falling means nothing in terms of interpreting our influence on the world. And T-Rex only exists on paper anyway. We experience an infinite number of temperatures at every location on Earth, not an average cobbled together from white boxes and good intentions. If things are changing in the climate, we have to look at them directly in all their particularity and complexity, rather than trying to reduce the problem to the artificial simplicity of a one-dimensional character in a B-movie.

As to the third question, the climate is not a clock we can take apart and put back together or manipulate at will. Even if the world were as sensitive to CO_2 as the Big Panel has tried to argue, all the king's horses and all the king's men have not succeeded in coming up with anything better than the Kyoto Protocol, which, with its compromises at Bonn and prostrations at Marrakech, make it pointless with regard to any effect on atmospheric carbon dioxide concentrations. T-Rex will do what it was going to do despite it.

The world is full of surprises. "Surprises" was precisely the word used to describe it in Section 14.2.2 in PUN's scientific report. But too few actually read the actual scientific report. There the real scientists were able to send us a message in a bottle through the sea of politics to tell us that they too do not know. For all we know, we may yet be on the verge of an ice age, in which case we won't think global warming such a bad thing.

No one knows why some versions of T-Rex are rising and others are not. Temperatures rise and fall in response to changes in solar output, fluctuations in the earth's orbit and the lunar pull on tides, volcanic activity, long-term ocean dynamics and many other factors we do not understand, including every flap of a seagull's wings. And sometimes it rises and falls for no external reason at all, just because nonlinear, turbulent and chaotic dynamics can generate large sudden uncaused changes in the state of a system. That is what they said in 14.2.2 and that is what we have shown in this book. The presumption that every marked change has a cause is simply wrong. On Earth, these things are beyond our understanding, let alone our control.

As to the question of whether we would *want* to change the climate even if we could, we must look at the costs and benefits of small steps. The steps that involve carbon dioxide emission reductions (such as Kyoto) cost more than any benefits they are likely to generate. So they are not worth taking.

That said, some kinds of economic changes, driven by the internal comparison of costs and benefits, lead to emission reductions. For instance, there is a continuing drive toward energy efficiency in industrial economies. Every year we get a bit more efficient in how we use fuels, because it is economical to do so. But we cannot point to those actions and say that *therefore* all carbon dioxide emission reduction proposals are warranted. The question is not whether beneficial emission reductions are warranted: They will be undertaken with or without a push from government. The question is whether costly emission reductions are warranted, and the answer is no.

Even so, some people will rebel against this counsel. Over the past couple of years, as we have talked to people in and out of government, we keep hearing the message, *Yes but we have to do something.* Whether the rationale is the "precautionary principle" (which doesn't apply in this case) or some other dressing up of a simple gut reaction, there is a prevailing assumption out there that it is unacceptable to simply "do nothing."

At some point, we hit the wall in terms of what reason can accomplish. For those who really feel that way, despite all we have argued, we have run out of arguments. We could keep rephrasing what we have already written, but we won't. What is needed is for the reader to climb over the psychological hurdle of admitting that we should not burden society with carbon dioxide emission reduction requirements.

Regardless of how much effort and ingenuity has been expended developing the Kyoto Accord, it is the wrong thing to do. The right thing to do is to muddle along, focusing on basic priorities like economic development, wealth creation, education and the spread of freedom. We know these are good things to do. Policies like Kyoto do nobody any good and only take resources away from those real priorities.

So the best policy on global warming is to make sure science is free to investigate it, without having to prove constantly that this or that is relevant to policy issues. This has created a climate in the research community that has systematically pushed out the absolutely essential basic research that is all but disappearing because of it. Adequate funds must be made available to those who are studying it just because the research is good, not because it is producing fodder for political action.

Otherwise, the best policy is to do nothing unless future information indicates otherwise. This has been regarded as a political position. "More study" does not appeal to impatient political ideologues. But it is the only course that reflects a proper humility in the face of the enormity of the scientific problem we are facing. . . .

POSTSCRIPT

Is the Threat of Global Warming Real?

The issue of global warming is to this decade what acid rain was to the 1970s. Just as the blighted trees and polluted lakes of Scandinavia captured the hearts of the then newly emerging group of environmentalists, the issue of global warming has been front page news for over a decade and fodder for environmentalists and policymakers everywhere.

In a sense, the issue of global warming is a prototype of the contemporary issue making its way onto the global agenda. Recall that a global issue is characterized by disagreement over the extent of the condition, disagreement over the causes of the condition, disagreement over desirable future alternatives to the present condition, and disagreement over appropriate policies to reach desired end states.

All of these characteristics are present in global warming. Both sides of the issue can find a substantial number of scientists, measured in the thousands, to support their case that the Earth is or is not warming. Both sides can find hundreds of experts who will attest that the warming is either a cyclical phenomenon or the consequence of human behavior. Both sides can find a substantial number of policymakers and policy observers who will say that the Kyoto Treaty is humankind's best hope to reverse the global warming trend or that the treaty is seriously flawed with substantial negative consequences for the United States. It is an issue whose potential solutions on the table will impact different sectors of the economy and different countries differently.

The IPCC 2001 report concludes the following. First, the global average surface temperature has risen over the twentieth century by about 0.6°C. Two, temperatures have risen during the past four decades in the lowest 8 kilometers of the atmosphere. Third, snow cover and ice extent have decreased. Fourth, global average sea level has risen, and ocean heat has increased. Fifth, changes have occurred in other important aspects of climate. Sixth, concentrations of atmospheric greenhouse gases have continued to increase as a result of human activity. Seventh, there is new and stronger evidence that most of the warming observed over the last 50 years is attributable to human activities. Eighth, human influences will continue to change atmospheric composition throughout the twenty-first century. Ninth, global average temperature and sea level are projected to rise under all IPCC scenarios. And tenth, climate change will persist for many centuries. The list is impressive.

Yet others argue the opposite position. In their comprehensive book on the subject, Essex and McKitrick lay out a series of nine statements that comprise the "Doctrine" of global warming. In their analysis of these components, they conclude that the evidence simply does not warrant the global community's undertaking policy making at this time to address a problem that may not exist or may have been caused by some other phenomena.

156

This view is a shared one. Brian Tucker's 1997 article in *The National Interest* ("Science Fiction: The Politics of Global Warming") suggests that science "does not support the conclusion that calamitous effects from global warming are high upon us." He continues: "There is no scientific justification for such a view." Tucker raises the stakes by asserting that the global warming controversy is much more than a debate about the causes and extent of the phenomenon. It is also about global development, power, and morality in the struggle between the rich and poor countries, with population "control" a central issue.

This view is echoed by John R. Christy in "The Global Warming Fiasco" (*Global Warming and Other Eco-Myths*, Ronald Bailey, Forum, 2002). He accuses those who suggest global warming is a major problem of adhering to the science of "calamitology" rather than the science of "climatology." His bottom line assessment is that "No global climate disaster is looming."

The Heartland Institute (www.heartland.org/studies/ieguide.htm) suggests "seven facts" to counteract observations such as those of the recent and earlier IPCC studies. First, "most scientists do not believe human activities threaten to disrupt the earth's climate." Second, "the most reliable temperature data show no global warming trend." Third, "global computer models are too crude to predict future climate changes." Fourth, the IPCC did not prove that human activities are causing global warming" (a reference to the 1995 study). Fifth, "a modest amount of global warming, should it occur, would be beneficial to the natural world and to human civilization." Sixth, "reducing our greenhouse gas emissions would be costly and would not stop global warming." And seventh, "the best strategy to pursue is one of 'no regrets'." The latter refers to the idea that it is it is not better to be safe than sorry (suggested by the other side), as immediate action will not make us safer, just poorer.

Three other studies make valuable reading, each one of which takes a different position (one each at the extreme ends of the debate and a third one that suggests moderate climate change). S. Fred Singer's *Hot Talk; Cold Science: Global Warming's Unfinished Debate* (Independent Institute, 1998) enhances the author's reputation as one of the leading opponents of global warming's adverse consequences. At the other extreme, John Houghton in *Global Warming: The Complete Briefing*, 2d ed. (Cambridge University Press, 1997) accepts global warming as a significant concern and describes how it can be reversed in the future. S. George Philander, in *Is the Temperature Rising?* (1998), concludes that the global temperatures will rise 2°C over several decades, creating the prospect of some regional climate changes, with major consequences. Finally, Roy W. Spencer's "How Do We Know the Temperature of the Earth?" (Ronald Bailey, ed., *Earth Report 2000*, 2000), presents a basic argument with evidence that the popular perception of global warming as an environmental catastrophe cannot be supported with evidence. Finally, an objective analysis of the issue can be found in Chapter 5 ("Is the Earth Warming?") of Jack M. Hollander's *The Real Environmental Crisis* (University of California Press, 2003).

What are we to make of all of this? Simply put, whether or not global warming exists and is caused by human behavior, the issue will remain on the front page until agreement can be reached on these two fundamental questions.

ISSUE 9

Is the Threat of a Global Water Shortage Real?

YES: Mark W. Rosegrant, Ximing Cai, and Sarah A. Cline, from "Global Water Outlook to 2025: Averting an Impending Crisis," A Report of the International Food Policy Research Institute and the International Water Management Institute (September 2002)

NO: Bjørn Lomborg, from *The Skeptical Environmentalist: Measuring the Real State of the World* (Cambridge University Press, 2001)

ISSUE SUMMARY

YES: Rosegrant and colleagues conclude that if current water policies continue, farmers will find it difficult to grow sufficient food to meet the world's needs.

NO: Water is not only plentiful but is a renewable resource that, if properly treated as valuable, should not pose a future problem.

Water shortages and other water problems are occurring with greater frequency, particularly in large cities. Some observers have speculated that the situation is reminiscent of the fate that befell ancient glorious cities like Rome. Recognition that the supply of water is a growing problem is not new. As early as 1964, the United Nations Environmental Programme (UNEP) indicated that close to a billion people were at risk from desertification. At the Earth Summit in Rio in 1992, world leaders reaffirmed that desertification was of serious concern.

Moreover, in conference after conference and in study after study, increasing population growth and declining water supplies and quality are being linked together, as is the relationship between the planet's ability to meet its growing food needs and available water. Lester R. Brown, in "Water Deficits Growing in Many Countries: Water Shortages May Cause Food Shortages," *Eco-Economy Update 2002–11* (August 6, 2002), sums up the problem this way: "The world is incurring a vast water deficit. It is largely invisible, historically recent, and growing fast." The World Water Council's study, "World Water Actions Report, Third Draft" (October 31, 2002), describes the problem in much the same way: "Water is no longer taken for granted as a plentiful resource that will always be available when we need it." The report continues with the observation that increasing numbers of people in more and more countries are discovering that water is a "limited resource that must be managed for the benefit of people and the environment, in the present and in the future." In short, water is fast becoming both a food-related issue and a health-related problem. Some scholars are now arguing that water shortage is likely to become the twenty-first century's analog to the oil crisis of the last half of the previous century. The one major difference, as scholars are quick to point out, is that water is not like oil; there is no substitute.

Proclamations of impending water problems abound. Peter Gleick, in *The World's Water 1998–99: The Biennial Report on Freshwater Resources* (Island Press, 1998), reports that the demand for freshwater increased six-fold between 1900 and 1995, twice the rate of population growth. The UN study "United Nations Comprehensive Assessment of Freshwater Resources of the World" (1997) suggests that one-third of the world's population live in countries that have medium to high water stress. One 2001 headline reporting the release of a new study proclaimed that "Global thirst 'will turn millions into water refugees'" (The Millennium Environment Debate). News reports released by the UN Food and Agricultural Organization in conjunction with World Food Day 2002 asserted that water scarcity could result in millions of people having inadequate access to clean water or sufficient food. And the World Meteorological Organization predicts that two out of every three people will live in water-stressed conditions by 2050 if consumption patterns remain the same.

Sandra Postel, in *Pillar of Sand: Can the Irrigation Miracle Last?* (W.W. Norton, 1999), suggests another variant of the water problem. For her, the time-tested method of maximizing water usage in the past, irrigation, may not be feasible as world population marches toward seven billion. She points to the inadequacy of surface water supplies, increasing depletion of groundwater supplies, the salinization of the land, and the conversion of traditional agricultural land to other uses as reasons for the likely inability of irrigation to be a continuing panacea. Yet the 1997 UN study concluded that annual irrigation use would need to increase 30 percent for annual food production to double, necessary for meeting food demands of 2025.

The issue of water quality is also in the news. The World Health Organization reports that in some parts of the world, up to 80 percent of all transmittable diseases are attributable to the consumption of contaminated water. Also, a UNEP-sponsored study, *Global Environment Outlook 2000*, reported that 200 scientists from 50 countries pointed to the shortage of clean water as one of the most pressing global issues.

In the following selection, Mark W. Rosegrant, Ximing Cai, and Sarah A. Cline project that by 2025, water scarcity will result in annual global losses of 350 million metric tons, equivalent to approximately the entire U.S. crop. In the second selection, Bjørn Lomborg takes issue with the prevailing wind in the global water debate. His argument can be summed up in his simple quote: "Basically we have sufficient water." Lomborg maintains that water supplies rose during the twentieth century and that we have gained access to more water through technology.

Mark W. Rosegrant, Ximing Cai,
and Sarah A. Cline

→ **YES**

Global Water Outlook to 2025: Averting an Impending Crisis

Introduction

Demand for the world's increasingly scarce water supply is rising rapidly, challenging its availability for food production and putting global food security at risk. Agriculture, upon which a burgeoning population depends for food, is competing with industrial, household, and environmental uses for this scarce water supply. Even as demand for water by all users grows, groundwater is being depleted, other water ecosystems are becoming polluted and degraded, and developing new sources of water is getting more costly.

A Thirsty World

Water development underpins food security, people's livelihoods, industrial growth, and environmental sustainability throughout the world. In 1995 the world withdrew 3,906 cubic kilometers (km^3) of water for these purposes (Figure 1). By 2025 water withdrawal for most uses (domestic, industrial, and livestock) is projected to increase by at least 50 percent. This will severely limit irrigation water withdrawal, which will increase by only 4 percent, constraining food production in turn.

About 250 million hectares are irrigated worldwide today, nearly five times more than at the beginning of the 20th century. Irrigation has helped boost agricultural yields and outputs and stabilize food production and prices. But growth in population and income will only increase the demand for irrigation water to meet food production requirements (Figure 2). Although the achievements of irrigation have been impressive, in many regions poor irrigation management has markedly lowered groundwater tables, damaged soils, and reduced water quality.

Water is also essential for drinking and household uses and for industrial production. Access to safe drinking water and sanitation is critical to maintain health, particularly for children. But more than 1 billion people across the globe lack enough safe water to meet minimum levels of health and income. Although the domestic and industrial sectors use far less water than agriculture, the growth in water consumption in these sectors has been rapid. Globally, withdrawals for

Figure 1

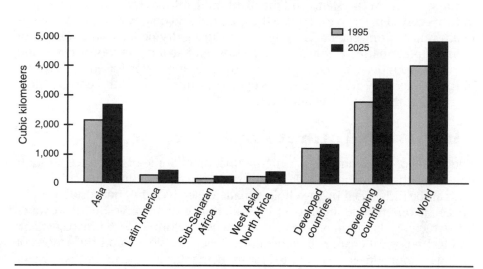

Total Water Withdrawal by Region, 1995 and 2025

Source: Authors' estimates and IMPACT-WATER projections, June 2002.
Note: Projections for 2025 are for the business as usual scenario.

Figure 2

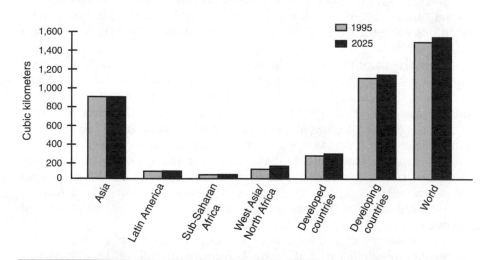

Total Irrigation Water Consumption by Region, 1995 and 2025

Source: Authors' estimates and IMPACT-WATER projections, June 2002.
Note: Projections for 2025 are for the business as usual scenario.

domestic and industrial uses quadrupled between 1950 and 1995, compared with agricultural uses, for which withdrawals slightly more than doubled.[1]

Water is integrally linked to the health of the environment. Water is vital to the survival of ecosystems and the plants and animals that live in them, and in turn ecosystems help to regulate the quantity and quality of water. Wetlands retain water during high rainfall, release it during dry periods, and purify it of many contaminants. Forests reduce erosion and sedimentation of rivers and recharge groundwater. The importance of reserving water for environmental purposes has only recently been recognized: during the 20th century, more than half of the world's wetlands were lost.[2]

Alternative Futures for Water

The future of water and food is highly uncertain. Some of this uncertainty is due to relatively uncontrollable factors such as weather. But other critical factors can be influenced by the choices made collectively by the world's people. These factors include income and population growth, investment in water infrastructure, allocation of water to various uses, reform in water management, and technological changes in agriculture. Policy decisions—and the actions of billions of individuals—determine these fundamental, long-term drivers of water and food supply and demand.

To show the very different outcomes that policy choices produce, we present three alternative futures for global water.[3] . . .

Business As Usual Scenario

In the business as usual scenario current trends in water and food policy, management, and investment remain as they are. International donors and national governments, complacent about agriculture and irrigation, cut their investments in these sectors. Governments and water users implement institutional and management reforms in a limited and piecemeal fashion. These conditions leave the world ill prepared to meet major challenges to the water and food sectors.

Over the coming decades the area of land devoted to cultivating food crops will grow slowly in most of the world because of urbanization, soil degradation, and slow growth in irrigation investment, and because a high proportion of arable land is already cultivated. Moreover, steady or declining real prices for cereals will make it unprofitable for farmers to expand harvested area. As a result, greater food production will depend primarily on increases in yield. Yet growth in crop yields will also diminish because of falling public investment in agricultural research and rural infrastructure. Moreover, many of the actions that produced yield gains in recent decades, such as increasing the density of crop planting, introducing strains that are more responsive to fertilizer, and improving management practices, cannot easily be repeated.

In the water sector, the management of river basin and irrigation water will become more efficient, but slowly. Governments will continue to transfer management of irrigation systems to farmer organizations and water-user associations. Such transfers will increase water efficiency when they are built upon existing patterns of cooperation and backed by a supportive policy and legal environment. But these conditions are often lacking.

In some regions farmers will adopt more efficient irrigation practices. Economic incentives to induce more efficient water management, however, will still face political opposition from those concerned about the impact of higher water prices on farmers' income and from entrenched interests that benefit from existing systems of allocating water. Water management will also improve slowly in rainfed agriculture as a result of small advances in water harvesting, better on-farm management techniques, and the development of crop varieties with shorter growing seasons.

Public investment in expanding irrigation systems and reservoir storage will decline as the financial, environmental, and social costs of building new irrigation systems escalate and the prices of cereals and other irrigated crops drop. Nevertheless, where benefits outweigh costs, many governments will construct dams, and reservoir water for irrigation will increase moderately.

With slow growth in irrigation from surface water, farmers will expand pumping from groundwater, which is subject to low prices and little regulation. Regions that currently pump groundwater faster than aquifers can recharge, such as the western United States, northern China, northern and western India, Egypt, and West Asia and North Africa, will continue to do so.

The cost of supplying water to domestic and industrial users will rise dramatically. Better delivery and more efficient home water use will lead to some increase in the proportion of households connected to piped water. Many households, however, will remain unconnected. Small price increases for industrial water, improvements in pollution control regulation and enforcement, and new industrial technologies will cut industrial water use intensity (water demand per $1,000 of gross domestic product). Yet industrial water prices will remain relatively low and pollution regulations will often be poorly enforced. Thus, significant potential gains will be lost.

Environmental and other interest groups will press to increase the amount of water allocated to preserving wetlands, diluting pollutants, maintaining riparian flora and other aquatic species, and supporting tourism and recreation. Yet because of competition for water for other uses, the share of water devoted to environmental uses will not increase.

The Water Situation

Almost all users will place heavy demands on the world's water supply under the business as usual scenario. Total global water withdrawals in 2025 are projected to increase by 22 percent above 1995 withdrawals, to 4,772 km^3 (see Figure 1).[4] Projected withdrawals in developing countries will increase 27 percent over the 30-year period, while developed-country withdrawals will increase by 11 percent.[5]

Together, consumption of water for domestic, industrial, and livestock uses—that is, all nonirrigation uses—will increase dramatically, rising by 62 percent from 1995 to 2025. Because of rapid population growth and rising per capita water use, total domestic consumption will increase by 71 percent, of which more than 90 percent will be in developing countries. Conservation and technological improvements will lower per capita domestic water use in developed countries with the highest per capita water consumption.

Industrial water use will grow much faster in developing countries than in developed countries. In 1995 industries in developed countries consumed much more water than industries in the developing world. By 2025, however, developing-world industrial water demand is projected to increase to 121 km^3, 7 km^3 greater than in the developed world. The intensity of industrial water use will decrease worldwide, especially in developing countries (where initial intensity levels are very high), thanks to improvements in water-saving technology and demand policy. Nonetheless, the sheer size of the increase in the world's industrial production will still lead to an increase in total industrial water demand.

Direct water consumption by livestock is very small compared with other sectors. But the rapid increase of livestock production, particularly in developing countries, means that livestock water demand is projected to increase 71 percent between 1995 and 2025. Whereas livestock water demand will increase only 19 percent in the developed world between 1995 and 2025, it is projected to more than double in the developing world, from 22 to 45 km^3.

Although irrigation is by far the largest user of the world's water, use of irrigation water is projected to rise much more slowly than other sectors. For irrigation water, we have computed both potential demand and actual consumption. Potential demand is the demand for irrigation water in the absence of any water supply constraints, whereas actual consumption of irrigation water is the realized water demand, given the limitations of water supply for irrigation. The proportion of potential demand that is realized in actual consumption is the irrigation water supply reliability index (IWSR).[6] An IWSR of 1.0 would mean that all potential demand is being met.

Potential irrigation demand will grow by 12 percent in developing countries, while it will actually decline in developed countries by 1.5 percent. The fastest growth in potential demand for irrigation water will occur in Sub-Saharan Africa, with an increase of 27 percent, and in Latin America, with an increase of 21 percent. Each of these regions has a high percentage increase in irrigated area from a relatively low 1995 level. India is projected to have the highest absolute growth in potential irrigation water demand, 66 km^3 (17 percent), owing to relatively rapid growth in irrigated area from an already high level in 1995. West Asia and North Africa will increase by 18 percent (28 km^3, mainly in Turkey), while China will experience a much smaller increase of 4 percent (12 km^3). In Asia as a region, potential irrigation water demand will increase by 8 percent (100 km^3).

Water scarcity for irrigation will intensify, with actual consumption of irrigation water worldwide projected to grow more slowly than potential consumption, increasing only 4 percent between 1995 and 2025. In developing countries a declining fraction of potential demand will be met over time. The IWSR for developing countries will decline from 0.81 in 1995 to 0.75 in 2025, and in dry river basins the decline will be steeper. For example, in the Haihe River Basin in China, which is an important wheat and maize producer and serves major metropolitan areas, the IWSR is projected to decline from 0.78 to 0.62, and in the Ganges of India, the IWSR will decline from 0.83 to 0.67.

In the developed world, the situation is the reverse: the supply of irrigation water is projected to grow faster than potential demand (although certain
basins will face increasing water scarcity). Increases in river basin efficiency will
more than offset the very small increase in irrigated area. As a result, after initially declining from 0.87 to 0.85 in 2010, the IWSR will improve to 0.90 in
2025 thanks to slowing growth of domestic and industrial demand (and actual
declines in total domestic and industrial water use in the United States and
Europe) and more efficient use of irrigation water. . . .

Water Crisis Scenario

A moderate worsening of many of the current trends in water and food policy
and in investment could build to a genuine water crisis. In the water crisis scenario, government budget problems worsen. Governments further cut their
spending on irrigation systems and accelerate the turnover of irrigation systems
to farmers and farmer groups but without the necessary reforms in water rights.
Attempts to fund operations and maintenance in the main water system, still
operated by public agencies, cause water prices to irrigators to rise. Water users fight
price increases, and conflict spills over to local management and cost-sharing
arrangements. Spending on the operation and maintenance of secondary and
tertiary systems falls dramatically, and deteriorating infrastructure and poor management lead to falling water use efficiency. Likewise, attempts to organize river
basin organizations to coordinate water management fail because of inadequate
funding and high levels of conflict among water stakeholders within the basin.

National governments and international donors will reduce their investments in crop breeding for rainfed agriculture in developing countries, especially
for staple crops such as rice, wheat, maize, other coarse grains, potatoes, cassava,
yams, and sweet potatoes. Private agricultural research will fail to fill the investment gap for these commodities. This loss of research funding will lead to further
declines in productivity growth in rainfed crop areas, particularly in more marginal areas. In search of improved incomes, people will turn to slash-and-burn
agriculture, thereby deforesting the upper watersheds of many basins. Erosion
and sediment loads in rivers will rise, in turn causing faster sedimentation of reservoir storage. People will increasingly encroach on wetlands for both land and
water, and the integrity and health of aquatic ecosystems will be compromised.
The amount of water reserved for environmental purposes will decline as unreg-
ulated and illegal withdrawals increase.

The cost of building new dams will soar, discouraging new investment in
many proposed dam sites. At other sites indigenous groups and nongovernmental organizations (NGOs) will mount opposition, often violent, over the environmental and human impacts of new dams. These protests and high costs will
virtually halt new investment in medium and large dams and storage reservoirs.
Net reservoir storage will decline in developing countries and remain constant in
developed countries.

In the attempt to get enough water to grow their crops, farmers will
extract increasing amounts of groundwater for several years, driving down
water tables. But because of the accelerated pumping, after 2010 key aquifers in
northern China, northern and northwestern India, and West Asia and North

Africa will begin to fail. With declining water tables, farmers will find the cost of extracting water too high, and a big drop in groundwater extraction from these regions will further reduce water availability for all uses.

As in the business as usual scenario, the rapid increase in urban populations will quickly raise demand for domestic water. But governments will lack the funds to extend piped water and sewage disposal to newcomers. Governments will respond by privatizing urban water and sanitation services in a rushed and poorly planned fashion. The new private water and sanitation firms will be undercapitalized and able to do little to connect additional populations to piped water. An increasing number and percentage of the urban population must rely on high-priced water from vendors or spend many hours fetching often-dirty water from standpipes and wells.

The Water Situation
The developing world will pay the highest price for the water crisis scenario. Total worldwide water consumption in 2025 will be 261 km^3 higher than under the business as usual scenario—a 13 percent increase—but much of this water will be wasted, of no benefit to anyone. Virtually all of the increase will go to irrigation, mainly because farmers will use water less efficiently and withdraw more water to compensate for water losses. The supply of irrigation water will be less reliable, except in regions where so much water is diverted from environmental uses to irrigation that it compensates for the lower water use efficiency.

For most regions, per capita demand for domestic water will be significantly lower than under the business as usual scenario, in both rural and urban areas. The result is that people will not have access to the water they need for drinking and sanitation. The total domestic demand under the water crisis scenario will be 162 km^3 in developing countries, 28 percent less than under business as usual; 64 km^3 in developed countries, 7 percent less than under business as usual; and 226 km^3 in the world, 23 percent less than under business as usual.

Demand for industrial water, on the other hand, will increase, owing to failed technological improvements and economic measures. In 2025 the total industrial water demand worldwide will be 80 km^3 higher than under the business as usual scenario—a 33 percent rise—without generating additional industrial production.

With water diverted to make up for less efficient water use, the water crisis scenario will hit environmental uses particularly hard. Compared with business as usual, environmental flows will drop significantly by 2025, with 380 km^3 less environmental flow in the developing world, 80 km^3 less in the developed world, and 460 km^3 less globally. . . .

Sustainable Water Scenario

A sustainable water scenario would dramatically increase the amount of water allocated to environmental uses, connect all urban households to piped water, and achieve higher per capita domestic water consumption, while maintaining food production at the levels described in the business as usual scenario. It would achieve greater social equity and environmental protection through both careful reform in the water sector and sound government action.

Governments and international donors will increase their investments in crop research, technological change, and reform of water management to boost water productivity and the growth of crop yields in rainfed agriculture. Accumulating evidence shows that even drought-prone and high-temperature rainfed environments have the potential for dramatic increases in yield. Breeding strategies will directly target these rainfed areas. Improved policies and increased investment in rural infrastructure will help link remote farmers to markets and reduce the risks of rainfed farming.

To stimulate water conservation and free up agricultural water for environmental, domestic, and industrial uses, the effective price of water to the agricultural sector will be gradually increased. Agricultural water price increases will be implemented through incentive programs that provide farmers income for the water that they save, such as charge-subsidy schemes that pay farmers for reducing water use, and through the establishment, purchase, and trading of water use rights. By 2025 agricultural water prices will be twice as high in developed countries and three times as high in developing countries as in the business as usual scenario. The government will simultaneously transfer water rights and the responsibility for operation and management of irrigation systems to communities and water user associations in many countries and regions. The transfer of rights and systems will be facilitated with an improved legal and institutional environment for preventing and eliminating conflict and with technical and organizational training and support. As a result, farmers will increase their on-farm investments in irrigation and water management technology, and the efficiency of irrigation systems and basin water use will improve significantly.

River basin organizations will be established in many water-scarce basins to allocate mainstream water among stakeholder interests. Higher funding and reduced conflict over water, thanks to better water management, will facilitate effective stakeholder participation in these organizations.

Farmers will be able to make more effective use of rainfall in crop production, thanks to breakthroughs in water harvesting systems and the adoption of advanced farming techniques, like precision agriculture, contour plowing, precision land leveling, and minimum-till and no-till technologies. These technologies will increase the share of rainfall that goes to infiltration and evapotranspiration.

Spurred by the rapidly escalating costs of building new dams and the increasingly apparent environmental and human resettlement costs, developing and developed countries will reassess their reservoir construction plans, with comprehensive analysis of the costs and benefits, including environmental and social effects, of proposed projects. As a result, many planned storage projects will be canceled, but others will proceed with support from civil society groups. Yet new storage capacity will be less necessary because rapid growth in rainfed crop yields will help reduce rates of reservoir sedimentation from erosion due to slash-and-burn cultivation.

Policy toward groundwater extraction will change significantly. Market-based approaches will assign rights to groundwater based on both annual withdrawals and the renewable stock of groundwater. This step will be combined with stricter regulations and better enforcement of these regulations. Groundwater overdrafts will be phased out in countries and regions that previously pumped groundwater unsustainably.

Domestic and industrial water use will also be subject to reforms in pricing and regulation. Water prices for connected households will double, with targeted subsidies for low-income households. Revenues from price increases will be invested to reduce water losses in existing systems and to extend piped water to previously unconnected households. By 2025 all households will be connected. Industries will respond to higher prices, particularly in developing countries, by increasing in-plant recycling of water, which reduces consumption of water.

With strong societal pressure for improved environmental quality, allocations for environmental uses of water will increase. Moreover, the reforms in agricultural and nonagricultural water sectors will reduce pressure on wetlands and other environmental uses of water. Greater investments and better water management will improve the efficiency of water use, leaving more water instream for environmental purposes. All reductions in domestic and urban water use, due to higher water prices, will be allocated to instream environmental uses.

The Water Situation

In the sustainable water scenario the world consumes less water but reaps greater benefits than under business as usual, especially in developing countries. In 2025 total worldwide water consumption is 408 km^3, or 20 percent, lower under the sustainable scenario than under business as usual. This reduction in consumption frees up water for environmental uses. Higher water prices and higher water use efficiency reduces consumption of irrigation water by 296 km^3 compared with business as usual. The reliability of irrigation water supply is reduced slightly in the sustainable scenario compared with business as usual, because this scenario places a high priority on environmental flows. Over time, however, more efficient water use in this scenario counterbalances the transfer of water to the environment and results in an improvement in the reliability of supply of irrigation water by 2025.

This scenario will improve the domestic water supply through universal access to piped water for rural and urban households. Globally, potential domestic water demand under the sustainable water scenario will decrease 9 percent compared with business as usual, owing to higher water prices. However, potential per capita domestic demand for connected households in rural areas will be 12 percent higher than that under business as usual in the developing world, and 5 percent higher in the developed world. This increase is accomplished by expanding universal access to piped water in rural areas even with higher prices for water. And in urban areas, potential per capita water consumption for poor households sharply improves through connection to piped water, while the initially connected households reduce consumption in response to higher prices and improved water-saving technology.

Through technological improvements and effective economic incentives, the sustainable water scenario will reduce industrial water demand. In 2025 total industrial water demand worldwide under the sustainable scenario will be 85 km^3, or 35 percent, lower than under business as usual.

The environment is a major beneficiary of the sustainable water scenario, with large increases in the amount of water reserved for wetlands, instream flows, and other environmental purposes. Compared with the business as usual scenario, the sustainable scenario will also result in an increase in the environmental flow of 850 km^3 in the developing world, 180 km^3 in the developed world, and 1,030 km^3 globally. This is the equivalent of transferring 22 percent of global water withdrawals under business as usual to environmental purposes.

Notes

1. W. J. Cosgrove and F. Rijsberman, *World Water Vision: Making Water Everybody's Business* (London: World Water Council and World Water Vision and Earthscan, 2000); I. A. Shiklomanov, "Electronic Data Provided to the Scenario Development Panel, World Commission on Water for the 21st Century" (State Hydrological Institute, St. Petersburg, Russia, 1999), mimeo.

2. E. Bos and G. Bergkamp, "Water and the Environment," in *Overcoming Water Scarcity and Quality Constraints*, 2020 Focus 9, ed. R. S. Meinzen-Dick and M. W. Rosegrant (Washington, D.C.: International Food Policy Research Institute, 2001).

3. The business as usual, crisis, and sustainable scenarios are compared using average 2025 results generated from 30 hydrologic scenarios. The other scenarios are compared with business as usual based on a single 30-year hydrologic sequence drawn from 1961–90, and results are shown as the average of the years 2021–25.

4. Water demand can be defined and measured in terms of withdrawals and actual consumption. While water withdrawal is the most commonly estimated figure, consumption best captures actual water use, and most of our analysis will utilize this concept.

5. The global projection is broadly consistent with other recent projections to 2025, including the 4,580 km^3 in the medium scenario of J. Alcamo, P. Döll, F. Kaspar, and S. Sieberg, *Global Change and Global Scenarios of Water Use and Availability: An Application of Water GAP 1.0* (Kassel, Germany: Center for Environmental System Research, University of Kassel, 1998), the 4,569 km^3 in the "business-as-usual" scenario of D. Seckler, U. Amarasinghe, D. Molden, S. Rhadika, and R. Barker, *World Water Demand and Supply, 1990 to 2025: Scenarios and Issues*, Research Report Number 19 (Colombo, Sri Lanka: International Water Management Institute, 1998), and the forecast of 4,966 km^3 (not including reservoir evaporation) of Shiklomanov, "Electronic Data."

6. Compared with other sectors, the growth of irrigation water potential demand is much lower, with 12 percent growth in potential demand between 1995 and 2025 in developing countries and a slight decline in potential demand in developed countries.

 **NO**

Water

There is a resource which we often take for granted but which increasingly has been touted as a harbinger of future trouble. Water.

Ever more people live on Earth and they use ever more water. Our water consumption has almost quadrupled since 1940. The obvious argument runs that "this cannot go on." This has caused government agencies to worry that "a threatening water crisis awaits just around the corner." The UN environmental report *GEO 2000* claims that the water shortage constitutes a "full-scale emergency," where "the world water cycle seems unlikely to be able to cope with the demands that will be made of it in the coming decades. Severe water shortages already hamper development in many parts of the world, and the situation is deteriorating."

The same basic argument is invoked when WWF [World Wildlife Fund] states that "freshwater is essential to human health, agriculture, industry, and natural ecosystems, but is now running scarce in many regions of the world." *Population Reports* states unequivocally that "freshwater is emerging as one of the most critical natural resource issues facing humanity." Environmental discussions are replete with buzz words like "water crisis" and "time bomb: water shortages," and *Time* magazine summarizes the global water outlook with the title "Wells running dry." The UN organizations for meteorology and education simply refer to the problem as "a world running out of water."

The water shortages are also supposed to increase the likelihood of conflicts over the last drops—and scores of articles are written about the coming "water wars." Worldwatch Institute sums up the worries nicely, claiming that "water scarcity may be to the nineties what the oil price shocks were to the seventies—a source of international conflicts and major shifts in national economies."

But these headlines are misleading. True, there may be *regional* and *logistic* problems with water. We will need to get better at using it. But basically we have sufficient water.

How Much Water in the World?

Water is absolutely decisive for human survival, and the Earth is called the Blue Planet precisely because most of it is covered by water: 71 percent of the Earth's surface is covered by water, and the total amount is estimated at the unfathomably large 13.6 billion cubic kilometers. Of all this water, oceans make up 97.2 percent and the polar ice contains 2.15 percent. Unfortunately sea water is too saline for direct human consumption, and while polar ice contains potable water it is hardly within easy reach. Consequently, humans are primarily dependent on the last 0.65 percent water, of which 0.62 percent is groundwater.

Fresh water in the groundwater often takes centuries or millennia to build up—it has been estimated that it would require 150 years to recharge all of the groundwater in the United States totally to a depth of 750 meters if it were all removed. Thus, thoughtlessly exploiting the groundwater could be compared to mining any other non-renewable natural resource. But groundwater is continuously replenished by the constant movement of water through oceans, air, soil, rivers, and lakes in the so-called hydrological cycle. The sun makes water from the oceans evaporate, the wind moves parts of the vapor as clouds over land, where the water is released as rain and snow. The precipitated water then either evaporates again, flows back into the sea through rivers and lakes, or finds its way into the groundwater.

The total amount of precipitation on land is about 113,000 km^3, and taking into account an evaporation of 72,000 km^3 we are left with a net fresh water influx of 41,000 km^3 each year or the equivalent of 30 cm (1 foot) of water across the entire land mass. Since part of this water falls in rather remote areas, such as the basins of the Amazon, the Congo, and the remote North American and Eurasian rivers, a more reasonable, geographically accessible estimate of water is 32,900 km^3. Moreover, a large part of this water comes within short periods of time. In Asia, typically 80 percent of the runoff occurs from May to October, and globally the flood runoff is estimated at about three-quarters of the total runoff. This leaves about 9,000 km^3 to be captured. Dams capture an additional 3,500 km^3 from floods, bringing the total accessible runoff to 12,500 km^3. This is equivalent to about 5,700 liters of water for every single person on Earth *every single day*. For comparison, the average citizen in the EU uses about 566 liters of water per day. This is about 10 percent of the global level of available water and some 5 percent of the available EU water. An American, however, uses about three times as much water, or 1,442 liters every day.

Looking at global water consumption, as seen in Figure 1, it is important to distinguish between water withdrawal and water use. Water withdrawal is the amount of water physically removed, but this concept is less useful in a discussion of limits on the total amount of water, since much of the withdrawn water is later returned to the water cycle. In the EU and the US, about 46 percent of the withdrawn water is used merely as cooling water for power generation and is immediately released for further use downstream. Likewise, most industrial uses return 80–90 percent of the water, and even in irrigation 30–70 percent of the water runs back into lakes and rivers or percolates into aquifers, whence it can be reused. Thus, a more useful measure of water consumption is the amount of water this consumption causes to be irretrievably lost through evaporation or transpiration from plants. This is called water use.

Over the twentieth century, Earth's water use has grown from about 330 km^3 to about 2,100 km^3. As can be seen from Figure 1 there is some uncertainty about the future use and withdrawal (mainly depending on the development of irrigation), but until now most predictions have tended to overestimate the

Figure 1

Global, Annual Water Withdrawal and Use, in Thousand km³ and Percentage of Accessible Runoff, 1900–95, and Predictions for 2025

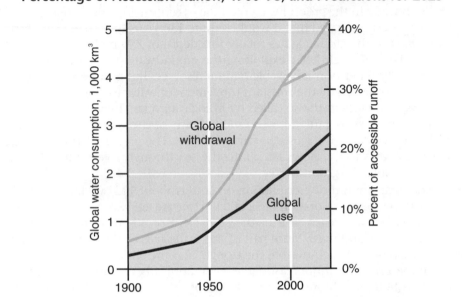

Source: Shiklomanov 2000:22 (high prediction), World Water Council 2000:26 (low prediction).

actual water consumption by up to 100 percent. Nevertheless, total use is still less than 17 percent of the accessible water and even with the high prediction it will require just 22 percent of the readily accessible, annually renewed water in 2025.

At the same time, we have gained access to more and more water, as indicated in Figure 2. Per person we have gone from using about 1,000 liters per day to almost 2,000 liters over the past 100 years. Particularly, this is due to an approximately 50 percent increase in water use in agriculture, allowing irrigated farms to feed us better and to decrease the number of starving people. Agricultural water usage seems, however, to have stabilized below 2,000 liters per capita, mainly owing to higher efficiency and less water consumption in agriculture since 1980. This pattern is also found in the EU and the US, where consumption has increased dramatically over the twentieth century, but is now leveling off. At the same time, personal consumption (approximated by the municipal withdrawal) has more than quadrupled over the century, reflecting an increase in welfare with more easily accessible water. In developing countries, this is in large part a question of health—avoiding sickness through better access to clean drinking water and sanitation, whereas in developed countries higher water use is an indication of an increased number of domestic amenities such as dishwashers and better-looking lawns.

So, if the global use is less than 17 percent of the readily accessible and renewable water and the increased use has brought us more food, less starvation, more health and increased wealth, why do we worry?

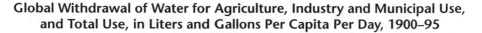

Figure 2

Global Withdrawal of Water for Agriculture, Industry and Municipal Use, and Total Use, in Liters and Gallons Per Capita Per Day, 1900–95

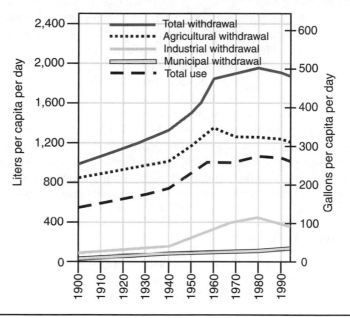

Source: Shiklomanov 2000:24.

The Three Central Problems

There are three decisive problems. First, precipitation is by no means equally distributed all over the globe. This means that not all have equal access to water resources and that some countries have much less accessible water than the global average would seem to indicate. The question is whether water shortages are already severe in some places today. Second, there will be more and more people on Earth. Since precipitation levels will remain more or less constant this will mean fewer water resources for each person. The question is whether we will see more severe shortages in the future. Third, many countries receive a large part of their water resources from rivers; 261 river systems, draining just less than half of the planet's land area, are shared by two or more countries, and at least ten rivers flow through half a dozen or more countries. Most Middle Eastern countries share aquifers. This means that the water question also has an international perspective and—if cooperation breaks down—an international conflict potential.

Beyond these three problems there are two other issues, which are often articulated in connection with the water shortage problem, but which are really conceptually quite separate. One is the worry about water pollution, particularly of potable water. While it is of course important to avoid water pollution in part because pollution restricts the presently available amount of freshwater, it is not related to the problem of water shortage *per se*. . . .

The second issue is about the shortage of *access* to water in the Third World. . . . This problem, while getting smaller, is still a major obstacle for global welfare. In discussing water shortage, reference to the lack of universal access to drinking water and sanitation is often thrown in for good measure, but of course this issue is entirely separate from the question of shortages. First, the cause is *not* lack of water (since human requirements constitute just 50–100 liters a day which any country but Kuwait can deliver, cf. Table 1) but rather a lack of investment in infrastructure. Second, the solution lies not in cutting back on existing consumption but actually in increasing future consumption.

Finally, we should just mention global warming . . . and its connection to water use. Intuitively, we might be tempted to think that a warmer world would mean more evaporation, less water, more problems. But more evaporation also means more precipitation. Essentially, global climate models seem to change *where* water shortages appear (pushing some countries above or below the threshold) but the total changes are small (1–5 percent) and go both ways.

Not Enough Water?

Precipitation is not distributed equally. Some countries such as Iceland have almost 2 million liters of water for each inhabitant every day, whereas Kuwait must make do with just 30 liters. The question, of course, is when does a country not have *enough* water.

It is estimated that a human being needs about 2 liters of water a day, so clearly this is not the restrictive requirement. The most common approach is to use the so-called *water stress index* proposed by the hydrologist Malin Falkenmark. This index tries to establish an approximate minimum level of water per capita to maintain an adequate quality of life in a moderately developed country in an arid zone. This approach has been used by many organizations including the World Bank, in the standard literature on environmental science, and in the water scarcity discussion in *World Resources*. With this index, human beings are assessed to need about 100 liters per day for drinking, household needs and personal hygiene, and an additional 500–2,000 liters for agriculture, industry and energy production. Since water is often most needed in the dry season, the water stress level is then set even higher—if a country has less than 4,660 liters per person available it is expected to experience periodic or regular water stress. Should the accessible runoff drop to less than 2,740 liters the country is said to experience chronic water scarcity. Below 1,370 liters, the country experiences absolute water scarcity, outright shortages and acute scarcity.

Table 1 shows the 15 countries comprising 3.7 percent of humanity in 2000 suffering chronic water scarcity according to the above definition. Many of these countries probably come as no surprise. But the question is whether we are facing a serious problem.

How does Kuwait actually get by with just 30 liters per day? The point is, it doesn't. Kuwait, Libya and Saudi Arabia all cover a large part of their water demand by exploiting the largest water resource of all—through desalination of

Table 1

Countries With Chronic Water Scarcity (Below 2,740 Liters Per Capita Per Day) in 2000, 2025, and 2050, Compared to a Number of Other Countries

Available water, liters per capita per day	2000	2025	2050
Kuwait	30	20	17
United Arab Emirates	174	129	116
Libya	275	136	92
Saudi Arabia	325	166	118
Jordan	381	203	145
Singapore	471	401	403
Yemen	665	304	197
Israel	969	738	644
Oman	1,077	448	268
Tunisia	1,147	834	709
Algeria	1,239	827	664
Burundi	1,496	845	616
Egypt	2,343	1,667	1,382
Rwanda	2,642	1,562	1,197
Kenya	2,725	1,647	1,252
Morocco	2,932	2,129	1,798
South Africa	2,959	1,911	1,497
Somalia	3,206	1,562	1,015
Lebanon	3,996	2,971	2,533
Haiti	3,997	2,497	1,783
Burkina Faso	4,202	2,160	1,430
Zimbabwe	4,408	2,830	2,199
Peru	4,416	3,191	2,680
Malawi	4,656	2,508	1,715
Ethiopia	4,849	2,354	1,508
Iran, Islamic Rep.	4,926	2,935	2,211
Nigeria	5,952	3,216	2,265
Eritrea	6,325	3,704	2,735
Lesotho	6,556	3,731	2,665
Togo	7,026	3,750	2,596
Uganda	8,046	4,017	2,725
Niger	8,235	3,975	2,573
Percent people with chronic scarcity	**3.7%**	**8.6%**	**17.8%**
United Kingdom	3,337	3,270	3,315
India	5,670	4,291	3,724
China	6,108	5,266	5,140
Italy	7,994	8,836	10,862
United States	24,420	20,405	19,521
Botswana	24,859	15,624	12,122
Indonesia	33,540	25,902	22,401
Bangladesh	50,293	35,855	29,576
Australia	50,913	40,077	37,930
Russian Federation	84,235	93,724	107,725
Iceland	1,660,502	1,393,635	1,289,976

Source: WRI 1998a.

sea water. Kuwait in fact covers more than half its total use through desalination. Desalting requires a large amount of energy (through either freezing or evaporating water), but all of these countries also have great energy resources. The price today to desalt sea water is down to 50–80 ¢/m^3 and just 20–35 ¢/m^3 for brackish water, which makes desalted water a more expensive resource than fresh water, but definitely not out of reach.

This shows two things. First, we can have sufficient water, if we can pay for it. Once again, this underscores that *poverty* and not the environment is the primary limitation for solutions to our problems. Second, desalination puts an upper boundary on the degree of water problems in the world. In principle, we could produce the Earth's entire present water consumption with a single desalination facility in the Sahara, powered by solar cells. The total area needed for the solar cells would take up less than 0.3 percent of the Sahara.

Today, desalted water makes up just 0.2 percent of all water or 2.4 percent of municipal water. Making desalination cover the total municipal water withdrawal would cost about 0.5 percent of the global GDP. This would definitely be a waste of resources, since most areas have abundant water supplies and all areas have some access to water, but it underscores the upper boundary of the water problem.

Also, there's a fundamental problem when you only look at the total water resources and yet try to answer whether there are sufficient supplies of water. The trouble is that we do not necessarily know *how* and *how wisely* the water is used. Many countries get by just fine with very limited water resources because these resources are exploited very effectively. Israel is a prime example of efficient water use. It achieves a high degree of efficiency in its agriculture, partly because it uses the very efficient drip irrigation system to green the desert, and partly because it recycles household wastewater for irrigation. Nevertheless, with just 969 liters per person per day, Israel should according to the classification be experiencing absolute water scarcity. Consequently, one of the authors in a background report for the 1997 UN document on water points out that the 2,740 liters water bench-mark is "misguidedly considered by some authorities as a critical minimum amount of water for the survival of a modern society."

Of course, the problem of faulty classification increases, the higher the limit is set. The European Environmental Agency (EEA) in its 1998 assessment somewhat incredibly suggested that countries below 13,690 liters per person per day should be classified as "low availability," making not only more than half the EU low on water but indeed more than 70 percent of the globe. Denmark receives 6,750 liters of fresh water per day and is one of the many countries well below this suggested limit and actually close to EEA's "very low" limit. Nevertheless, national withdrawal is just 11 percent of the available water, and it is estimated that the consumption could be almost doubled without negative environmental consequences. The director of the Danish EPA has stated that, "from the hand of nature, Denmark has access to good and clean groundwater far in excess of what we actually use."

By far the largest part of all water is used for agriculture—globally, agriculture uses 69 percent, compared to 23 percent for industry and 8 percent for households. Consequently, the greatest gains in water use come from cutting down on agricultural use. Many of the countries with low water availability therefore compensate by importing a large amount of their grain. Since a ton of grain uses about 1,000 tons of water, this is in effect a very efficient way of importing water. Israel imports about 87 percent of its grain consumption, Jordan 91 percent, Saudi Arabia 50 percent.

Summing up, more than 96 percent of all nations have at present sufficient water resources. On all continents, water accessibility has *increased* per person, and at the same time an ever higher proportion of people have gained access to clean drinking water and sanitation. While water accessibility has been getting *better* this is not to deny that there are still widespread shortages and limitations of basic services, such as access to clean drinking water, and that local and regional scarcities occur. But these problems are primarily related not to physical water scarcity but to a lack of proper water management and in the end often to lack of money—money to desalt sea water or to increase cereal imports, thereby freeing up domestic water resources.

Will It Get Worse in the Future?

The concerns for the water supply are very much concerns that the current problems will become worse over time. As world population grows, and as precipitation remains constant, there will be less water per person, and using Falkenmark's water stress criterion, there will be more nations experiencing water scarcity. In Figure 3 it is clear that the proportion of people in water stressed nations will increase from 3.7 percent in 2000 to 8.6 percent in 2025 and 17.8 percent in 2050.

Figure 3

Share of Humanity With Maximum Water Availability in the Year 2000, 2025, and 2050, Using UN Medium Variant Population Data

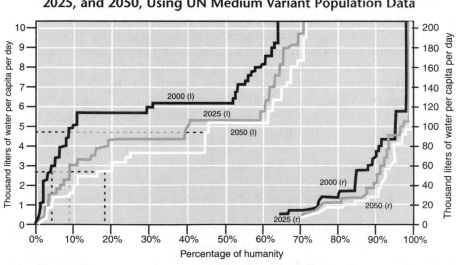

The left side uses the left axis, the right side the right axis.
Source: WRI 1998a.

It is typically pointed out that although more people by definition means more water stress, such "projections are neither forecasts nor predictions." Indeed, the projections merely mean that if we do not improve our handling of water resources, water will become more scarce. But it is unlikely that we will not become better at utilizing and distributing water. Since agriculture takes up the largest part of water consumption, it is also here that the largest opportunities for improving efficiency are to be found. It is estimated that many irrigation systems waste 60–80 percent of all water. Following the example of Israel, drip irrigation in countries as diverse as India, Jordan, Spain and the US has consistently been shown to cut water use by 30–70 percent while increasing yields by 20–90 percent. Several studies have also indicated that industry almost without additional costs could save anywhere from 30 to 90 percent of its water consumption. Even in domestic distribution there is great potential for water savings. EEA estimates that the leakage rates in Europe vary from 10 percent in Austria and Denmark up to 28 percent in the UK and 33 percent in the Czech Republic.

The problem of water waste occurs because water in many places is not well priced. The great majority of the world's irrigation systems are based on an annual flat rate, and not on charges according to the amount of water consumed. The obvious effect is that participants are not forced to consider whether all in all it pays to use the last liter of water—when you have first paid to be in, water is free. So even if there is only very little private utility from the last liter of water, it is still used because it is free. . . .

This is particularly a problem for the poor countries. The poorest countries use 90 percent of their water for irrigation compared to just 37 percent in the rich countries. Consequently, it will be necessary to redistribute water from agriculture to industry and households, and this will probably involve a minor decline in the potential agricultural production (i.e. a diminished increase in the actual production). The World Bank estimates that this reduction will be very limited and that water redistribution definitely will be profitable for the countries involved. Of course, this will mean increased imports of grain by the most water stressed countries, but a study from the International Water Management Institute indicates that it should be possible to cover these extra imports by extra production in the water abundant countries, particularly the US.

At the same time there are also large advantages to be reaped by focusing on more efficient household water consumption. In Manila 58 percent of all water disappears (lost in distribution or stolen), and in Latin America the figure is about 40 percent. And on average households in the Third World pay only 35 percent of the actual price of water. Naturally, this encourages overconsumption. We know that pricing and metering reduces demand, and that consumers use less water if they have to pay for each unit instead of just paying a flat rate.

Actually, it is likely that more sensible pricing will not only secure future water supplies but also increase the total social efficiency. When agriculture is given cheap or even free water, this often implies a hidden and very large subsidy—in the United States the water subsidy to farmers is estimated to be above 90 percent or $3.5 billion. For the developing countries this figure is even larger: it is estimated that the hidden water subsidy to cities is about $22 billion, and the hidden subsidy to agriculture around $20–25 billion.

Thus, although an increasing population will increase water demands and put extra water stress on almost 20 percent of humanity, it is likely that this scarcity can be solved. Part of the solution will come from higher water prices, which will cut down on inefficient water use. Increased cereal imports will form another part of the solution, freeing up agricultural water to be used in more valuable areas of industry or domestic consumption. Finally, desalting will again constitute a backstop process which can produce virtually unlimited amounts of drinking water given sufficient financial backing.

POSTSCRIPT

Is the Threat of a Global Water Shortage Real?

The authors of the two selections agree that something must be done—that is, major public policy making must occur—if the water future is going to be acceptable. They disagree, however, on the urgency of the task and the level of optimism (or pessimism) that they bring to their analysis of likely success. Rosegrant and his colleagues approach the global water problem from the perspective of its role in food production. They argue that water is among one of the main factors that will limit food production in the future. Competition for water comes from many quarters, and farmers find themselves in an increasingly competitive situation as they attempt to keep pace with agricultural needs.

Rosegrant and his colleagues' view of past water trends—whether for irrigation or for drinking and household uses—leads them to conclude that the future of water is "highly uncertain," at best. The authors' pessimism stems from the failure of governments to address the issue adequately to date. Their model used to analyze future trends posits three basic scenarios: business as usual, water crisis, and sustainable water scenarios.

In the business-as-usual scenario, farmers will be unable to grow food as quickly as in the past. The cost of supplying water to users will increase, as heavier demands are placed on the global water supply. Industrial water use in the developing world is expected to grow faster than in the developed world as the developing countries push toward industrialization. Water availability for irrigation in the poorer sectors of the globe will decline significantly. Sub-Saharan Africa will, as it has for three decades, be the region of the globe hardest hit by the business-as-usual scenario. The water crisis scenario posits that "increasing water scarcity, combined with poor water policies and inadequate investment in water, has the potential to generate sharp increases in cereal food prices over the coming decades." Finally, the sustainable water scenario would result in a dramatic increase in water availability, improve environmental uses, lead to higher per capita domestic water consumption, and maintain food production at the business-as-usual scenario levels.

Lomborg argues that our ability to find more water has resulted in a global usage rate of 17 percent of the readily accessible and renewable water. The consequence, according to Lomborg, is that the world has "more food, less starvation, more health and increased wealth [so] why do we worry?" He does suggest that there are significant problems and identifies three: (1) the unequal distribution of precipitation throughout the globe, (2) increasing global population, and (3) the fact that many countries receive their water through shared river systems. Additionally, water pollution and the shortage of access to water in the developing world are issues, but of a different sort.

Both readings point out the need for aggressive policy action on the part of governments and other actors in the global water regime. This is not to suggest, however, that the world's leaders have been idle. At least 10 major international conferences since 1977 have addressed water issues, resulting in significant action-oriented proposals. For example, the 1992 International Conference on Water and Environment in Dublin established four basic ideas (known as the Dublin Principles). In addition, the 1992 Earth Summit highlighted water as an integral part of the ecosystem.

The current cry is for aggressive global water management. Every major study related to water concludes with the observation that governments need to do much more if future water crises are to be avoided. The Web is replete with public (and private) interest groups urging more global action. The Third World Traveler site titles one such plea from the International Forum on Globalization "The Failure of Governments" in a discussion of the crisis. The World Water Council's recent draft of its *World Water Actions Report* presents an overview of global actions to improve water management and to spell out priorities for future efforts. The report centers on a new paradigm for looking at water. No longer to be viewed as a physical product, the paradigm calls for water to be seen as an ecological process "that connects the glass of water on the table with the upper reaches of the watershed." The key ideas are "scarcity, conservation, and awareness of water's life cycle from rain to capture, consumption, and disposal."

The good news for the World Water Council is that the international community has broad consensus, forged at the 10 conferences mentioned above, about the need for policy action. The creation of the World Water Council itself in 1996 and the Global Water Partnership the same year is an example of such institutional action. But it was the Second World Water Forum, held at The Hague in The Netherlands in March 2000, that was a landmark event in raising global water consciousness. Among the many research reports and documents that emerged from the conference, two stand out: *Vision for Water, Life, and the Environment in the 21st Century* and *Towards Water Security: Framework for Action.*

Vision proposed five key actions: "Involve all stakeholders in integrated management, move towards full-cost pricing of all water services, increase public funding for research and innovation in the public interest, increase cooperation in international water basins, (and) massively increase investments in water." The *Framework for Action* document addresses the question of where do we go from here, suggesting four basic steps: "generating water wisdom . . ., expanding and deepening dialogue among diverse stakeholders, strengthening the capabilities of the organizations involved in water management, and ensuring adequate financial resources to pay for the many actions required."

A final important source for understanding the global water problem, particularly its relationship to food and the environment, is *Dialogue on Water, Food and Environment* (2002), published by 10 important actors in the field (FAO, GWP, ICID, IFAP, IWMI, IUCN, UNEP, WHO, WWC, and WWF). Perhaps it is fitting to end this discussion with the quote from UN Secretary General Kofi Annan on the cover of this report: "We need a blue revolution in agriculture that focuses on increasing productivity per unit of water—more crop per drop."

Globalization: Threat or Opportunity?

This Web site contains the article "Globalization: Threat or Opportunity?" by the staff of the International Monetary Fund (IMF). This article discusses such aspects of globalization as current trends, positive and negative outcomes, and the role of institutions and organizations. "The Challenge of Globalization in Africa," by IMF acting managing director Stanley Fischer, and "Factors Driving Global Economic Integration," by Michael Mussa, IMF's director of research, are also included on this site.

http://www.imf.org/external/np/exr/ib/2000/
041200.htm

Center for Democracy & Technology

The Center for Democracy & Technology (CDT) works to preserve and enhance free expression on the Internet as well as to protect privacy. Democratic values are promoted through public policy advocacy, online grassroots organizing, and public education campaigns. Explore this site to learn about the CDT's proposals for achieving its goals as well as its thoughts on other Internet issues, such as digital authentication, bandwidth, terrorism, and access to government information.

http://www.cdt.org

Popular Culture Association/American Culture Association

The Popular Culture Association/American Culture Association Web site, a member of the H-NET Humanities OnLine initiative, fosters intellectual discussion of popular culture. View the discussion logs by month or by specific topic. Also included on this site is a list of links to sites on specific aspects of American culture.

http://www2.h-net.msu.edu/~pcaaca/

United Nations Office on Drugs and Crime

Established in 1997, this UN organization assists members in their struggle against illicit drugs. It focuses on research, assistance with treaties, and field-based technical assistance. It is headquartered in Vienna with 21 field offices.

http://www.unodc.org

Friends of the Earth

Founded in 1969 in the United States, this organization is now an international network of grassroots groups in 70 countries who act as watchdogs on, among other topics, international financial institutions.

http://wwwfoe.org/camps/intl/institutions/

PART 3

Expanding Global Forces and Movements

*O*ur ability to travel from one part of the globe to another in a short amount of time has expanded dramatically since the Wright brothers first lifted an airplane off the sand dunes of North Carolina's Outer Banks. The decline of national borders has also been made possible by the explosion of global technology.

One consequence is the international flow of illegal drugs. Another is the flow of information, ideas, and money across the globe. The devices that allow people to connect with the globe at all of these levels also allow the rest of the world to connect with them. Fiber-optic lines and satellites carry information, products, money, ideas, and values across cultural and territorial divides as never before. The impact of these new patterns of access have yet to be calculated, but we know that is both uplifting and unsettling to tens of millions.

- Can the Global Community "Win" the Drug War?

- Is Globalization a Positive Development for the World Community?

- Will the Digital/Computer World Pose a Threat to Our Individual Privacy?

- Is the World a Victim of American Cultural Imperialism?

- Do International Financial Institutions and Multinational Corporations Exploit the Developing World?

- Is the Transnational Media Hurting Global Society?

ISSUE 10

Can the Global Community "Win" the Drug War?

YES: Federico Mayor in collaboration with Jérôme Bindé, from "Winning the Fight Against Drugs: Education, Development and Purpose," *The World Ahead: Our Future in the Making* (UNESCO, 2001)

NO: Harry G. Levine, from "The Secret of Worldwide Drug Prohibition," *The Independent Review* (Fall 2002)

ISSUE SUMMARY

YES: Mr. Mayor suggests that drug trafficking and consumption "constitute one of the most serious threats to our planet" and the world must dry up the demand and attack the financial power of organized crime.

NO: Harry G. Levine, professor of sociology at Queens College, City University of New York, argues that the emphasis on drug prohibition should be replaced by a focus on "harm reduction," creating mechanisms to address tolerance, regulation, and public health.

In 1999, the United Nations pegged the world illicit drug trade at $400 billion, about the size of the Spanish economy. Such activity takes place as part of a global supply chain that "uses everything from passenger jets that can carry shipments of cocaine worth $500 million in a single trip to custom-built submarines that ply the waters between Colombia and Puerto Rico." In June 2004, the United Nations Office on Drugs and Crime (UNODC) reported that 3 percent of the world population (and 4.7 percent of those aged 15 to 64), approximately 185 million people, had "abused drugs during the previous 12 months." This includes people from virtually every country on the planet and from every walk of life. The *2004 World Drug Report* suggested that opiates remained the most serious problem in terms of treatment, with 67 percent of the drug treatment in Asia, 61 percent in Europe, and 47 percent in Oceana. Cocaine is still the drug of choice in the Americas, and the number of admissions for treatment is now higher for cocaine in the United States. In Africa, cannabis tops the treatment list (65 percent). One-fifth of the world's population (and 29 percent of those above age 15) also use tobacco.

While the UN report suggests that the spread of drug abuse may be losing some momentum, the use of cannabis is growing at an accelerated rate. After cannabis, the highest increases in the past decade have been ATS (mainly ecstasy), then cocaine and then opiates. The amount of drugs seized has also increased over the last decade, rising from 14 billion doses in 1990 to 26 billion in 2000. The bottom line statement from the United Nations report is that "the second half of the twentieth century . . . witnessed an epidemic of illicit drug use."

The report also included some good news. Especially important has been the emergence of a consensus among governments and global public opinion that the current levels of illegal drug use is unacceptable. In two drug-producing regions, declines in production have actually occurred. In Southeast Asia, opium poppy cultivation continues to drop in Myanmar and Laos. In the Andean region, coca cultivation has declined for four straight years in the three leading producing countries (Colombia, Peru, and Bolivia).

The illegal movement of drugs across national borders is accompanied by the same kind of movement for illegal weapons. They go hand in hand with one another. The UN estimates that only 3 percent of such weapons (18 million of a total of 550 million in circulation) are used by government, the military, or police.

This increase in drug use has occurred despite a rather long history of government attempts to control the illegal international drug trade. Beginning in 1961, such efforts have been part of governments' worldwide social policies. Precisely because drug policy crosses over into social policy, policymakers and scholars have been at odds over how best to deal with this ever-growing problem, whether talking about national policy or international policy. Simply stated, the debate has centered on legalization vs. prohibition, and treatment vs. prevention.

Policies of the United States have always had the goal of drug use reduction and punishment for abusers, resulting in less attention to treatment. This includes a number of important elements, as outlined in a Congressional Research Service Brief for Congress (2003): "(1) eradication of narcotic crops, (2) interdiction and law enforcement activities in drug-producing and drug-transmitting countries, (3) international cooperation, (4) sanctions/economic assistance, and (5) institution development." Many have charged the United States and those other countries that share its fundamental philosophy of drug wars, of using the issue to expand its national power in other domains. On the other hand, other countries, particularly those in Western Europe, have been shifting attention for some time away from "repressive policies" and toward those associated with harm reduction and treatment.

The following two selections contribute to the debate over the proper approach to "winning" the drug war. Federico Mayor understands the need for creating a declining market for drugs. He argues that although the control of narcotics is preferable, it is not likely to happen. The major reason for such pessimism is due to the overwhelming power of criminal organizations. Harry G. Levine suggests that drug prohibition has had wide acceptability throughout the world because of pressure from the United States. These countries have embraced this position because it is useful for other purposes associated with national power.

➔ **YES**

Winning the Fight Against Drugs: Education, Development and Purpose

A Rapidly Expanding Market

While many economists are rejoicing at the sustained growth of the world economy, there is one market in particular that is undergoing uninterrupted expansion throughout the world: the drug market, the cause of the most radical marginalization of human beings, since drugs abolish all notions of self or of other human beings.[1] According to the United Nations, profits derived from drug trafficking amount to $400 billion annually, that is to say 8% of world trade or, at a rough estimate, the equivalent of international trade in textiles in 1994,[2] or 1% of world GNP, or the GNP of the whole of Africa.[3] The production and consumption of drugs are rising constantly. According to the 1996 report of the International Narcotics Control Board (INCB), 'in spite of increased repression, production of and trafficking in drugs, together with drug addiction, have now reached hitherto preserved regions of the world.' As emphasized by an expert, 'Narco-states and narco-democracies, narco-terrorism and narco-guerrilla activities, narco-tourism and narco-dollars, are all signs that drugs have penetrated every sphere of political, economic and social life. The expansion of drug trafficking now goes hand in hand with the globalization of the economy and free market democracy.'[4] The highly efficient organization of drug trafficking, through a worldwide network based on extremely flexible and constantly changing units, has rendered any control of narcotics particularly difficult.[5] Production as such is still highly concentrated, as 90% of the illicit production of narcotics derived from opium worldwide originates from two major areas: the 'Golden Crescent' (Afghanistan, Iran and Pakistan) and the 'Golden Triangle' (Laos, Burma and Thailand) and 98% of world supplies of cocaine come from the Andean countries (Peru, Colombia and Bolivia).[6] But new connections have developed, production and trafficking areas have expanded and new synthetic drugs have appeared on the market. Drug trafficking is the hidden face of globalization. It also happens to be one of its main beneficiaries, owing to increasingly porous national frontiers, the volatile nature of financial operations and the contagion of lifestyles, and even what might be called 'death styles.' INTERPOL estimates that only 5% to 15% of banned drugs are actually seized, which means that at least 85% of narcotics escape repression and circulate freely, in a clandestine market controlled by criminals.[7]

According to OECD estimations, $85 billion derived from profits from such trafficking are laundered every year on the financial markets—that sum is greater than the GNP of three-quarters of the 207 economies in the world, according to a group of G7 experts.[8] The wealth accumulated by drug traffickers over the last 10 to 15 years could amount to as much as 'several trillion dollars.'[9] The drugs trade first and foremost 'benefits' the industrialized countries, if we may venture to say so: 90% of these sums is thought to be reinvested in the Western countries.[10] Many experts are increasingly concerned at the growing expansion of 'grey areas' within the world economy, which enable major organized crime networks to penetrate the very heart of some strategic spheres of the international economy such as the major world financial exchanges.[11] As pointed out by an expert, 'in every country, the banking system is actively involved in recycling drugs revenue, particularly through subsidiaries and correspondents established in the worldwide constellation of tax havens,' which means, sometimes, that 'laundered money from drugs enables debt instalments to be paid or funds structural adjustment plans.'[12] The growing sophistication of financial operations, the globalization of the banking system, which is no longer hampered by frontiers and is able to operate round the clock, and the rapid emergence of unrecorded 'cyber-payments' require increased vigilance on the part of regulatory bodies and the extension of their partnership to the whole range of world financial institutions. The USA, for its part, recently advocated regulation of the non-banking financial sector, ranging from currency exchange and brokerage houses to casinos, as well as express delivery services, insurance firms and the precious-metal trade.[13] Multinational corporations and transnational finance companies should abide by codes of conduct in order to prevent the laundering of money derived from crime, whether it comes from drug trafficking, arms dealing or any form of criminal trade or mafia activity (embezzlement of public funds, racketeering, prostitution, illicit gambling, etc.).

The influence, whether overt or covert, of major criminal organizations, seemingly on the increase in many countries in both North and South, means that serious dangers are threatening economic ethics and the rule of law. According to the UNDP, expenditure on consumption of narcotics in the USA alone exceeds the accumulated GDP of more than eighty developing countries. Furthermore, organized crime has considerably extended its geographical areas of influence thanks to globalization and the development of drug trafficking, which often occurs in symbiosis with other criminal activities (arms trafficking, prostitution and the slave trade, the embezzlement of public funds, illegal gambling and penetration of the casino network by the mafia, etc.). The gigantic scale of illicit profits from drugs, together with the 'penetration' of entire sectors of the legal economy now controlled through money-laundering, could ultimately lead, through the dynamics of accumulation and concentration observed during the last two decades, to an irreversible situation whereby no state or organized force would be in a position to react as, through the laundering process, a substantial part of the economy and of pressure groups, in both North and South, would fall under the influence of drug trafficking. Keeping quiet in this matter amounts to observing the very principle that underlies the power of criminal organizations, namely the law of silence.

In our opinion, drug trafficking and consumption constitute one of the most serious threats to our planet, with disastrous consequences for health, development and society. We are all the more sensitive to this problem as we know all too well what effects drug addiction has on the brain's receptors and how irreversible lesions are caused above a certain level of consumption.[14] To these evils are added the effects of the spread of AIDS among drug addicts who absorb drugs intravenously. Young people, education and human values are affected first but drug addiction makes life unbearable for the whole family of the addict and, sooner or later, it is democracy itself that is threatened and, with it, peace. As pointed out by a specialist in this field, 'there is virtually no local conflict today that is not linked to a greater or lesser extent to drug trafficking.'[15] Drugs have become a form of violence not only towards the individual but also towards the whole of society. . . .

Eliminate Supply or Dry up Demand?

If effective action is to be taken against drugs, we must first of all open our eyes wider and open those of others. We need to discuss the problem with greater scientific rigour and critical awareness within the institutions that disseminate knowledge, namely schools, universities and all the channels of mass communication. Our task must be to show both clearly and unceasingly the real damage caused by the various drugs, the moral and physical servitude and the destruction of mind and body, attitudes and values, of which they are the cause. We must shed light on the harm caused to both society—starting with the immense suffering inflicted on the addict's family—and the individual. We must cease to portray drugs as malevolent yet attractive, and thus avoid demonizing them, as is too often the case, unaware that, in so doing, we turn them into a symbol of the urge to transgress social rules. If we cry 'Wolf!' too often, we are more likely to push young people towards drugs than to put them off. A survey recently conducted by a French opinion poll agency showed that for 52% and 44% respectively of drug-takers, pleasure and curiosity were the prime motive.[16] What we need to do is to demystify drugs by explaining to children that they are first and foremost a denial of existence, as stressed by Rita Levi-Montalcini, Nobel prizewinner in medicine, in her fine essay dedicated to young people, aptly entitled 'Your future.' Drugs will stop attracting adolescents once they have understood that they constitute, above all, a 'lessening of the power to act,' as Spinoza might have put it, when he defined sadness in such terms in his work *Ethics*.

Some people believe that drugs should no longer be banned. According to these liberal-minded opponents of prohibition, it is through re-establishing control of narcotics by the legal economy that we can best fight the plagues which are the result of the illegal nature of the market, namely the enrichment of dealers and middlemen (which stimulates expansion of this trade), crime, violence, the marginalization of addicts, arms trafficking, terrorism and the suspicion of corruption that in many countries weighs on wide areas of public and political life. Nevertheless, we are among those who believe that we must firmly oppose the legalization of the non-therapeutic use of drugs. In that regard, the report of the International Narcotics Control Board (INCB) for 1997 deplored the existence of 'an overall climate of acceptance that is favourable to or at least tolerant of drug abuse.'[17] We can no more play around with narcotics than we can with weapons or medicinal drugs. The risk of a real surge in consumption is far too great. That is why drugs cannot be left to market forces, whether legal or illegal. In this domain, we cannot afford to be sorcerers' apprentices.

Conversely, the idea that it is possible to eliminate the production and consumption of drugs, however commendable it may be, strikes us as being scarcely credible. Many commentators are now advocating the replacement of prohibition policies by measures to reduce the harmful effects of drugs, in the belief that 'drugs are here to stay and we have no choice but to learn to live with them so that they cause the least possible harm.'[18] The policy of all-round repression can fail because, by making drugs scarcer, it merely makes drug trafficking more lucrative for new networks that have replaced earlier ones. While we thought that we were fighting organized crime, we were actually strengthening its financial power and capacity to corrupt, and dragging down large sections of society into delinquency.

A recent report states that 'Despite regional successes supply suppression is not a prescription for solving the world's illicit-drug problem. It is a prescription for funding drug mafias, peasant growers, petty traffickers and smugglers.'[19] There is naturally a price to pay for such a policy: several tens of billions of dollars are spent every year on repression, with results that, to put it mildly, are hardly convincing: delinquency in a growing part of society (the inner cities, ghettos, minorities, the younger generation, the interaction between consumption and 'petty trafficking,' etc.), and the corresponding excessive growth of the repressive and penitentiary system, which ultimately penalizes addicts rather than traffickers. Everything, therefore, needs to be changed. On the one hand, an efficient judicial and penitentiary system must be set up to deal with drug dealers. On the other hand, preventive and curative measures should be introduced on a large scale. We must stress that addicts need to be given treatment—and adequate funds should be allocated for that purpose—by bringing into play all the means required, whether they be medical, scientific or of another nature.

We must therefore learn how to deal with the problems of drug addiction in our societies while reducing this phenomenon to a minimum and avoiding the criminalization of addicts. In terms of public health alone, realistic public policies 'for reducing the harmful effects' are required to respond to the fact that the total absence of control over the drugs market is a powerful vector for the development of AIDS and other epidemics: the distribution of free syringes helped in several countries to lower the contamination of drug addicts by HIV. Moreover, we believe the time has come to contemplate an international agreement whereby it would be possible, under medical supervision, to distribute a limited quantity of drugs to addicts who are not able to break out of addiction.[20] Drug addicts should be treated as patients rather than delinquents. As such, patients have a right to benefit from medical supervision and social assistance in much the same way as any human being suffering from a curable pathology. What is more, such a measure could reduce violence and delinquency and contribute to dismantling the illegal drugs market and, therefore, the major source of profits for organized crime. The difficulty underlying such a policy can be summarized as follows: any agreement would have to be international, as policies for fighting against drugs can no longer be conceived in purely national terms. Harmonizing the policies of various states would be the key to effectiveness in this field where interdependence is particularly marked.[21]

Clearly, such a policy will have to be accompanied by an in-depth survey on the specific harmfulness of drugs by the scientific and medical community, in close cooperation with the relevant national and international authorities. Such concertation might, as suggested by a French consultative body, make it possible to establish regulations for each substance, 'taking account of its toxicity, the risks of dependence relating to its consumption, the danger of desocialization it might entail and the risks to which its consumption might expose other people.'[22]

Prevention by educating and informing the public would also be indispensable for this project. We need the help of the media, as well as municipal and local authorities, to foster appropriate awareness, commitment and participation and to ensure that drug addiction does not become commonplace, the lame excuse of a society that tolerates the degeneration and distress of those who symbolize its future, namely, its younger generations. The Youth Charter for a Twenty-first Century Free from Drugs (1997), which received the support of UNESCO and the United Nations Programme for International Narcotics Control (UNPINC), rightly states that 'the first experiences with drugs are often motivated by curiosity, idleness, lack of self-confidence, indifference and violence in our immediate surroundings, but also by the difficulties and trials of everyday life.'[23]

Many experts and institutions nevertheless continue to give priority to reducing the supply of drugs. One of the solutions for reducing the production and therefore the supply of drugs would be to develop sufficiently lucrative alternative crops and new markets for the peasant farmers whose livelihood depends on poppy and coca cultivation. To do that, the cooperation of the peasant farmers concerned is all the more important in the choice of new crops as the cultivation of toxic plants is often related to cultural traditions. They need to be made aware of the dangers of drugs for their health and for the life of their community, as well as for the well-being of the whole of humanity. Unfortunately, policies for the eradication of plantations and help for the substitution of illegal crops have often failed and have had adverse effects through ignorance of the cultural factors of development, the local social environment, the requirements of sustainable development, as well as through anthropocentric naivety.

In such circumstances, the establishment of a scheme for subsidies and guaranteed prices for new crops bringing into play national and international resources would seem indispensable. In this type of situation, the international community, by lending its financial support, could invest in the future with success, benefit and a sense of long-term vision.

In addition to these difficulties, replacing drugs with alternative crops, in the absence of accompanying structural measures and international support, would appear virtually condemned to failure for four main reasons. No government would seem ready to pay the very high price of replacing crops worldwide, if the operation had to be subsidized. No agricultural production, on the basis of market prices, would be competitive with the price of base plants. Furthermore, the economy of many producing countries, which are often very poor, is now increasingly based on drugs. It is no secret, as illustrated by various reports produced by the United Nations, that the drug industry accounts for as much as 20% of GNP for some countries.[24] Worse still, once crops have actually been eradicated in a particular country, the outcome is disappointing and deceptive as production moves elsewhere, often to a neighbouring country.[25]

A global threat requires a global solution. A large-scale threat calls for large-scale solutions. Many countries are members of major international alliances that guarantee their borders and security. If we wish to mark the advent of the new century by making a fresh start, we must then have comparable alliances at our disposal for combating drugs, as would be required for combating global catastrophes of all kinds. The drugs problem should not be confined to protracted, trivial discussions on the respective responsibility of the producer and consumer countries: such a dispute is all the more futile as the frontier between the former and the latter has become increasingly blurred since the explosion of consumption in the South. Let us endeavour instead to fight the causes of supply and demand by offering acceptable living conditions and a better future to the peasant farmers attracted by illicit but more profitable crops, to the middlemen who, in many underprivileged countries or communities, often have no alternative income and to the consumers who, through lack of education and information, are ignorant of the dangers of drugs or who feel excluded from a society in which they are not able to fashion their own lives. Let us give hope and a future to them all. The terrible effects of drug addiction on human dignity constitute a powerful illustration of the importance of preventive action, which involves educating young people as early as possible.

Let us recall that international action against drug trafficking began some 80 years ago when the opium trade came under international jurisdiction. Since then, the multilateral system has devised many conventions and plans of action for combating this traffic which, at the highest levels of responsibility, may be considered to be a crime against humanity. In fact, the United Nations General Assembly proclaimed the 1990s as the 'United Nations Decade against Drugs.'

The most efficient means of fighting against drug trafficking is, as stated by the Italian judge, Giovanni Falcone, shortly before he was assassinated by the Mafia, 'the destruction of the financial power of organized crime, which would presuppose powerful international collaboration.'[26] This alone can help to prevent the emergence of the 'chain of connivance' composed of obscure acts of corruption and unavowed links described by the great Sicilian writer, Leonardo Sciascia, in his novel *The Context*, published in 1971. It is that chain of connivance that undermines democratic institutions and threatens their legitimate representatives.[27]

Judge Falcone added that it was necessary, with that aim in mind, to encourage and coordinate 'efforts aimed at identifying and confiscating wealth of illegal origin,' which requires 'adapting international laws and achieving constant international collaboration.' Giovanni Falcone advocated 'first and foremost, the elimination of tax havens which, up to now, have countered the most serious attempts of various countries to identify financial flows originating from illegal trafficking.' According to Judge Falcone, 'this is a fight that concerns all members of the international community because its outcome will determine whether organized crime is destroyed or at least limited in such a way as to be no longer a serious peril for society.' The advice and sacrifice of Judge Falcone would not seem to have been totally pointless: since then, the Italian magistracy has intensified the seizure of assets of illegal origin, while the profits of the four major Italian criminal organizations (estimated by the Italian Anti-Mafia Investigatory Department at 10 trillion lire, or 30% of a turnover estimated at 30 trillion in 1994), seem to have shrunk massively in the same year.[28]

Furthermore, the production of narcotics is by no means limited to substances of natural origin. In its 1996 and 1997 reports, the INCB highlighted the preoccupying expansion throughout the world of synthetic drugs, particularly amphetamines or by-products, such as ecstasy, produced in clandestine laboratories. They supply a very lucrative illicit market for dealers and meet with alarming success among young people. The US Department of State believes that amphetamines, on account of their simple manufacturing process and the sudden growth in demand, are about to become 'the drug-control nightmare of the next century.'[29] In the face of this growing threat, which, paradoxically, has benefited from progress achieved in the field of pharmaceutical research, new measures for control, information, research and education are required, particularly for the benefit of young people.

There is therefore no miracle solution to the drugs problem. As long as there is demand, there will be supply. As noted by the UNDP in a lucid report, the real solution requires tackling the causes of drug addiction and eliminating the poverty that leads farmers to become involved in producing narcotics.[30]

We are particularly concerned with the consequences of drug consumption on the fate of street children who, today, number more than 100 million and who are fighting every day to survive in conditions of total deprivation. These children are those who are most threatened by violence, sexual and economic exploitation, AIDS, hunger, solitude and the scourges of exclusion, illiteracy and drugs. They are the 'golden fish' referred to by the French novelist Le Clézio which the ill-intentioned fishermen in search of innocent prey attempt to catch in their nets. Everything must be done to ensure that these children are fully integrated into society, that they learn to live in it, that they have access to education and that they are no longer manipulated by criminals who make them serve their evil purposes. The latter deserve to be punished all the more severely as by destroying innocence, they attempt to eradicate faith and confidence in the future.

The fight against drug addiction—like that against AIDS or against the collective shame represented by street children and children who are sexually or professionally exploited—will not be truly effective unless it is based on a major alliance between all countries, translated practically into a political will not to abandon the cause, just as we defend our country when national sovereignty is in jeopardy. In fact, in all the cases we have just referred to, it is national dignity that is threatened, which cannot be defended simply through charity or by organizing tombolas and galas. The best way of celebrating human rights, the fiftieth anniversary of the declaration of which we celebrated in 1998, would be an internationally reached decision aimed at ensuring its effective exercise by all human beings. The rest is no more than ceremony and empty rhetoric. The United Nations International Drug Control Programme (UNDCP) should be one of the most powerful in the entire United Nations system, in terms both of its authority and of its resources. This should also be the case, in a different field, for the United Nations of Environment Programme (UNEP). The limited means available to these programmes reflect a lack of political will and of public awareness as to what is required for combating drugs or preserving the global environment. We are, to varying degrees, all responsible for this twofold deficiency.

To fight against drug-related problems, the causes of marginalization and exclusion have to be tackled by investing in the welfare of young people, particularly through sports and training activities. It is UNESCO's responsibility to fight against the demand for drugs through education, and more especially preventive education. While education may be the main victim of drugs, it is also its best antidote. In fact, it is thanks to education that young people can become aware of the real dangers of narcotics, that they can escape 'the blues' and find their true path in society, and that they can acquire the knowledge and ethical attitudes that will enable them to assert their own personality and take their destiny in hand. Instead of paying the price for war and over-investing in armed defence, let us invest in the peaceful defence of individuals and young people, in cultural security and in genuine spiritual freedom which access to the world of knowledge and freedom from any servitude, can provide. Once education is widely perceived as having the objective of 'ensuring that people have control over their lives,' then it is through education that any form of dependency can be combated, such as dependency on alcohol, tobacco, drugs and sects, etc. Through education we can learn to be free and responsible.

More than ever before, the vital issue is the political will of governments to agree on effective solutions and on implementing them. More than ever, UNESCO has a major role to play in the context of its fields of competence: education and information against drug abuse, communication activities among the populations, and the contribution of the social sciences and scientific research in order to fine tune action plans and national and international strategies, and to assess the specific harmfulness of drugs, which are a subject of debate.[31]

To be perfectly frank, however, education and information alone cannot, even in the long term, be the only solutions in this field, nor can development if it is reduced to mere economic prosperity. Without referring to the great minds who succumbed to 'artificial paradises,' it is striking to observe the number of people with a high level of education and a comfortable income, who consume drugs in a number of countries. Psychological malaise as reflected by drug addiction cannot be cured merely by knowledge as, to paraphrase Henri Michaux, the poet, knowledge itself may lead to an abyss. If the twenty-first century is to win the battle against drug addiction (which some experts in science fiction doubt, imagining, on the contrary, the expansion of a form of 'addiction to soft drugs' controlled by the neurosciences and pharmacology), it will have to win the battle against nihilism, consumerism, and the fruitless pursuit of intoxication and ecstasy. We shall have to bring about a 'global mobilization' of governments, parliaments, the media, industry and society as a whole against drugs and addiction. The next century will have to give a new meaning to life.

Education, economic development and material well-being will probably not suffice to eliminate drugs even if they are the major instruments of prevention. To think so would be to imagine that human beings can be prevented from walking along the edge of precipices, from seeking to experience ecstasy or trance or, quite simply, from wanting to poison themselves. What therefore has to be done is to construct humanity's defences in people's minds, all the more so as drugs tempt the minds as much as they do the body. Thus new forms of wisdom and ethics will have to develop. What must also be done is to build humanity's defences by investing in human dignity, by reducing dire poverty, racism and exclusion. Fighting against drugs, the source of destruction, suffering and war, also means responding to the aims of the founders of UNESCO. It simply means building peace and development on the basis of the intellectual and moral solidarity of humanity. Fighting against drugs, in a united effort, with human and financial means that correspond to the scale of the plague, means protecting young people, our children and our future. It also means speeding up the transition from a culture of violence, war and indignity to a culture of peace, non-violence and dignity for all.

Pointers and Recommendations

- Reduce the demand for drugs in consumer countries, particularly through education, prevention and treatment.
- Educate and inform children and young people about the risks of drug consumption.
- Mobilize the international community against the main causes of drug consumption which are marginalization and poverty, in both urban and rural environments.
- Develop specific machinery, on international, regional and national scales, for fighting against corruption, the laundering of money from drugs and organized crime. Encourage ratification and implementation of international treaties related to narcotics control and the conclusion of international agreements aimed at destroying the financial power of organized crime.
- Help drug addicts to overcome their dependency and to adopt a sustainable lifestyle without drug consumption, through appropriate educational, rehabilitation and vocational training programmes.
- Reduce the perverse effects of drug trafficking and consumption (financial development of organized crime, criminalization, delinquency and social pathologies) by studying the feasibility, on an international scale, of an agreement that, under medical supervision, would allow addicts who are in need and incapable of giving up their habit a limited supply of drugs.

- Give serious consideration to the adoption of 'reduction of the harmful effects' policies implemented in various countries under the terms of a policy and a world programme for narcotics control.
- Encourage scientific and medical research on the specific harmfulness of drugs and scientific research on the environmental impact of drug cultivation.
- Plan for the convening of a world summit on drugs, organized by the United Nations (United Nations International Drug Control Programme and the World Health Organization), which would take account of all aspects, whether old or new, of the problem of drug consumption and trafficking. Strengthen the means and authority of the UNDCP.
- Bring about a 'global mobilization' of governments, parliaments, the media, industry and society against drugs and addiction.

Notes

1. The term 'drugs' refers to narcotics and psychotropic substances as defined by the relevant international conventions.

2. UNDCP, *1997 World Drug Report,* Oxford University Press, New York, 1997; *Le Monde,* 27 June 1997.

3. Eric de la Maisonneuve, *La Violence qui vient,* Éditions Arléa, Paris, 1997.

4. Christian de Brie, 'La drogue dopée par le marché,' *Le Monde Diplomatique,* April 1996.

5. Hugues de Jouvenel, 'L'inextricable marché des drogues illicites,' *Futuribles,* No. 185, March 1994 (special issue 'Géopolitique et économie politique de la drogue').

6. UNDCP, *1997 World Drug Report.*

7. Address by the director-general of UNESCO at the 58th session of the United Nations International Narcotics Control Board (INCB), Vienna, 9 May 1995.

8. Quoted in UNDP, *1994 World Report on Human Development,* UNDP, Economica, Paris, 1994, p. 37; *UN Chronicle,* No. 3, 1996.

9. Alain Labrousse, *Les Idées en mouvement,* No. 35, January 1996.

10. Alain Labrousse, interview in *Le Nouvel Observateur,* 19–25 September 1996.

11. See 'The Mob on Wall Street,' *Business Week,* 16 December 1996; Laurent Zecchini, 'La "pieuvre" mafieuse prolifère à Wall Street,' *Le Monde,* 3 January 1997; Brie, 'La drogue dopée par le marché'; LaMond Tullis, *Unintended Consequences: Illegal Drugs and Drug Policies in Nine Countries,* Studies on the Impact of the Illegal Drug Trade, Vol. 4, series editor LaMond Tullis, United Nations University and United Nations Research Institute for Social Development, Lynne Rienner, Boulder, CO and London, 1995.

12. De Brie, *Monde Diplomatique* feature, *Le Monde Diplomatique,* February 1996.

13. US Department of State, *International Narcotics Control Strategy Report, 1996,* March 1997.

14. Address by the director-general of UNESCO at the 58th session of the United Nations International Narcotics Control Board (INCB), Vienna, 9 May 1995.

15. Labrousse, *Les Idées en mouvement.*

16. Patrick Piro, *Les Idées en mouvement,* No. 35, January 1996.

17. Report of the International Narcotics Control Board, 1997, para. 20.

18. Ethan A. Nadelmann, 'Commonsense drug policy,' *Foreign Affairs,* January/February 1998, p. 112; Anthony Lewis, 'The war on drugs is being lost,' *International Herald Tribune,* 6 January 1998.

19.   LaMond Tullis, *Unintended Consequences,* p. 183.

20.   See address by the director-general of UNESCO at the 58th session of the United Nations International Narcotics Control Board (INCB), Vienna, 9 May 1995.

21.   Kopp and Schiray, 'Les sciences sociales.'

22.   Conclusions of the Comité consultatif national d'éthique (CCNE), November 1994, quoted in Observatoire géopolitique des drogues, *Géopolitique des Drogues 1995,* Paris, 1995.

23.   Charter co-ordinated by the NGO Environnement sans frontière, with the support of UNESCO and the United Nations Programme for International Narcotics Control (UNPINC). The signatories also emphasized that 'drug trafficking and use are a threat to the development and progress of our societies, they invariably cause greater violence, crime, exploitation and other infringements of our rights' and that the fight against drugs hinges on guaranteeing 'peace, freedom, democracy, solidarity, justice, protection of the environment and access to employment'.

24.   UNDP, *World Report on Human Development,* 1994.

25.   Ibid.

26.   Giovanni Falcone, 'What is the mafia?' lecture to the Bundeskriminalamt (Wiesbaden), 1990, *Frankfurter Allgemeine Zeitung,* 27 May 1992, reproduced in *Esprit,* No. 185, October 1992, pp. 111–18.

27.   See *Le Monde,* 10 February 1998, 'Cadavres exquis.' 'Cadavres exquis' is the French title of Francesco Rosi's film *Cadaveri Eccellenti,* based on Leonardo Sciascia's novel *The Context.*

28.   *Libération,* 13 September 1995.

29.   US Department of State, *International Narcotics Control Strategy Report, 1996,* March 1997.

30.   UNDP, *World Report on Human Development,* 1994.

31.   See for example, 'Marijuana: special report,' *New Scientist,* 21 February 1998.

NO ↵

The Secret of Worldwide Drug Prohibition: The Varieties and Uses of Drug Prohibition

What percentage of countries in the world have drug prohibition? Is it 100 percent, 75 percent, 50 percent, or 25 percent? I recently asked many people I know to guess the answer to this question. Most people in the United States, especially avid readers and the politically aware, guess 25 or 50 percent. More suspicious individuals guess 75 percent. The correct answer is 100 percent, but *almost no one* guesses that figure. Most readers of this paragraph will not have heard that every country in the world has drug prohibition. Surprising as it seems, almost nobody knows about the existence of worldwide drug prohibition.

In the last decade of the twentieth century, men and women in many countries became aware of *national* drug prohibition. They came to understand that the narcotic or drug policies of the United Sates and some other countries are properly termed *drug prohibition*. Even as this understanding spread, the fact that drug prohibition covers the entire world remained a kind of "hidden-in-plain-view" secret. Now, in the twenty-first century, that situation, too, is changing. As "global drug prohibition" becomes more visible, it loses some of its ideological and political powers.

In this article, I briefly describe the varieties and uses of drug prohibition and the growing crisis of the worldwide drug prohibition regime. . . .

Drug Prohibition Is Useful to All Types of Governments

There is no doubt that governments throughout the world have accepted drug prohibition because of enormous pressure from the U.S. government and a few powerful allies, but U.S. power alone cannot explain the global acceptance of drug prohibition. Governments of all types, all over the world, have found drug prohibition *useful for their own purposes,* for several reasons.

The Article is reprinted with permission of the publisher from *The Independent Review: A Journal of Political Economy* (Fall 2002, vol. V11, no. 2). @ Copyright 2002, The Independent Institute, 100 Swan Way, Oakland, California 94621-1428 USA; info@independent.org; www.independent.org

The Police and Military Powers of Drug Prohibition

Drug prohibition has given all types of governments additional police and military powers. Police and military narcotics units can go undercover almost anywhere to investigate—after all, almost anybody might be in the drug business. More undercover police in the United States are in narcotics squads than are in any other branch of police work. Antidrug units within city, county, and state police departments are comparatively large and often receive federal subsidies. Police antidrug units have regular contact with informers; they can make secret recordings and photographs; they have cash for buying drugs and information. In the United States, police antidrug units sometimes are allowed to keep money, cars, houses, and other property that they seize. Top politicians and government officials in many countries may have believed deeply in the cause of drug prohibition, but other health-oriented causes could not have produced for them so much police, coast guard, and military power.

Government officials throughout the world have used antidrug squads to conduct surveillance operations and military raids that they would not otherwise have been able to justify. Many times these antidrug forces have been deployed against targets other than drug dealers and users—as was the case with President Richard Nixon's own special White House antidrug team, led by former CIA agents, which later became famous as the Watergate burglars. Nixon was brought down by his squad's mistakes, but over the years government antidrug forces all over the world have carried out countless successful nondrug operations.

The Usefulness of Antidrug Messages and of Drug Demonization

Drug prohibition also has been useful for governments and politicians of all types because it has required at least some antidrug crusades and what is properly called *drug demonization*. Antidrug crusades articulate a moral ideology that depicts "drugs" as extremely dangerous and destructive substances. Under drug prohibition, police departments, the media, and religious and health authorities tend to describe the risks and problems of drug use in extreme and exaggerated terms. "Drugs" are dangerous enemies. "Drugs" are evil, vile, threatening, and powerfully addicting. Politicians and governments crusade against "drugs," declare war on them, and blame them for many unhappy conditions and events. Antidrug crusades and drug scares popularize images of "drugs" as highly contagious, invading evils. Words such as *plague, epidemic, scourge,* and *pestilence* are used to describe psychoactive substances, drug use, and moderate, recreational drug users.

Government officials, the media, and other authorities have found that almost anyone at any time can blame drug addiction, abuse, and even use for long-standing problems, recent problems, and the worsening of almost anything. Theft, robbery, rape, malingering, fraud, corruption, physical violence, shoplifting, juvenile delinquency, sloth, sloppiness, sexual promiscuity, low productivity, and all around irresponsibility—anything at all—can be and has been blamed on "drugs." Almost any social problem is said to be made worse—often much worse—by "drugs."

In a war on "drugs," as in other wars, defining the enemy necessarily involves defining and teaching about morality, ethics, and the good things to be defended. Since the temperance or antialcohol campaigns of the nineteenth century, antidrug messages, especially those aimed at children and their parents, have had recognizable themes. Currently in the United States, these antidrug messages stress individual responsibility for health and economic success, respect for police, resistance to peer-group pressure, the value of God or a higher power in recovering from drug abuse, parents' knowledge of where their children are, sports and exercise as alternatives to drug use, drug testing of sports heroes, low grades as evidence of drug use, abstinence as the cause of good grades, and the need for parents to set good examples for their children. Almost anyone—police, politicians, schools, medical authorities, religious leaders—can find some value that can be defended or taught while attacking "drugs." (See the U.S. government-sponsored antidrug Web site at www.theantidrug.com).

In the United States, newspapers, magazines, and other media have long found that supporting antidrug campaigns is good for public relations. The media regularly endorse government antidrug campaigns and favorably cover antidrug efforts as a "public service." For doing so, they receive praise from government officials and prominent organizations. No doubt many publishers and editors deeply believe in the "war on drugs" and in defending the criminalized, prison-centered tradition of U.S. drug policies. But few of the other causes that people in the media support can be turned so easily into stories that are good for public relations and, simultaneously, that are very good for attracting customers and business.

Since at least the 1920s, top editors in the news media have recognized as an economic fact of their business that an alarming front-page antidrug story will likely increase sales of magazines and newspapers, especially when it is about a potential drug epidemic threatening to destroy middle-class teenagers, families, and neighborhoods. Editors know that a frightening story about a new, tempting, addicting drug attracts more TV viewers and radio listeners than most other kinds of news stories, including nonscary drug stories. In short, whatever their personal values, publishers, editors, and journalists give prominent space to scary antidrug articles because they know the stories attract customers.

Consider the case of crack cocaine and the still active U.S. war on drugs. In the 1980s, the media popularized the image of crack cocaine as "the most addicting drug known to man." Politicians from both parties then used that image to explain the deteriorating conditions in America's impoverished city neighborhoods and schools. Front-page stories in the *New York Times* and other publications warned that crack addiction was rapidly spreading to the suburbs and the middle class. In the election years of 1986 and 1988, politicians from both parties enthusiastically voted major increases in funding for police, prisons, and the military to save America's children from crack cocaine.

Even if crack was as bad as Republicans, Democrats, and the media said, it still probably could not have caused all the enduring problems they blamed on it, but the truth about crack cocaine is as startling as the myths. Crack cocaine, "the most addicting drug known to man," turned out to be a drug that very few people used continuously for long. Many Americans tried crack, but not many kept on using it heavily and steadily for a long time, mainly because most people cannot physically tolerate, much less enjoy, frequent encounters with crack's brutally brief and extreme ups and downs. Nor has crack become popular anywhere else in the world. Heavy, long-term crack smoking appeals only to a small number of deeply troubled people, most of whom are also impoverished. Because frequent bingeing on the drug is so thoroughly unappealing, it was extremely unlikely that an epidemic or plague of crack addiction would spread across America to the middle class and the suburbs.

Nonetheless, the contradictions between the drug war's myths about crack and the reality of crack cocaine's very limited appeal have not undermined the credibility or usefulness of antidrug messages, news stories, or political statements. In this respect, drug war propaganda is like the propaganda from other wars: the claims often remain useful even though they are patently false or do not make logical sense. In the 1990s, when crack cocaine finally ceased to be a useful enemy, American politicians, media, and police did not acknowledge their exaggerations and falsehoods about crack cocaine. They simply claimed victory, stopped discussing crack, and focused on other scary drugs, most recently MDMA (ecstasy) and prescription narcotics.

Additional Political and Ideological Support for Drug Prohibition

In many countries, popular and political support for drug prohibition also has been rooted in the widespread faith in the capacity of the state to penetrate and police many aspects of daily life for the "common good." This romantic or utopian view of the coercive state became especially strong and pervasive in the twentieth century. Unlike, say, the dissenters who insisted on the U.S. Bill of Rights in the eighteenth century, and unlike the members of many nineteenth-century political movements, in the twentieth century liberals, conservatives, fascists, communists, socialists, populists, left-wingers, and right-wingers shared this vision of the benevolent national state—if *they* controlled it. Drug prohibition was one of the few things on which they could all agree. Drug prohibition has been part of what I think it is appropriate to call the twentieth century's "romance with the state."

Because politicians in many countries, from one end of the political spectrum to the other, have shared this positive view of the powerful, coercive state, they could all agree on drug prohibition as sound nonpartisan government policy. In the United States during the 1980s and 1990s, Democrats feared and detested Presidents Reagan and Bush, and Republicans feared and detested President Clinton, but the parties united to fight the war on drugs. They even competed to enact more punitive antidrug laws, build more prisons, hire more drug police, expand antidrug military forces, and fund many more government-sponsored antidrug messages and crusades for a "drug-free" America. Opposing political parties around the world have fought about many things, but until recently they have often united in endorsing efforts to fight "drugs."

Finally, drug prohibition has enjoyed widespread support and legitimacy because the United States has used the UN as the international agency to create, spread, and supervise worldwide prohibition. Other than the U.S. government, the UN has done more to defend and extend drug prohibition than any other organization in the world. The UN currently identifies a "drug-free world" as the goal of its antidrug efforts (http://www.odccp.org/adhoc/gass/ga9411.htm). . . .

The Place of Harm Reduction Within Drug Prohibition

Since the early 1980s, harm-reduction workers and activists in Europe and increasingly throughout the world have sought to provide drug users and addicts with a range of services aimed at reducing the harmful effects of drug use. In the United States, conservative pundits and liberal journalists have accused harm-reduction advocates of being "drug legalizers" in disguise, but in most other countries many prominent politicians, public-health professionals, and police officials who are strong defenders of drug prohibition also have supported harm-reduction programs as practical public-health policies. Even the UN agencies that supervise worldwide drug prohibition have come to recognize the public-health benefits of harm-reduction services *within* current drug prohibition regimes.

A better understanding of the varieties and scope of worldwide drug prohibition helps us to see better the place of the "harm-reduction movement" within the history of drug prohibition. I suggest that harm reduction is a movement within drug prohibition that shifts drug polices from the criminalized and punitive end to the more decriminalized and openly regulated end of the drug policy continuum. Harm reduction is the name of the movement within drug prohibition that in effect (though not always in intent) moves drug policies away from punishment, coercion, and repression and toward tolerance, regulation, and public health. Harm reduction is not inherently an enemy of drug prohibition. However, in the course of pursuing public-health goals, harm reduction necessarily seeks to reduce the criminalized and punitive character of U.S.–style drug prohibition.

Consider the many programs identified as part of harm reduction: needle exchange and distribution, methadone maintenance, injection rooms, heroin clinics, medical use of marijuana by cancer and AIDS patients, truthful drug education aimed at users, drug-testing services at raves, and so on. Harm-reduction programs have pursued all these ways to increase public health and to help users reduce the harms of drug use. In order to carry out their stated objectives, these programs have often required laws, policies, or funding that reduce the harshness of drug prohibition. The reforms seek to reduce the punitive character of drug prohibition without necessarily challenging drug prohibition itself.

Harm-reduction advocates' stance toward drug prohibition is exactly the same as their stance toward drug use. Harm reduction seeks to reduce the harmful effects of drug use without requiring users to be drug free. It also seeks to reduce the harmful effects of drug prohibition without requiring governments to be prohibition free. Harm-reduction organizations say to drug users: "We are not asking you to give up drug use; we just ask you to do some things (such as using clean syringes) to reduce the harmfulness of drug use (including the spread of AIDS) to you and the people close to you." In precisely the same way, these organizations say to governments: "We are not asking you to give up drug prohibition;

we just ask you to do some things (such as making clean syringes and metha-done available) to reduce the harmfulness of drug prohibition."

Harm reduction offers a radically tolerant and pragmatic approach to both drug use and drug prohibition. It assumes that neither is going away soon and suggests therefore that reasonable and responsible people try to persuade both those who use drugs and those who use drug prohibition to minimize the harms that their activities produce. . . .

The Future of Global Drug Prohibition

Global drug prohibition is in crisis. The fact that it is at long last becoming visible is one symptom of that crisis. In the long run, the more criminalized and puni-tive forms of drug prohibition almost certainly are doomed. In the short run, the ever-growing drug-law and drug-policy reform movements make it likely that criminalized drug prohibition will find itself confronted with new opponents. (This prediction is already becoming a reality, in Switzerland, Australia, Germany, Portugal, Canada, the Netherlands, Spain, the United Kingdom, the United States, and other counties.)

In the twentieth century, for specific practical and ideological reasons, the nations of the world constructed a global system of drug prohibition. In the twenty-first century, because of the spread of democracy, trade, and information and for other practical and ideological reasons, the peoples of the world will likely dismantle and end worldwide drug prohibition.

It is important to understand that *this process of dismantling global drug prohibi-tion will not end local drug prohibition.* The end of *global* drug prohibition will not (and cannot) be the end of *all* national drug prohibition. Advocating the end of worldwide drug prohibition is not the same as advocating worldwide drug legaliza-tion. Long after the demise of the UN's Single Convention, communities, regions, and some democratic nations will choose to retain forms of drug prohibition. Many places in the world will also continue to support vigorous antidrug crusades.

However, as accurate information about drug effects and alternative drug policies becomes more widespread, increasing numbers of countries, especially democratic ones, will likely choose not to retain full-scale criminalized drug prohibition. Most places eventually will develop their own varied local forms of regulated personal cultivation and use of the once-prohibited plants and substances. Many places also eventually will allow some forms of commercial production and sale—of cannabis, first of all and above all, because it is by far the most widely grown, traded, sold, and used illegal drug in the world.

These changes will take time. Prohibitionists and drug warriors in every country will fight tenaciously to maintain their local regimes, and enormous power will be employed to prevent the Single Convention of 1961 and its related treaties from being repealed or even modified. As a result, in coming years, all around the world there will be even greater public discussion and debate about drug prohibition, about criminalized drug policies, and about the worldwide movement within drug prohibition to decriminalize the possession and use of cannabis, cocaine, heroin, and other substances.

As part of that process of conversation and debate, many more people will discover—often with considerable astonishment—that they have lived for decades within a regime of worldwide drug prohibition. That growing understanding will itself push worldwide drug prohibition closer to its end. Here in the twenty-first century, it may turn out that the most powerful three holding global drug prohibition in place is the secret of its existence. . . .

POSTSCRIPT

Can the Global Community "Win" the Drug War?

The October 2003 Lisbon International Symposium on Global Drug Policy provided a forum for leading drug policymakers from national governments, senior representatives from various UN agencies, and other experts to address new ideas and innovative solutions. Speakers addressed such varied topics as an international framework for combating drugs, better public health policy, new approaches to the war on drugs, and a variety of new challenges facing the international community. Four key areas of division were spelled out by Martin Jelsma: (1) repression vs. protection; (2) zero tolerance vs. harm reduction; (3) the North-South or donors vs. recipients divide; and (4) demand vs. supply. The failure of nations of the world to reach agreement on these four major areas of contention has resulted, in the judgment of many, in the inability of the global community to address the drug problem successfully. Not enough funds are made available to international agencies like the United Nations, and money that is given is likely to have strings attached to it.

The conference was particularly timely because for the first time in the global war on drugs, policymakers did not debate the issue of whether the current policy was working. Instead, conference participants focused on how to organize a better drug control system. Honorary Secretary-General of Interpol, Raymond Kendall, echoed this view in his closing speech. While Kendall was pleased that debate had shifted to what he called "new levels," the conference did not develop a new plan of action. The issues outlined above were too great to overcome. Nonetheless, progress was made as national examples of successful alternative public policy programs were presented to the delegates.

The United Nations continued the theme of alternative approaches in its "2004 World Drug Report." Acknowledging that effective strategies are discovered through trial and error, the UN alluded to a number of recent developments that appear to have potential for helping the global war on drugs. Four major ideas were discussed under the rubric of a "holistic approach." The first is "Addressing the drug problem in a broader sustainable development context." On the one hand, the drug problem hinders development in poor countries and compromises peacemaking efforts in countries torn by civil strife. On the other hand, "poverty, strife and feeble governance are fertile ground for drug production, trafficking and abuse." These situations are interconnected and can only be addressed by a comprehensive approach that recognizes the causes as well as the symptoms of the drug problem.

The second idea is "providing an integrated response to the drugs and crime nexus." The connections between drug trafficking, organized crime, and even the financing of terrorism has meant that those responsible for addressing each of these scourges must work within the same multilateral system rather than in isolation. The third development is "addressing the drugs and crime nexus under the new paradigm of human security." Growing out of the 2000 UN Millennium Summit, the Commission on Human Security is developing a new approach to security that combines human development and human rights. The UN Report suggests that this could provide a critical link between drugs/crime control and sustainable development.

The fourth development is termed "a more synergistic approach." This simply means that not only must there be an integrated and balanced approach to the war on drugs, but that much more needs to be learned. For example, the structure and dynamics of drug markets at the national, regional, and global levels are a mystery beyond the simple belief that normal supply and demand principles are at work.

In the selection by Mayor, the basic question of eliminating the supply or drying up the demand is raised. Mayor believes it critical that a major emphasis must be placed on both sides of the equation. His motivation is based on the dual points that eliminating the supply is difficult but that the health issues associated with illegal drug use demand that we educate existing and potential users about the evils of such behavior. Addicts must be treated as patients, not criminals. At the same time, Mayor argues that the most effective way to fight the drug war is to destroy the financial power of organized crime. Both strategies are critically important.

Levine suggests that global drug prohibition and a focus on both punishing the supplier and the user has not worked very well. Instead, this approach must be reexamined with a view toward addressing in a much different way the plight of the drug user.

Is Globalization a Positive Development for the World Community?

YES: Timothy Taylor, from "The Truth About Globalization," *Public Interest* (Spring 2002)

NO: Jose Bove, from "Globalisation's Misguided Assumptions," *The OECD Observer* (September 2001)

ISSUE SUMMARY

YES: Mr. Taylor argues that globalization is not the boogeyman that some scholars and politicos contend but rather a complex array of economic trends that provides positive results for the vast majority of people.

NO: Mr. Bove contends that globalization and the capitalist principles that underlie it are misleading and create inequities and myths regarding progress and wealth creation that are not supported by an analysis of the facts.

Globalization is a phenomenon and a revolution. It is sweeping the world with increasing speed and changing the global landscape into something new and different. Yet, like all such trends, its meaning, development, and impact puzzle many. We talk about globalization and experience its effects, but few of us really understand the forces that are at work in the global political economy.

When people use their cell phones, log onto the Internet, view events from around the world on live television, and experience varying cultures in their own backyards, they begin to believe that this process of globalization is a good thing that will bring a variety of new and sophisticated changes to people's lives. Many aspects of this technological revolution bring fun, ease, and sophistication to people's daily lives. Yet the anti–World Trade Organization (WTO) protests in Seattle, Washington, in 1999 and Washington, D.C., in 2000 are graphic illustrations of the fact that not everyone believes that globalization is a good thing. Many Americans who have felt left out of the global economic boom, as well as Latin Americans, Africans, and Asians who feel that their job skills and abilities are being exploited by multinational corporations (MNCs) in a global division of labor, believe that this system does not meet their needs. Local cultures that believe that Wal-Mart and McDonald's bring cultural change and harm rather than inexpensive products and convenience criticize the process. In this way, globalization, like all revolutionary forces, polarizes people, alters the fabric of their lives, and creates rifts within and between people.

Many in the West, along with the prominent and elite—among MNCs, educators, and policymakers—seem to have embraced globalization. They argue that it helps to streamline economic systems, disciplines labor and management, brings forth new technologies and ideas, and fuels economic growth. They point to the relative prosperity of many Western countries and argue that this is proof of globalization's positive effects. They see little of the problems that critics identify. In fact, those who recognize some structural problems in the system argue that despite these issues, globalization is like an inevitable tide of history, unfortunate for some but unyielding and impossible to change. Any problems that are created by this trend, they say, can be solved.

Many poor and middle-class workers, as well as hundreds of millions of people across the developing world, view globalization as an economic and cultural wave that tears at the fabric of centuries-old societies. They see jobs emerging and disappearing in a matter of months, people moving across the landscape in record numbers, elites amassing huge fortunes while local cultures and traditions are swept away, and local youth being seduced by promises of American material wealth and distanced from their own cultural roots. These critics look past the allure of globalization and focus on the disquieting impact of rapid and system-wide change.

The irony of such a far-ranging and rapid historical process such as globalization is that both proponents and critics may be right. The realities of globalization are both intriguing and alarming. As technology and the global infrastructure expand, ideas, methods, and services are developed and disseminated to greater and greater numbers of people. As a result, societies and values are altered, some for the better and others for the worse.

In the selections that follow, the authors explore the positive and negative impacts of globalization and come to different conclusions. Timothy Taylor explores the economic results of globalization and contends that the benefits are clear and largely positive. He cites data in the areas of trade, wages, and investment to make the case for a growing global economy that is benefiting most people, provided local governments manage their affairs wisely. Bove argues that globalization is based on four falsehoods about capitalist economics that fool people into thinking that globalization provides for all. In reality, Bove contends, the world must look to other models if we are serious about improving the economic condition of the majority of humanity.

Timothy Taylor ➡ **YES**

The Truth About Globalization

To keep my economist union card, I am required every morning when I arise to place my hand on the leather-bound family heirloom copy of Adam Smith's *The Wealth of Nations* and swear a mighty oath of allegiance to globalization. I hereby do asseverate my solemn belief that globalization, taken as a whole, is a positive economic force and well worth defending. I also believe that the economic and social effects of globalization are exaggerated by both its detractors and supporters.

In media coverage of anti-globalization protests, "globalization" often becomes a catch-all term for capitalism and injustice. (Indeed, for some protestors, referring to capitalism *and* injustice would be redundant.) But economic globalization in fact describes a specific phenomenon: the growth in flows of trade and financial capital across national borders. The trend has consequences in many areas, including sovereignty, prosperity, jobs, wages, and social legislation. Globalization is too important to be consigned to buzzword status.

One World?

The degree to which national economies are integrated is not at all obvious. It depends on your choice of perspective. During the last few decades, international flows of goods and financial capital have certainly increased dramatically. One snap measure of globalization is the share of economic production destined for sale in other countries. In the U.S. economy, exports of goods and services were 4.9 percent of the gross domestic product (GDP) in 1965, but 10.8 percent of GDP in 2000. From a global perspective, exports rose from 12 percent of world GDP in 1965 to 22 percent of world GDP in 2000. In round numbers, international trade of goods and services has doubled in about four decades.

International financial markets are not tracked as easily as cross-border flows of goods and services. But by a variety of measures, they have also expanded considerably, especially in the last decade. Total assets held by U.S. investors in other nations nearly tripled from $2.3 trillion in 1991 to $6.2 trillion in 2000. Conversely, total foreign-owned assets in the U.S. economy quadrupled from $2 trillion in 1991 to $8 trillion in 2000. Annual global flows of "foreign direct investment"—that is, investment that creates a lasting management interest,

Reprinted with permission of the author from THE PUBLIC INTEREST, No. 147 (Spring 2002), pp. 24–22. @ 2002 by National Affairs, Inc.

often defined as more than 10 percent of voting stock in a company—rose from $200 billion in 1990 to nearly $900 billion in 1999. A 1998 survey by the Bank of International Settlements found that $1.5 trillion *per day* was traded on foreign-exchange markets. Since foreign-exchange trading has been growing at double-digit rates, its volume now must exceed $2 trillion per day.

For many countries, international financial markets feel more like an anvil on their toes than a parade of dry statistics. Argentina in 2002 has no remaining doubts about the size and power of international capital movements. A number of countries and regions have recently suffered through economic instability and recessions caused by rapid outflows of international capital: Russia in 1998, East Asia in 1997 and 1998, Mexico in 1995, and European countries in 1993, when international financial speculators blew down the mechanism for coordinating their exchange rates.

But while international flows of goods, services, and financial capital have increased dramatically in the last few decades, the term "globalization" implies more. It implies that the world is a single market—or nearly so. In a fully globalized economy, goods and investments would flow across national borders with no more difficulty than that encountered by a company from California selling its products or taking out a loan in New York. But studies testing the importance of national borders in determining the flows of goods and services have shown that we are still a long way from a single world market.

Indeed, national borders continue to play a significant role. For example, the Canadian province of Ontario is an equal distance from Washington state and the province of British Columbia. In a borderless world, one might expect Ontario's level of trade with Washington state and with British Columbia to be about the same, at least after adjusting for the size of the local economies. Yet this is not the case. The levels of trade have been measured between pairs of Canadian regions, pairs of U.S. regions, and pairs of U.S.-Canadian regions, and it turns out that trade between regions of Canada, and between regions of the United States, is commonly 12 times higher than trade between equivalent regions across the U.S.-Canadian border. In Europe, similar studies have found that trade between regions within countries is three to ten times higher than trade that crosses national borders, even after adjusting for factors like size of local economies and geographic distance. . . .

It's easy to identify reasons why national borders play such a substantial role in limiting international movements of goods and capital. Transportation and communications networks are often built by national governments and focus more on connections within a country than across national borders. Economic transactions across a border must face additional costs associated with different legal systems, regulations, institutions, cultures, and languages. Movements in exchange rates add a degree of risk to economic transactions across borders. Moreover, many countries retain tariff and nontariff barriers to trade.

Many discussions of globalization often begin with the assumption, buttressed by a few lively anecdotes, that international flows of goods and services have become so powerful that national borders are economically obsolete, or nearly so.

Don't believe it.

The Benefits of Trade

The circumstantial evidence that international trade provides economic benefits is overwhelming. Eras of expanding global trade, like recent decades, have generally been times of economic growth. Periods of contracting trade have often involved recession or worse. When a country's economy expands, its level of international trade typically increases. International trade and investment played an important role in the great economic success stories of the last four decades: Japan, East Asia, and now China.

A recent study by the World Bank grouped developing economies into two categories according to their participation in the global economy. In one group of two dozen countries that includes China, Mexico, and India—and a total population of 3 billion people—the ratio of exports to total GDP has doubled in the last two decades. In these countries, the level of per capita GDP rose 5 percent per year in the 1990s. In the second group of countries, which includes much of Africa, Russia, and the Middle East—and a population of 2 billion people—the average share of exports to GDP has fallen over the last two decades. In these countries, per capita GDP fell an average of 1 percent per year in the 1990s.

The fundamentals of economic theory explain how trade should benefit all parties. Trade allows countries to specialize in the products they have the greatest advantage in producing. Specialization encourages learning and innovation about these products and allows nations to take advantage of economies of scale. When countries specialize and trade, the world's productive resources of labor, physical resources, and time are used more efficiently. Trade allows consumers and businesses to seek out the best deal in a global market, giving producers an incentive to compete in this market. International flows of financial capital allow the world's savings to flow toward the most productive investment opportunities—even if those opportunities are halfway around the world. National economies can draw on international capital markets when they need funding and diversify their risks by investing their savings outside their own borders.

These textbook reasons as to how trade benefits an economy are fine, as far as they go, but I suspect they don't go far enough. Many of the benefits of globalization are not from the products that are shipped but from other ways in which trade shapes the economic environment.

The most valuable Japanese export to the United States over the last few decades may not have been cars or computers. Instead, it may have been the competitive push that Japanese firms gave U.S. automakers and high-tech firms to make better products in more efficient ways. Or it may be the idea of "just-in-time" inventory procedures that U.S. firms learned from Japanese producers, which has made manufacturing across many industries leaner and more efficient. Or it may be Japanese quality-control procedures. Trade often carries with it a wealth of skills and institutions: technology, training, management, accounting, outside monitoring of businesses, and even exposure to expertise in such areas as bank regulation, antitrust policies, and environmental protection.

These indirect benefits of trade are especially important for smaller and poorer economies. The United States has an enormous internal market, a remarkable capacity for innovation, and advanced capitalist institutions. It could limp along with a reduced level of external trade if it needed to do so. But if small economies like Belgium or Chile or Ivory Coast were forced to consume only what they produce domestically, their standards of living would fall dramatically. Without trade, small economies cannot sustain a high degree of specialization. Furthermore, they would have a difficult time generating a broad array of expertise in science, technology, management, and financial regulation. If small economies were forced to rely only on their domestic production, they would end up consuming a far smaller variety of products at far higher prices. . . .

Unemployment in the Global Village

During the debate over the North American Free Trade Agreement (NAFTA), Ross Perot infamously predicted that free trade with Mexico would cause a "giant sucking sound" as U.S. jobs fled south over the border. In fact, NAFTA passed into law in 1994 and was immediately followed by seven years of the lowest unemployment in the history of the U.S. economy. It is worth savoring the occasions when ignorance is so quickly and decisively revealed by events. There is no evidence at all that globalization increases overall unemployment rates. As globalization increased over the last few decades, global unemployment levels did not rise. Moreover, countries with higher levels of international trade do not typically have higher unemployment.

However, international competition can cause reallocations of jobs from one industry to another. For example, the number of unionized steelworkers in the U.S. economy fell from just over 1 million in 1975 to 572,000 in 1985, and the number of unionized autoworkers declined from 974,000 in 1985 to 751,000 by 1995. Increased competition from steel and auto imports surely played a role in these declines. This problem is not best described as unemployment, since total unemployment isn't rising, but rather as an increase in what economists call "displaced workers," workers who are pushed out of one job and thus forced to find another. For an economy as a whole, a flexible workforce that moves between different employers, different geographic areas, and even different occupations is an enormous strength. However, the individual workers involved in these labor-market movements can suffer high costs: lost wages due to unemployment or new employment with lower wages, costs of moving and retraining, and the psychological costs of uncertainty and a disrupted life. For present purposes, the question is how much globalization contributes to the churning of the labor market as people move in and out of jobs.

The U.S. labor market has more turnover than many people realize. Early in 2002, about 8 million Americans were unemployed. Based on historical experience, we can estimate that more than half of these workers will find another job within a couple of months, even in a slow economy. But the number of unemployed will not fall by half in the next two months, because the churning of the labor market will cause millions of other workers to become unemployed over that same period. Even in the strong U.S. economy of the late 1990s, when overall unemployment was very low, mass layoffs totaled about 1.1 million per year. Longer-term studies of U.S. labor markets found that in a given year in the 1990s, about 10 percent of all jobs disappeared during the year. But the unemployment rate did not rise, because an even larger number of new jobs were being created. Indeed, workers often moved fairly seamlessly from old jobs to new ones without a period of unemployment in between. . . .

But I see no reason why efforts to reduce the cost of job mobility for individuals should focus on globalization. It's hard to estimate how much of the churning in America's labor market is due to increased foreign trade, but compared to all the other factors of a dynamic market economy, globalization is a minor contributor. Moreover, from the view of public policy, it doesn't matter why a worker lost a job. A worker who is out of a job because of tough domestic competition or poor management is no less deserving of support in the transition to a new job than one laid off because of foreign competition. Public policy should neither demonize globalization as the primary cause of job loss, nor pretend that job dislocation won't happen in a dynamic market-oriented economy. Rather, it should build institutions and programs to help workers make the transition to new employment.

The Wages of Globalization

The most important factor determining the average level of wages in an economy is the average productivity of its workers. Since globalization contributes to productivity, it also increases the average wage level. However, even as globalization increases the overall level of wages, it may also increase inequality. The gap between high and low incomes might expand within a developed economy like the United States, or within a developing economy like Mexico. Also, the gap between average incomes in high-income and low-income economies around the world can expand. Let's take each of these cases in turn.

It is conceivable that low-wage workers in high-income economies like that of the United States suffer from increased competition with low-wage workers in other countries. But most studies have attributed the growth in inequality in high-income economies to the remarkable advances in information and communication technology over recent decades. These advances have meant that the labor market pays a higher premium for skilled workers who are better equipped to use the new technologies. The share of income going to the top fifth of U.S. households rose from 40.7 percent in 1975 to 46.5 percent in 1995. Most of that increase was accounted for by the income share of the top 5 percent of households, which rose from 14.9 percent in 1975 to 20 percent in 1995. Since then, inequality in the U.S. income distribution has not risen further, but neither has it receded.

Mainstream studies of the rise in U.S. inequality have attributed relatively little of it—perhaps one-fifth or so—to higher levels of trade. Why does trade play a relatively small role in shaping the wage distribution? One reason is that more than half of America's imports come from other industrialized countries—Canada, Japan, the European Union—and there is no reason why trade with these other high-wage economies should much affect U.S. income inequality. Only about one-quarter of America's imports come from the truly low-wage economies of the world such as China. So the argument that trade with low-income countries increases domestic wage inequality supposes that a relatively small tail of overall U.S. trade wags the entire U.S. wage distribution.

In addition, most U.S. workers, including low-wage workers, are in service industries. Traditionally, imports have not competed much with services. Imported goods like textiles can directly affect the jobs and wages of U.S. textile workers. But the U.S. economy cannot import hotel cleaning services, restaurant labor, yard work, or many of the other services provided by low-wage workers, so the impact of inexpensive imported products on the wages of low-income workers in these industries is muted and indirect.

There are other reasons to doubt that global trade is a primary shaper of the income distribution. The time frame of the rise in income inequality, the mid 1970s to the mid 1990s, doesn't match the time frame of rising globalization, which started in the 1950s and has continued since 1995. Moreover, the rise in globalization has been worldwide, but the U.S. economy saw a larger rise in inequality than other high-income countries. Patterns of education, unionization, minimum wages, taxation, and executive stock options appear to have a more powerful effect in shaping income distributions than other economic influences such as new technologies and globalization. . . .

It is sometimes popular to assert that the high-income levels in the United States and Europe are achieved through the exploitation of Africa, Latin America, and other low-income areas of the world. There have been enough episodes of malign behavior by colonial powers and multinational corporations to give surface plausibility to such a claim. Nonetheless, the idea that global trade is a primary source of the gap between high-income and low income nations is based on an outdated view of the wealth of nations. The model of economic growth that prevailed in much of the world up to the twentieth century was based on conquest: A nation became richer by controlling a greater quantity of land, people, or resources. Yet in the modern world, economic welfare is based not on far-flung political control but on skilled workers taking advantage of recent technology in a market-oriented environment. Global trade is part of this environment but not the most important part. Certainly, Belgium once benefited from its colonial control over rubber, timber, and other resources from the Congo. But Belgium's current economic performance as a country with income levels above the European Union average is based on the capacities of its citizens to produce goods and services inside Belgium. Modern economic growth is fundamentally built upon voluntary transactions made for mutual benefit, not upon coercion.

The global tide of economic growth over the last century has not raised all economic ships. But globalization is an avenue through which high-income nations can reach out to low-income ones. Expecting the poorest people in the world to pull themselves up by their bootstraps, without access to foreign investment, training, technical skills, or markets, verges on indifference or cruelty. Foreign aid has its place, but as a matter of practical politics, it will never arrive in sufficient quantities, nor be spent with sufficient wisdom, to raise overall standards of living dramatically in low-income countries. Only a combination of institutional reforms within low-income countries, coupled with much closer connections to the extraordinary resources and buying power of international markets, offers a realistic chance of substantially improving the plight of the poorest people in the world.

Race to the Bottom?

The "race to the bottom" is the darkly logical scenario of environmental destruction and worker exploitation. In this story, political jurisdictions are pressured by profit-seeking businesses to reduce their costly environmental rules and labor standards—or else lose jobs and economic clout as businesses relocate to jurisdictions that lack such protections. The result is a race to the bottom in which the selfish, soulless forces of capitalism run roughshod over the social good.

This is a forbidding scenario, but one that does not appear to be happening. Consider trade across the United States. States and localities vary widely in their taxes, welfare benefits, school expenditures, environmental rules, workmen's compensation policies, and much else. Tax rates and social benefits in the highest states are often double those in the lowest states. But these differences have existed for a long time, and free trade across the United States has not forced a race to the bottom. Businesses in the United States have not fled en masse to low-regulation, low-tax states like North Dakota or Mississippi. State and local taxes and regulations have on average been rising, not dropping, over recent decades. . . .

It is impossible to disprove the thesis that environmental and worker protections might have increased even further and faster in recent decades in the absence of globalization. But it seems more plausible that the rise in globalization is helping to improve environmental and labor standards in low-income countries, rather than hindering them. Globalization spreads knowledge about the rest of the world, including the fact that air and water need not become ever dirtier as a country develops its economy. Globalization illustrates that improving the skills of workers, and treating them as if their morale and motivation matter, is a more effective route to economic success than treating them as dumb muscle-power. Globalization also brings with it international public pressure for implementing basic standards and tourists who will not tolerate a filthy environment or a surly workforce. But the ultimate protection for the environment and for workers in low-income countries is an involved citizenry with an income level high enough to lift its sights above basic needs and allow other social values to take their rightful place.

Perpetual Peace?

Before September 11, 2001, it was common to hear blithe predictions that as countries became intertwined with one another through global trade, they would move toward domestic tranquility and peaceful international relations. The world now seems a more dangerous and unstable place.

But well before September 11, economic historians looked skeptically at the notion that globalization assures peace and international stability. The world experienced an extraordinary surge of international trade in the second half of the nineteenth century, but this did not lead to a peaceful twentieth century. During this boom, exports rose from 5 percent of global GDP in 1870 to 8.7 percent by 1913. The speed of communications improved by orders of magnitude as the telegraph, which has been nicknamed the "Victorian Internet," linked information about prices and business conditions around the world. Messages that used to take weeks to cross the Atlantic by ship were now received in minutes by wire. The railroad and the steamship made transportation much faster and cheaper. Waves of immigration reallocated labor across the world. The gold standard provided a reasonably fixed exchange rate that facilitated international movements of financial capital. Indeed, the net sums that national economies borrowed and loaned abroad, expressed as a share of GDP, were often higher in the years around 1900 than they are today.

This earlier wave of globalization was brought to a halt by two world wars and the Great Depression. Clearly, countries that traded with one another were quite willing to fight each other. But even without the disruptions of war and depression, the first phase of globalization faced a severe political backlash in the early decades of the twentieth century. Before the infamous Smoot-Hawley tariffs of the Great Depression (which probably deepened the depression, although they were not a primary cause of it), protectionist barriers were going up on both sides of the Atlantic. European agricultural interests wanted protection from U.S. farm products, and U.S. manufacturing firms wanted insulation from European competitors. Barriers to immigration arose as well. International finance was shaken by a series of bank runs and panics, and the gold standard began to look very shaky. Both situations discouraged flows of international capital in the same way that financial panics and exchange-rate volatility discourage flows of international capital today.

Violence and war often parade behind grievances that are expressed in economic terminology. But economic grievances alone do not seem either sufficient or necessary to cause war. The vast majority of poor people around the world are not advocates or practitioners of violence. Most wars of recent memory—for example, in the former Yugoslavia or in Rwanda—have not been about unfair trade relations. The terrorist attacks of September 11 were primarily motivated by conflicts over political power, culture, and religion, not by U.S. negotiating positions at the World Trade Organization. Severe civil strife or outright warfare is most often generated by nationalist or ethnic sentiment, fueled and fanned by demagogic leadership. Globalization can make the world a more peaceful place by creating connections between peoples that make it harder to demonize others and by giving powerful political interests a stake in peaceful relations. But economic globalization is far from being an inoculation against international violence. . . .

Worth Defending

Economic globalization is a powerful trend, driven by a combination of technological developments, profit-seeking businesses, and generally supportive public policy. But globalization is also less pervasive and more fragile than is widely believed by both its supporters and its opponents. The demise of the first wave of globalization in the early decades of the twentieth century vividly demonstrates that it is not an irreversible trend. National borders and policies continue to play a role in limiting and directing international transactions. Discontent with globalization is widespread enough to be troubling. The most extreme and active of globalization's opponents caricature it as the source of all that is wrong in the modern world, including poverty, injustice, inequality, violence, and war.

Those of us who believe in globalization need to defend it. We need to stick up forthrightly for the benefits it has provided and will continue to provide—and to emphasize not just the actual goods, services, and capital that flow across international borders, but the associated trade in ideas, skills, and institutions as well. When accompanied by sensible, market-oriented public policies, globalization can be a great boon to national wealth and social development.

An additional line of defense, perhaps just as important, is to be honest about the meaning and the limits of globalization. Globalization is not a magic cure-all for what ails a nation's economy, nor is it a plot by profit-hungry megacorporations to exploit workers and despoil the environment. Globalization is not the return of colonialism, nor is it the arrival of world government. At the most fundamental level, globalization simply means an expansion of the range of possible commercial activities. Acts of buying, selling, producing, borrowing, and lending that used to be ruled out by geographic, technological, or legal barriers have now become practical. Seeking out and sorting through the possibilities opened up by globalization will require a daunting amount of effort, flexibility, and change, precisely because globalization embodies such a vast and marvelous array of new economic opportunities.

NO ↵

Globalisation's Misguided Assumptions

Markets, competition and free trade are all essential for healthy globalisation
to take place. Or are they? Critics argue that while globalisation has the
potential to become a positive force for economic growth, too many of the
benefits go to well-off countries, while the costs of adjusting markets and
institutions are being borne by millions of already poor workers worldwide.
Others go further still and blame free trade for many of the woes of the poor
everywhere, and farmers in particular. José Bové of France's Confédération
Paysanne (Small Farmers' Federation) is one such anti-globalisation activist.
In this article he explains what he considers to be the "false assumptions"
underlying the arguments of the free-market camp. The article originally
appeared in the French newspaper *Le Monde*.*

Humanity is grappling with a formidable creed, which, like so many others, is
totalitarian and planetary in scope, namely free trade. The gurus and zealous servants of this doctrine ("responsible" people) are saying that the Market is the only
god, and that those who want to combat it are heretics ("irresponsible" people).
So we find ourselves faced with a modern-day obscurantism—a new opium on
which the high priests and traffickers are sure they can make populations dependent. Recent articles in the international press supporting new trade rounds and
the like are quite clear on the dogma that some people would like to impose on the
men and women of this planet.

More and more people are coming out against the free market credo advocated by the WTO, the damage inflicted by it being so plain to see, and the falsehoods on which it is based so blatant.

The first falsehood is the market's self-regulating virtues, which form the
basis of the dogma, but this ideological mystification is belied by the facts. In
the field of agriculture, for example, since 1992 the major industrialised countries have embraced global markets with open arms—the United States enacted
the FAIR (Federal Agriculture Improvement and Reform) Act, a policy instrument that did away with direct production subsidies, instead "decoupling" aid
and allowing farmers to produce with no restrictions whatsoever—but this has
done nothing to calm the wild swings in the markets.

From *The Observer* (Guardian Unlimited) by Jose Bove, no. 228, September 2001, pp. 17–19.
Copyright © 2001 by Guardian Publications. Reprinted by permission.

It has, in fact, done quite the opposite, since markets have experienced unprecedented instability since the trade agreements signed in Marrakech in 1995. The most spectacular effect of this American "decoupling" has been the explosion of emergency direct subsidies to offset declining prices. These subsidies reached a record high of more than $23 billion in 2000 (four times more than the amount budgeted in the 1996 Farm Act).

So, contrary to free-marketers' assertions, markets are inherently unstable and chaotic. Government intervention is needed to regulate markets and adjust price trends, to guarantee producers' incomes and thus ensure that farming activity is sustained.

The second blatant untruth is that competition generates wealth for everyone. Competition is meaningful only if competitors are able to survive. This is especially true for agriculture, where labour productivity varies by a factor of a thousand to one between a grain farmer on the plains of the Middle West and a spade-wielding peasant in the heart of the Sahel.

To claim that the terms of competition will be healthy and fair, and thus tend towards equilibrium if farm policy does not interfere with the workings of a free market, is hypocritical. How can there be a level playing field in the same market between a majority of 1.3 billion farm workers who harvest the land with their hands or with harnessed animals, and a tiny minority of 28 million mechanised farmers formidably equipped for export? How can there be "fair" competition when the most productive farmers of rich countries receive emergency subsidies and multiple guarantees against falling prices on top of their direct and indirect export bonuses?

The third falsehood is that world market prices are a relevant criterion for guiding output. But these prices apply to only a very small fraction of global production and consumption. The world wheat market accounts for only 12% of overall output and international trade takes place at prices that are determined not by aggregate trade, but by the prices of the most competitive exporting country.

The world price of milk and dairy products is determined by production costs in New Zealand, while New Zealand's share of global milk production averaged only 1.63% between 1985 and 1998. The world price of wheat itself is pegged to the price in the United States, which accounted for only 5.84% of aggregate world output from 1985 to 1998.

What is more, these prices are nearly always tantamount to dumping (i.e., to selling below production costs in the producing and importing countries) and are only economically viable for the exporters thanks to the substantial aid they receive in return.

The fourth falsehood is that free trade is the engine of economic development. For free marketers, customs protection schemes are the root of all evil: they claim that such systems stifle trade and economic prosperity, and even hinder cultural exchanges and vital dialogue between peoples. Yet who would dare to claim that decades of massive northbound coffee, cocoa, rice and banana exports have enriched or improved the living standards of farmers in the south? Who would dare to make such a claim, looking these poverty-stricken farmers straight in the eyes? And who would dare to tell African breeders, bankrupted by competition from subsidised meat from Europe, that it is for their own good that customs barriers are falling?

To achieve their ends, the proponents of free trade exploit science in the name of so-called "modernism," asserting that the development of any scientific discovery constitutes progress—as long as it is economically profitable. They cannot bear the idea that life can reproduce on its own, free of charge, whence the race for patents, licences, profits and forcible expropriation.

Obviously, when talking about agriculture, it is impossible not to evoke the farce of GMOs. Nobody is asking for them, yet they must be the answer to everyone's dreams! There is pressure on us to concede that genetically modified rice (cynically dubbed "golden rice") is going to nourish people who are dying of hunger and protect them from all sorts of diseases, thanks to its new Vitamin A-enriched formula. But this will not solve the problems of vitamin deficiencies, because a person would have to eat three kilograms of dry rice every day, whereas the normal ration is no more than 100 grams.

The way to fight malnutrition, which affects nearly a third of humanity, is to diversify people's diets. This entails rethinking the appalling state of society, underpinned by free market economics which strives to keep wages in southern countries as low as possible in order to maximise profits. It is therefore a good idea to throw some vitamins into the rice that is sold to poor people, so that they don't die too quickly and can continue working for low wages, rather than helping them build a freer and fairer society. Jacques Diouf, Director-General of FAO, recently pointed out that "to feed the 8003 million people who are hungry, there is no need for GMOs" (*Le Monde*, 10 May). No wonder the Indian farmers of Via Campesina, an international small farmers' movement, destroy fields of genetically modified rice.

The FAO is not the only international institution to question some of the certainties and radical WTO positions regarding the benefits of free markets. The highly free-market OECD acknowledges in a recent report entitled *The Well-being of Nations* that the preservation and improvement of government services (healthcare, education) are a key factor underlying the economic success of nations.

We therefore have every reason to oppose the dangerous myth of free trade. Judging by the substantial social and environmental damage free trade has inflicted, before anything else, it is necessary for all of us—farmers and non-farmers alike—to make it subject to three fundamental principles: food sovereignty—the right of peoples and of countries to produce their food freely, and to protect their agriculture from the ravages of global "competition"; food safety—the right to protect oneself from any threat to one's health; and the preservation of bio-diversity.

Along with adherence to these principles must come a goal of solidarity-based development, via the institution of economic partnership areas among neighbouring countries, including import protection for such groups of countries having uniform structures and levels of development.

The WTO wants to take its free-market logic even further. Next November, in the seclusion of a monarchy that outlaws political parties and demonstrations—Qatar—it will attempt to attain its goals. But if major international institutions are becoming increasingly critical and are casting doubt on these certainties, then mobilised citizens can bring their own laws to bear on trade.

Between the absolute sovereign attitude of nationalists and the proponents of free trade, there are other roads. To echo the theme of the World Social Forum that took place in Porto Alegre last January, "another world is possible!"—a world that respects different cultures and the particularities of each, in a spirit of openness and understanding. We are happy and proud to be part of its emergence.

POSTSCRIPT

Is Globalization a Positive Development for the World Community?

It is hard to argue that this kind of revolution is all positive or all negative. Many will find the allure of technological growth and expansion too much to resist. They will adopt values and ethics that seem compatible with a materialistic Western culture. And they will embrace speed over substance, technical expertise over knowledge, and wealth over fulfillment.

Others will reject this revolution. They will find its promotion of materialism and Western cultural values abhorrent and against their own sense of humanity and being. They will seek enrichment in tradition and values rooted in their cultural pasts. This resistance will take many forms. It will be political and social, and it will involve actions ranging from protests and voting to division and violence.

Trying to determine whether a force as dominant and all-encompassing as globalization is positive or negative is like determining whether the environment is harsh or beautiful. It is both. One can say that in the short term, globalization will be destabilizing for many millions of people because the changes that it brings will cause some fundamental shifts in beliefs, values, and ideas. Once that period is past, it is conceivable that a more stable environment will result as people come to grips with globalization and either learn to embrace it, cope with it, or keep it at bay.

The literature on globalization is growing rapidly, much of it centering on defining its parameters and evaluating its impact. One important work that presents a deeper understanding of globalization is Friedman's best-seller *The Lexus and the Olive Tree* (Farrar, Straus & Giroux, 1999), which provides a positive view of the globalization movement. Two counterperspectives are William Greider's *One World Ready or Not: The Manic Logic of Global Capitalism* (Simon & Schuster, 1997) and David Korten's *When Corporations Rule the World* (Kumarian Press, 1996). Also see Jan Nederveen Pieterse, ed., *Global Futures: Shaping Globalization* (Zed Books, 2001).

The literature that will help us to understand the full scope of globalization has not yet been written. Also, the determination of globalization's positive or negative impact on the international system has yet to be decided. For certain, globalization will bring profound changes that will cause people from America to Zimbabwe to rethink assumptions and beliefs about how the world works. And equally certain is the realization that globalization will empower some people but that it will also leave others out, helping to maintain and perhaps exacerbate the divisions that already exist in global society.

ISSUE 12

Will the Digital/Computer World Pose a Threat to Our Individual Privacy?

YES: Jerry Kang, from "Cyberspace Privacy: A Primer and Proposal," *Human Rights* (Winter 1999)

NO: Renata Salecl, from "The Exposure of Privacy in Today's Culture," *Social Research* (Spring 2002)

ISSUE SUMMARY

YES: Professor of law Jerry Kang argues that privacy, which is fundamental to the concept of freedom, is severely threatened by the current love of cyberspace. He suggest that the U.S. Congress and, by extension, individual states should take action to ensure individual privacy in cyberspace.

NO: Senior researcher Renata Salecl contends that modern technology does not hinder freedom and privacy but rather enhances what is only a modern concept.

Scanning the Internet for products, services, friends, lovers, and information has become an obsession for some people, a necessity for others, and a desire for still others. More than 60 percent of all U.S. households have a computer. Most of these have Internet access. Experts estimate that by the end of the decade, over 80 percent of U.S. households will have a computer and Internet access. In an increasingly wireless world, these numbers will explode in the years to come. New technologies, such as the all-in-one device (computer, television, and telephone), will solidify the presence of the computer in the Western home. In economic terms, the largest growth businesses are Web-based, with estimates running from a conservative $25 billion to an astounding $1 trillion in financial impact over the next decade.

The ease with which people can find information, purchase products, and seek out communication and companionship makes the computer and cyberspace appealing. From information storage and retrieval to communicating around the globe, the computer is fast becoming a necessary tool for many.

Educators and students with computer access can search and retrieve information, explore new and heretofore inaccessible data, and make rare the age-old complaint heard by many professors that "I can't write on this topic because there is not enough information on the subject."

Yet not all people have access to Internet services, and those who do often do not know how to utilize the technology to empower themselves. In fact, most of the world does not have widespread access to this technology and are therefore left out of its potential benefits. Many who do have access become caught in a growing web of information, making their personal needs common knowledge and, indeed, public record. As people surf the Internet, they leave trails of information about their money, buying habits, personal tastes, and thoughts. In essence, people leave volumes of personal material that can be collected, stored, and used to market, investigate, or monitor those on the information superhighway.

The digital/computer age brings with it a variety of interesting and conflicting experiences. One can scan the Internet for products and items not found elsewhere. A person can obtain valuable information, research his or her ancestry, and experience virtual tours of faraway places. A computer-active person can speak with people in far-off lands, talk to his or her leaders, and solicit information from a growing variety of sources. One can also have one's bank accounts searched and drained, personal information collected and distributed, private messages recorded and read, and rights as an individual violated without one's knowledge. In some ways, logging on and being wired in have become a form of gambling with information, privacy, and freedom.

Privacy, which is fundamental to the conduct of political, economic, and social relations, is at risk in a cyberworld with little law and numerous technological ways of peeping into others' lives. The incredible speed of the growth of this technology has outstripped laws and ethics about its use and misuse. As a result, cyberspace is like a Wild West of information, ideas, and activities, where one can make a fortune or have one's life torn asunder with a few keystrokes. People's conception of individual freedom and expression will be severely tested in this new world.

In the following selections, Jerry Kang contends that the nature of cyberspace makes it a dangerous highway where millions of people are capable of peering into others' bank accounts, journals, and files; compiling information; and making people hostage to unseen forces. Kang maintains that these forces can manipulate individuals' buying habits, examine their political views, and track their whereabouts, thus inhibiting privacy and freedom.

Renata Salecl argues that the concept of privacy is a legal construct, relatively new and thus not an inalienable civil right. Consequently, new cyber technologies do not take away what we actually don't really possess. Further, she contends that new technologies have created our modern concept of privacy and thus new technologies will merely help us redefine it in the coming decades.

Jerry Kang

Cyberspace Privacy:
A Primer and Proposal

Human ingenuity has provided us a great gift, cyberspace. This blooming network of computing-communication technologies is quickly changing the world and our behavior in it. Already, it has become cliche to catalog cyberspace's striking benefits, its endless possibilities. But great gifts often come with a great price. Congress thinks that the price will be sexual purity due to easy access to pornography. Industry thinks it will be our economy due to easy copying of Hollywood's and Silicon Valley's programs. I worry that it will be our privacy.

When I mention "privacy," lawyers naturally think of privacy as used in the historic case, *Roe v. Wade*. Others think of the privacy of their own homes and backyards, largely in territorial terms. I use "privacy" differently. Instead of emphasizing privacy in a decisional or spatial sense, I mean it in an information sense. Information privacy is an individual's claim to control the terms under which personal information is acquired, disclosed, and used.

My thesis is that cyberspace threatens information privacy in extraordinary ways, and without much thought or collective deliberation, we may be in the process of surrendering our privacy permanently as we enter the next century.

Why Care About Privacy?

Some people do not understand what the big deal is about privacy. They assume that privacy is important only for those who have something to hide. This view is misguided. Let me articulate why individuals should enjoy meaningful control over the acquisition, disclosure, and use of their personal information.

Use of personal data. Personal data are often misused. For example, personal data can be used to commit identity theft, in which an impostor creates fake financial accounts, runs up enormous bills, and disappears leaving only a wrecked credit report behind. Personal data, such as home addresses and telephone numbers, can be used to harass and stalk. Personal data, such as one's sexual orientation, can be used to deny employment because of unwarranted prejudice.

Disclosure of personal data.   Sometimes, even if such data will not be "used" against us, its mere disclosure may lead to embarrassment. In any culture, certain conditions are embarrassing even when they are not blameworthy. Take impotency for example. In most cases, impotency will not affect whether one receives a job, a loan, or a promotion. In this sense, the data will not be misused in the allocation of rewards and opportunities. However, the mere disclosure of this medical condition would cause intense embarrassment for most men.

In addition to causing embarrassment, the inability to control the disclosure of personal data can hamper the building of intimate relationships. We construct many intimacies not only by sharing experiences but also by sharing secrets about ourselves, details not broadcast to a mass audience. If we have information privacy, we can regulate the outflow of such private information to others. By reducing this flow to a trickle (for example, to your boss), we maintain aloofness; by releasing a more telling stream (for example, to your former college roommate living afar) we invite and affirm intimacy. If anyone could find out anything about us, secrets would lose their ability to help construct intimacy.

Acquisition of personal data.   Finally, consider the fact that personal information is acquired by observing who we are and what we do. When such observation is nonconsensual and extensive, we have what amounts to surveillance, which is in tension with human dignity. Human beings have dignity because they are moral persons—beings capable of self-determination, with the capacity to reflect upon and choose personal and political projects. Extensive, undesired observation interferes with this exercise of choice because we act differently when we are being watched. Simply put, surveillance leads to self-censorship. When we do not want to be surveilled, it disrespects our dignity to surveil us nonetheless, unless some important social justification exists. This insult to individual dignity has social ramifications. It chills out-of-the-mainstream behavior. It corrodes private experimentation and reflection. It threatens to leave us with a bland, unoriginal polity, characterized by excessive conformity.

The Difference That Cyberspace Makes

So now that we know why privacy matters, we must ask what difference does cyberspace make? My claim is that cyberspace makes broad societal surveillance possible and, if we do nothing, likely. To see the greater threat that cyberspace poses, imagine the following two visits to a mall—one in real space, the other in cyberspace.

In real space, you drive to a mall, walk its corridors, peer into numerous shops, and stroll through the aisles of inviting stores. You walk into a bookstore and flip through a few magazines. Finally, you stop at a clothing store and buy a friend a scarf with a credit card. In this narrative, numerous persons interact with you and collect information along the way. For instance, while walking through the mall, fellow visitors visually collect information about you, if for no other reason than to avoid bumping into you. But such information is general, e.g., it does not pinpoint the geographical location and time of the sighting, is not in a format that can be processed by a computer, is not indexed to your name or any unique identifier, and is impermanent, residing in short-term human memory. You remain a barely noticed stranger. One important exception is the credit card purchase.

By contrast, in cyberspace, the exception becomes the norm: Every interaction may soon be like the credit card purchase. The best way to grasp this point is to take seriously, if only for a moment, the metaphor that cyberspace is an actual place, a virtual reality. In this alternate universe, you are invisibly stamped with a bar code as soon as you venture outside your home. There are entities called "road providers" (your Internet Service Provider), who supply the streets and ground you walk on, who track precisely where, when, and how fast you traverse the lands, in order to charge you for your wear on the infrastructure. As soon as you enter the cyber-mall's domain, the mall tracks you through invisible scanners focused on your bar code. It automatically records which stores you visit, which windows you browse, in which order, and for how long. The specific stores collect even more detailed data when you enter their domain. For example, the cyberbookstore notes which magazines you skimmed, recording which pages you have seen and for how long, and notes the pattern, if any, of your browsing. It notes that you picked up a health magazine featuring an article on Viagra, read for seven minutes a newsweekly detailing a politician's sex scandal, and flipped ever-so-quickly through a tabloid claiming that Elvis lives. Of course, whenever any item is actually purchased, the store, as well as the credit, debit, or virtual cash company that provides payment through cyberspace, takes careful notes of what you bought—in this case, a silk scarf, red, expensive, a week before Valentine's Day.

All these data generated in cyberspace are detailed, computer-processable, indexed to the individual, and permanent. While the mall example may not concern data that appear especially sensitive, the same extensive data collection can take place as we travel through other cyberspace domains—for instance, to research health issues and politics; to communicate to friends, businesses, and the government; and to pay our bills and manage our finances. Moreover, the data collected in these various domains can be aggregated to produce telling profiles of who we are, as revealed by what we do and with whom we associate. The very technology that makes cyberspace possible also makes detailed, cumulative, invisible observation of ourselves possible. One need only sift through the click streams generated by our cyber-activity.

It turns out that few laws limit what can be done with this data collected in cyberspace. Unlike Europe, the United States has no omnibus privacy law covering the private sector's processing of personal information. Instead, U.S. law features a legal patchwork that regulates different types of personal information in different ways, depending on how it is acquired, by whom, and how it will be used. To be sure, there are numerous statutes that govern specific sectors, such as consumer credit, education, cable programming, electronic communications, videotape rentals, motor vehicle records, and the recently enacted Children's Online Privacy Protection Act. But it turns out that in toto, information collectors can largely do what they want.

The Market Solution

Let me restate the problem. All cyberactivity, even simply browsing a Webpage, involves a "transaction" between an individual and potential information collectors. These collectors not only include the counterparty to the transaction but also intermediaries (transaction facilitators) that support the electronic communications (telephone company, cable company, Internet service provider) and sometimes payment (credit card company, electronic cash company). In these transactions, personal information is inevitably generated as either necessary or incidental by-products. Privacy enthusiasts insist that the individual owns this data; information collectors vigorously disagree. What shall be done?

Perhaps the market might solve the problem. One might reasonably view personal information as a commodity, whose pricing and consumption can and should be governed by the laws of supply and demand. Through offers and counteroffers between individual and information collector, the market will move the correctly priced personal data to the party that values it most. Economists love this approach because it appears to be economically efficient. The private sector loves this approach because it staves off regulation. Regulators love this approach for the same reason in the current antiregulatory environment.

The problem is that in practice, individuals and information collectors do not negotiate express privacy contracts before engaging in each transaction. Although privacy notices have become more frequent on Webpages, it is a stretch to say that there is a "meeting of the minds" on privacy terms each time an individual browses a Webpage. What is necessary, then, is a clear articulation of the default rules governing personal data collected in a cyberspace transaction, when parties have not agreed explicitly otherwise.

There are two default rules that society might realistically adopt. First, there is the status quo's "plenary" default rule: Unless the parties agree otherwise, the information collector may process the personal data anyway it likes. Second, there is the "functionally necessary" default rule: Unless the parties agree otherwise, the information collector may process the personal data only in functionally necessary ways. This rule allows the information collector to process personal data on a need-only basis to complete the transaction in which the information was originally collected.

A one-size-fits-all default rule is efficient for some transactions but inefficient for others. Those parties for whom the default is inefficient will either contract around the rule—"flip"—or they will "stick" with the rule and accept the inefficiencies. Thus, the social cost of a default rule equals the sum of the transaction costs of contracting around the rule—the "flip cost"—plus the inefficiency cost of not contracting around the rule even when it would be more efficient to do so—the "stick cost." We seek the rule that minimizes social costs.

If we implement the plenary rule, most parties will stick because it is hard for a consumer to flip out of the default rule. First, she would face substantial research costs to determine what information is being collected and how it is used. That is because individuals today are largely clueless about how personal information is processed through cyberspace. Sometimes, they are deceived by the information collectors themselves, as the Federal Trade Commission recently charged against the Internet Service Provider, Geocities. Second, the individual would run into a collective action problem. Realistically, the information collector would not entertain one person's idiosyncratic request to purchase back personal information because the costs of administering such an individually tailored program would be prohibitive. Therefore, to make it worth the firm's while, the individual would have to band together with other like-minded individuals to renegotiate the privacy terms of the underlying transaction. These individuals would suffer the collective action costs of locating each other, then coming to a mutually acceptable proposal to deliver to the information collector—all the while discouraging free riders.

By contrast, the "functionally necessary" rule would not be sticky at all. With this default, if the firm valued personal data more than the individual, then the firm would have to buy permission to process the data in functionally unnecessary ways. Note, however, two critical differences in contracting around this default. First, unlike the individual who has to find out how information is processed, the collector need not bear such research costs since it already knows what its information practices are. Second, the collector does not confront collective action problems. It need not seek out other like-minded firms and reach consensus before coming to the individual with a request. This is because an individual would gladly entertain an individualized, even idiosyncratic, offer to purchase personal information.

Now, the task is to compare the costs of the equilibrium generated by each default rule. For the "plenary" equilibrium, the cost of the default rule is approximately the stick cost because few parties will flip. By contrast, for the "functionally necessary" equilibrium, the cost of the default rule is approximately the flip cost; almost all parties who care to flip will flip. Which cost is higher? We lack the data to be confident in our answer. However, we do know that given how seriously many individuals feel about their privacy, the stick cost of the plenary rule will not be trivial. Many individuals who care deeply about their privacy will not be able to get it. By contrast, the flip cost of the functionally necessary rule will be small because cyberspace makes communications cheap. The information collector can ask in a simple dialog box whether the individual will allow some unnecessary use of personal data, in exchange for some benefit. What is more, this inequality will increase over time. As information processing becomes more sophisticated, people will feel less in control of their personal information; accordingly they will value control more (making the cost of "sticking" greater). Simultaneously, the cost of communication will decrease as cyberspace improves (making the cost of "flipping" less).

In conclusion, I think it is more likely than not that a functionally necessary rule will be less costly to society than the plenary rule we currently have. Putting economic efficiency aside, the functionally necessary rule also better respects human dignity by respecting an individual's desire not to be surveilled.

A Modest Proposal

Congress should adopt a Cyberspace Privacy Act (the "Act"), that implements the "functionally necessary" default rule for all personal information collected in cyberspace. This rule is more efficient and more respectful of human dignity.

Parties are, of course, free to contract around the default rule. The full proposed statute is available online at `http://www.law.ucla.edu/faculty/kang/scholars`, but here is a quick summary:

- First, a person who acquires personal data in the course of a cyberspace transaction must provide clear notice about what will be done with that information.
- Second, a person will not process personal information in a manner functionally unnecessary to the transaction without the prior consent of the individual.
- Third, an individual will have reasonable access to and rights of correction of personal data.
- Fourth, personal data that is no longer functionally necessary to the cyberspace transaction will be generally destroyed unless there is some legitimate pending request or the individual has given consent otherwise.
- Fifth, if compelled by court order or a dire emergency to the individual's own welfare, personal data may be disclosed as necessary.
- Finally, a person who violates this Act may be sued in federal court for civil damages. Moreover, the Federal Trade Commission will have administrative authority to enforce the Act.

Politically moderate, the proposed legislation should enjoy broad appeal. The private sector should not oppose the Act because it does not choke off electronic commerce in cyberspace.

Although the Act constrains certain forms of advertising based on detailed data collection, and the sharing of data with third parties, these constraints can be lifted simply by obtaining the customer's consent. Moreover, the Cyberspace Privacy Act would promote consumer confidence in—and thereby encourage— electronic commerce. Multinational corporations working in Europe might have an independent reason to accept the Act. By applying the Act to data received from the European Union, these corporations could credibly assert that they have begun to adopt adequate privacy protections necessary to maintain transborder flows under the recent European Union Data Protection Directive. Finally, the Act does not violate the First Amendment. In structure, the proposed Act does not differ materially from the privacy provisions of the Cable Act or the Video Privacy Protection Act. Neither act has been successfully challenged on First Amendment grounds.

Conclusion

A vision protective of information privacy in cyberspace will be singularly hard to maintain. Cyberspace's essence is the processing of information in ways and at speeds unimaginable just years ago. To retard this information processing juggernaut in the name of privacy seems antitechnology, even antiprogress. It cuts against the hackneyed cyber-proclamation that information wants to be free. Nevertheless, this intentional application of friction to personal information flows is warranted. If profit-seeking organizations are instituting such friction in the name of intellectual property, individuals should not be chastised for doing the same in the name of privacy.

Historically, privacy issues have been an afterthought. Technology propels us forward, and we react to the social consequences only after the fact. But the amount of privacy we retain is—to use a decidedly low-tech metaphor—a one-way ratchet. Once we ratchet privacy down, it will be extraordinarily difficult to get it back. More disturbingly, after a while, we might not mind so much. It may dawn on us too late that privacy should have been saved along the way.

NO ⏎

Renata Salecl

The Exposure of Privacy in Today's Culture

Today it seems to be a universally accepted thesis that people have less and less privacy and that new technologies allow new invasions into people's private lives. However, when we complain about our endangered privacy, we often forget that the idea of privacy is a distinctly modern phenomenon and that contemporary understanding about the protection of privacy has evolved not despite new technologies, but because of them. Historians who deal with the issue of privacy like to point out that writing, as one of the earliest technologies, enabled forms of private communication that did not exist before. And with the technological advances of the Industrial Revolution, the idea of privacy took on unprecedented dimensions. The prosperous middle class began to build houses with separate rooms for family members; telephone lines allowed for direct, private communication; radio and television brought entertainment into private homes; and the mass production of cars saw private travel become widely available. The newest technological advances in the forms of personal computers, the Internet, wireless devices, biological engineering radically expanded privacy while simultaneously also allowing for new forms of intrusions into it. Among the fastest growing businesses are those that provide new forms of privacy protection. If Orwell were to write *1984* today, his Big Brother would surely have found a twin sibling in a Big Protector. Or maybe Orwell would have simply taken the name of one of the existing privacy protection companies, such as Anonymizer, IDecide, Disappearing, Hushmail, Zip Lip, or Zero Knowledge.

However, if we now have great concerns about protection of privacy, we also have an expanded exposure of privacy. Some have found an entirely new form of enjoyment in being overtly concerned that someone is constantly watching; others are constantly trying to develop new devices to expose their private life to the public; and still others love to observe these exposures. Before analyzing these trends, let me first look at how psychoanalysis perceives the problem of respect of privacy. My concern will not be the privacy of the analytic session, but that something untouchable in the subject that has to be protected from public exposure.

From *Social Research* by Renata Salecl, vol. 69, no. 1, Spring 2002, pp. 1–9. Copyright © 2002 by Social Research. Reprinted by permission.

When we speak about privacy, we often claim that someone's privacy has to be respected or that one needs to honor the privacy of another. How would psychoanalysis explain this logic of honor and respect? Freud dealt with the issue of respect when he discussed the problem of female shyness, which he linked to the lack: the absence of a phallus in women. By being shy, a woman tries to cover up the lack and avert the gaze from it. However, this shyness has in itself a phallic character. So it can be said that the very lack of the phallic organ in a woman results in the phallicization of her whole body or a special part of the body; and covering this part of the body has a special seductive effect.

There is no significant difference between women's shyness and their honor: "The respect for women means that there is something that should not be seen or touched." Shyness and respect both concern the problem of castration, the lack that marks the subject. The insistence on respect is a demand for distance, which also means a special relation that the subject needs to have toward the lack in the other.

Freud thought that woman is the subject who actually lacks something, which means that in her case castration is effective. As a result of this, the woman has *Penisneid* (penis envy). In psychoanalytic practice, women's "deprivation" appears in many forms: as a fantasy of some essential injustice, as an inferiority complex, as a feeling of nonlegitimacy, as a lack of consistency or a lack of control, or even as a feeling of body fragmentation. The Freudian solution for this "deprivation" is motherhood.

For Jacques Lacan, women's relation toward the lack is much more complicated: femininity's trouble is not simply linked to having or not having a penis. The lack concerns the subject's very being—both a man and a woman are marked by lack, but they differently relate to this lack. A woman does not cover up the lack by becoming a mother, since for Lacan the problem of the lack cannot be solved on the level of having but on the level of being. Motherhood is not a solution to a woman's lack, since there is no particular object (not even a child) that can ever fill this lack.

Respect, therefore, concerns the subject's relation to the lack in the other, which also means that respect is just another name for the anxiety the subject feels with regard to this lack. The respect for the father, for example, needs to be understood as a way in which the subject tries to avoid the recognition that the father is actually impotent and powerless—that there is nothing behind his authority. Here we come again to the problem of castration. Lacan understands castration as something that is linked to the radical emptiness of the subject. The subject is nothing by him or herself; he or she gets all authority and power only from outside—from symbolic insignias. When we respect the father, we believe that the insignias have real power and thus we cover up the fact that the father is castrated, which means that he is himself an empty and powerless subject.

Respect is therefore an imaginary relationship that the subject has with another subject, or, better, with the symbolic status that this other subject temporarily assumes. (Of course, respect does not concern only our relationships with another subject but also with the big Other, the symbolic structure as such. Paying respect to our homeland, the flag, the law are all a subject's imaginary means of taking the big Other as consistent order.)

When we speak about the right to privacy we usually invoke the idea of respect. Privacy concerns some part of the subject's inner freedom, which the community or other people have no right to violate. If we take into account the aforementioned thesis that respect means the need for a distance toward the lack that marks the subject, the idea of respect that the right to privacy invokes also assumes another meaning. The inner freedom of the subject that this right protects concerns nothing other than the lack that marks the subject when he or she becomes a speaking being. When, for example, we respect the bodily integrity of the subject, this actually means that we avert our gaze from the fact that the subject actually does not have a naturally given bodily integrity, since this integrity comes into being only when the subject undergoes symbolic castration. Then the subject will be temporarily entrusted with a certain symbolic power, but the lack that pertains to his or her subjectivity will nonetheless remain.

If the right to privacy has to do with not being exposed to the gaze of the other, we can say that in conjunction with various attempts of today's lawyers and entrepreneurs to keep the gaze averted, we have an increasing desire to see what it supposed to be hidden. The Internet has become a place in which people are creating ever-new ways to expose their private lives. We have all heard about individuals who daily record their private lives and distribute images throughout the Net, and reality TV shows seem to be competing to expose the most embarrassing and obscene parts of everyday life. From fights and flirting on *Survivor* and *Big Brother* we have come to sex on *Temptation Island* and the ability to witness the suffering of military training on *Boot Camp*.

Reality TV has gained popularity recently, but the exposure of everyday life has been present in the arts for some time. Although we had in the last decades trends in art that showed everyday life as an art object, other trends try to depict the behind of things, the inside of the body. It appears that everything can be exposed and that nothing that is supposed to be behind the mask surprises us. For example, in the case of television reports on war, we can see all kinds of suffering exposed on the screen: the bodies being torn in front of our eyes, people killing each other and even recording their acts with video cameras. In the arts we have a similar trend, with violence against the body presented today as artwork. These trends have been especially evident in the show *Sensation*. The exposure of what is supposed to be the hidden inside can be seen in Damien Hirst's split animals, Monna Hatum's video of the intestines, Alain Miller's picture of the face behind the skin, Mark Quinn's skin without the body, Ron Mueck's *Dead Dad,* and even Chris Ofili's use of animal excreta. On the other hand, we have depiction of everyday life in the work of Tracey Emin's exposure of the names of all her lovers and Sara Lucas's mattress.

In contemporary society there appears to be no social antagonism anymore; that is, there is no lack. Everything today is visible—there seems to be no secret. However, this exposure of the inside and the revelation of the everyday life is nothing subversive. It goes hand in hand with the dominant ideology. We can find a number of examples of this logic of "there is no secret" in today's society.

In Germany, one of the country's most visited exhibitions has been "The World of the Body," which shows the inside of an actual human body. Anatomist Gunter von Hagens uses a special technique of plastification of real parts of corpses so one can see the skin without the body, the body without the skin. A number of artists in other countries also use actual dead bodies in their artwork. The most known are Joel-Peter Witkin, Stephen J. Shanabrook, and Ilja Cickin.

Examples of the exposure of the inside are also found in contemporary architecture. If you look at the designs of many new restaurants, you can see that the work process is supposed to be totally exposed to the public. Everywhere, one now finds restaurants in which, when a customer walks in, he or she sees low-paid workers preparing the food or washing the dishes. We observe these workers as decorative art objects and do not think about the hardship that these people might endure and the discomfort they might feel being exposed as if they were in a zoo.

Still other examples of this logic of exposing the secret can be found in today's election campaigns. Politicians in their television advertisements do not deliver a final product: a speech to convince the electorate. Often, the advertisement exposes the very preparation of the speech. We might see a politician while he shaves himself in the bathroom, sips his morning coffee, and talks to his advisers preparing his speech. If, in the past, a politician would have hidden the fact that it is not he who writes the speech, this very revelation is today used as a campaign advertisement. The message that this advertisement puts across is: we show you the truth; the politician is just the ordinary man like you, and he is very honest, since he even shows you how he is not writing his own speeches. . . .

Does the conclusion that with the advancement of technology people seek more solitude and privacy still hold? Alan Westlin, the renowned legal scholar on privacy, divides the population into three categories when it comes to the issue of privacy. While roughly a quarter of the American population falls into the category of "privacy fundamentalists," who are deeply concerned about privacy rights, a number of people (approximately 12 percent from Westlin's predictions) are unconcerned with privacy. The majority are part of the intermediate category of "privacy pragmatists" who, depending on what they get in return for their information, are willing to forsake different degrees of privacy protection. Privacy protection enterprises are fighting precisely over this group of "privacy pragmatists," when they are marketing their new protection devices. As one privacy businessman says: "People are being fed expectations—by the media, by politicians, by privacy advocates, by companies trying to sell privacy services—that they've never had before. . . . And as soon as someone tells you to worry about something, it is hard not to" (Lester, 2001). Another privacy businessman has a different agenda than selling protection of privacy—and this is selling privacy itself. Someone has determined that people would benefit most if they can store their data (such as age, sex, race, family, income, sexual orientation, and consumer preferences) in a protected data center to which companies will have access for advertising only if they pay a high fee, a portion of which will go to the people who stored the data. This system supposedly gives consumers exactly what they want; or, as the leader of this web-based enterprise says: "We work for you, the consumer, as an agent to extract the maximum value for your identity. We help you copyright your profile. Not only that, we take your click trail and consider that a unique work of authorship" (Lester, 2001).

Almost 20 years ago, the French artist Sophie Calle made from the exposure of privacy a unique work of art, but unfortunately she did not have consent from the person whose life she exposed. The art project began when Calle found on the streets of Paris an address book that belonged to a certain Paul V. Calle decided to learn as much as possible about this person by contacting people whose phone numbers were in Paul V.'s address book. Each day Calle met with one of these people, and for a whole month *Liberation* published her reports on these meetings. From people's recollections about Paul, we slowly learned quite a bit about his documentary filmmaking, his love passions, his odd life routines, and even the fact that he was now away at a film seminar in Norway. When Paul returned to Paris, he was shocked that he has been an object of such particular art project and wrote a furious response that was also published in *Liberation*. Paul took the project as an extreme form of violence, an utter intrusion into his private life. But he also stated that as a documentary filmmaker he believed that one should never try to comprehend another person's life by simply looking at him or her from the outside—that is, taking seriously other people's reflection on him or her—but should always give voice to the person in question. In this case, we have two very different views about privacy. Sophie Calle took the exposure of Paul's life as a work of art that tried to unravel the mystery of another person, and Paul radically disagreed with this approach.

Some expose privacy as a unique work of art, others as a unique work of authorship, and still others as a new form of public entertainment. In a world obsessed with the issue of identity and individuality, privacy is the hottest theme. However, both protectors of privacy and people who want to expose it sooner or later stumble upon the fact that social agreement on what is private is constantly changing. Lawyers and businessmen are now trying to determine if privacy can be bought and sold or if is it an inalienable human right. But even if we agree that it is an essential human right, we come to the point that what privacy protects is not the subject's identity but the very lack of it. If psychoanalysis teaches us anything, it is that the subject is constantly failing in his or her bid for an identity. And what the right to privacy at the end protects is not to be exposed in our failing.

Reference

Lester, Toby. "The Reinvention of Privacy." *Atlantic Monthly* 287:3 (March 2001).

POSTSCRIPT

Will the Digital/Computer World Pose a Threat to Our Individual Privacy?

The computer, like the television, telephone, and automobile before it, provides millions with access to a larger world around them. They can access information, ideas, people, places, and cultures beyond their own physical and economic limitations. The computer is a tool that can empower and inform beyond the tools of any previous generation. The digital/computer world is the next great technological revolution, and it may turn out to be the greatest in many respects. Information storage, retrieval, and access will be greater than at any other time in human history. Theoretically, billions will have the collected human experience at their fingertips and will be able to educate themselves about almost any subject or issue. The impact for education, economics, politics, and society is phenomenal.

However, the digital/computer world is also a commercial world. The technology and infrastructure that support this world is privately owned and run for profit. Therefore, the technology must serve the interests of those who own it in order for it to be implemented and mass-produced. To understand the full impact of this technology on individual freedom, one must understand the goals of those who control it. For example, the phone allows people to call out as well as others to call in. This means that external access demands internal access. Likewise, the digital/computer world allows for the cataloging of information—commercial, personal, private, and political. People's tastes, buying habits, desires, and beliefs are cataloged and stored so that more and more products can be marketed to individuals based on their tastes and so that more and more actors can know who those individuals are. Anonymity will die, and with it a degree of freedom.

Is it convenient that The Gap can know your tastes in clothes, buying habits, and even your friends' and family members' birthdays so that they can send you e-mail ads at key times offering products for you to buy? Or is that kind of knowledge intrusive? Is it good that the government can track your buying habits, private communications, and travel patterns down to the minute? Or is that potentially threatening to your freedom?

Three works of note that will help you to explore this issue in greater depth are Benjamin Barber, "Three Scenarios for the Future of Technology and a Strong Democracy," *Political Science Quarterly* (Winter 1998–1999); Michael Mehta and Eric Darier, "Virtual Control and Disciplining on the Internet: Electronic Governmentality in the New Wired World," *Information Society* (April/June 1998); and Albert H. Teich, *Technology and the Future* (St. Martin's Press, 1997).

ISSUE 13

Is the World a Victim of American Cultural Imperialism?

YES: Julia Galeota, from "Cultural Imperialism: An American Tradition," *The Humanist* (May/June 2004)

NO: Philippe Legrain, from "In Defense of Globalization," *The International Economy* (Summer 2003)

ISSUE SUMMARY

YES: Julia Galeota, the seventeen-year-old winner of the 2004 Humanist Essay Contest, contends that American cultural imperialism is a reality promoted by commercial images presented through the media and the selling of American products across the globe.

NO: Philippe Legrain is a British economist who presents two views of cultural imperialism and argues that the notion of American cultural imperialism "is a myth" and that the spreading of cultures through globalization is a positive, not negative, development.

In 1989 the Berlin Wall collapsed. Two years later the Soviet Union ceased to exist. With this relatively peaceful and monumental series of events, the cold war ended, and with it one of the most contentious and conflict-ridden periods in global history. It is easy to argue that in the wake of those events the United States is in ascendancy. The United States and its Western allies won the cold war, defeating communism politically and philosophically. Since 1990 democracies have emerged and largely flourished as never before across the world stage. According to a recent study, over 120 of the world's 190 nations now have a functioning form of democracy. Western companies, values, and ideas now sweep across the globe via airwaves, computer networks, and fiber-optic cables that bring symbols of U.S. culture and values (such as Michael Jordan and McDonald's) into villages and schools and cities around the world.

If American culture is embodied in the products sold by many multinational corporations (MNCs), such as McDonald's, Ford, IBM, The Gap, and others, then the American cultural values and ideas that are embedded in these products are being bought and sold in record numbers around the world. Globalization largely driven by MNCs and their control of technology brings with it values and ideas that are largely American in origin and expression. Values such as speed and ease of use, a strong emphasis on leisure time over work time, and a desire for increasing material wealth and comfort dominate the advertising practices of these companies. For citizens of the United States, this seems a natural part of the landscape. They do not question it; in fact, many Americans enjoy seeing signs of "home" on street corners abroad: a McDonald's in Tokyo, a Sylvester Stallone movie in Djakarta, or a Gap shirt on a student in Nairobi, for example.

While comforting to Westerners, this trend is disquieting to the hundreds of millions of people around the world who wish to partake of the globalizing system without abandoning their own cultural values. Many people around the globe wish to engage in economic exchange and develop politically but do not want to abandon their own cultures amidst the wave of values embedded in Western products. This tension is most pronounced in its effect on the youth around the world. Millions of impressionable young people in the cities and villages of the developing world wish to emulate the American icons that they see on soft drink cans or in movie theaters. They attempt to adopt U.S. manners, language, and modes of dress, often in opposition to their parents and local culture. These young people are becoming Americanized and, in the process, creating huge generational rifts within their own societies. Some of the seeds of these rifts and cultural schisms can be seen in the actions of the young Arab men who joined Al Qaeda and participated in the terrorist attacks of September 11, 2001.

Julia Galeota presents the case that American culture is seen in every corner of the globe. From McDonald's and Coke to MTV and CNN, America's values, ideals, and customs are presented to hundreds of millions of people as right and good and thus homogenizing world culture based on an American matrix. She contends that this imagery and its delivery systems need to be thwarted by "protective filters [that] should help to maintain the integrity of a culture in the face of cultural imperialism." Legrain posits two counter arguments, one by Thomas Freidman in favor of globalization and its impact, and one by Naomi Klein against critical of such impact to argue that people define themselves, not advertising and images. He argues that it is a myth to regard globalization as culturally imperialist and that living in an era of multiple forms of identity and diverse cultures and globalization does not hinder nor stifle any of those cultures.

YES ⤶

Julia Galeota

Cultural Imperialism: An American Tradition

Travel almost anywhere in the world today and, whether you suffer from habitual Big Mac cravings or cringe at the thought of missing the newest episode of MTV's *The Real World,* your American tastes can be satisfied practically everywhere. This proliferation of American products across the globe is more than mere accident. As a byproduct of globalization, it is part of a larger trend in the conscious dissemination of American attitudes and values that is often referred to as *cultural imperialism.* In his 1976 work *Communication and Cultural Domination,* Herbert Schiller defines cultural imperialism as:

> The sum of the processes by which a society is brought into the modern world system, and how its dominating stratum is attracted, pressured, forced, and sometimes bribed into shaping social institutions to correspond to, or even to promote, the values and structures of the dominant center of the system.

Thus, cultural imperialism involves much more than simple consumer goods; it involves the dissemination of ostensibly American principles, such as freedom and democracy. Though this process might sound appealing on the surface, it masks a frightening truth: many cultures around the world are gradually disappearing due to the overwhelming influence of corporate and cultural America.

The motivations behind American cultural imperialism parallel the justifications for U.S. imperialism throughout history: the desire for access to foreign markets and the belief in the superiority of American culture. Though the United States does boast the world's largest, most powerful economy, no business is completely satisfied with controlling only the American market; American corporations want to control the other 95 percent of the world's consumers as well. Many industries are incredibly successful in that venture. According to the *Guardian,* American films accounted for approximately 80 percent of global box office revenue in January 2003. And who can forget good old Micky D's? With over 30,000 restaurants in over one hundred countries, the ubiquitous

From *The Humanist* by Julia Galeota, vol. 64, no. 3, May/June 2004, pp. 22–24, 46. Copyright © 2004 by American Humanist Association. Reprinted by permission.

golden arches of McDonald's are now, according to Eric Schlosser's *Fast Food Nation,* "more widely recognized than the Christian cross." Such American domination inevitably hurts local markets, as the majority of foreign industries are unable to compete with the economic strength of U.S. industry. Because it serves American economic interests, corporations conveniently ignore the detrimental impact of American control of foreign markets.

Corporations don't harbor qualms about the detrimental effects of "Americanization" of foreign cultures, as most corporations have ostensibly convinced themselves that American culture is superior and therefore its influence is beneficial to other, "lesser" cultures. Unfortunately, this American belief in the superiority of U.S. culture is anything but new; it is as old as the culture itself. This attitude was manifest in the actions of settlers when they first arrived on this continent and massacred or assimilated essentially the entire "savage" Native American population. This attitude also reflects that of the late nineteenth-century age of imperialism, during which the jingoists attempted to fulfill what they believed to be the divinely ordained "manifest destiny" of American expansion. Jingoists strongly believe in the concept of social Darwinism: the stronger, "superior" cultures will overtake the weaker, "inferior" cultures in a "survival of the fittest." It is this arrogant belief in the incomparability of American culture that characterizes many of our economic and political strategies today.

It is easy enough to convince Americans of the superiority of their culture, but how does one convince the rest of the world of the superiority of American culture? The answer is simple: marketing. Whether attempting to sell an item, a brand, or an entire culture, marketers have always been able to successfully associate American products with modernity in the minds of consumers worldwide. While corporations seem to simply sell Nike shoes or Gap jeans (both, ironically, manufactured *outside* of the United States), they are also selling the image of America as the land of "cool." This indissoluble association causes consumers all over the globe to clamor ceaselessly for the same American products.

Twenty years ago, in his essay "The Globalization of Markets," Harvard business professor Theodore Levitt declared, "The world's needs and desires have been irrevocably homogenized." Levitt held that corporations that were willing to bend to local tastes and habits were inevitably doomed to failure. He drew a distinction between weak multinational corporations that operate differently in each country and strong global corporations that handle an entire world of business with the same agenda.

In recent years, American corporations have developed an even more successful global strategy: instead of advertising American conformity with blonde-haired, blue-eyed, stereotypical Americans, they pitch diversity. These campaigns—such as McDonald's new international "I'm lovin' it" campaign—work by drawing on the United State's history as an ethnically integrated nation composed of essentially every culture in the world. An early example of this global marketing tactic was found in a Coca Cola commercial from 1971 featuring children from many different countries innocently singing, "I'd like to teach the world to sing in perfect harmony/I'd like to buy the world a Coke to keep it company." This commercial illustrates an attempt to portray a U.S. goods as a product capable of transcending political, ethnic, religious, social, and economic differences to unite the world (according to the Coca-Cola Company, we can achieve world peace through consumerism).

More recently, Viacon's MTV has successfully adapted this strategy by integrating many different Americanized cultures into one unbelievably influential American network (with over 280 million subscribers worldwide). According to a 1996 "New World Teen Study" conducted by DMB&B's BrainWaves division, of the 26,700 middle-class teens in forty-five countries surveyed, 85 percent watch MTV every day. These teens absorb what MTV intends to show as a diverse mix of cultural influences but is really nothing more than manufactured stars singing in English to appeal to American popular taste.

If the strength of these diverse "American" images is not powerful enough to move products, American corporations also appropriate local cultures into their advertising abroad. Unlike Levitt's weak multinationals, these corporations don't bend to local tastes; they merely insert indigenous celebrities or trends to present the facade of a customized advertisement. MTV has spawned over twenty networks specific to certain geographical areas such as Brazil and Japan. These specialized networks further spread the association between American and modernity under the pretense of catering to local taste. Similarly, commercials in India in 2000 featured Bollywood stars Hrithik Roshan promoting Coke and Shahrukh Khan promoting Pepsi (Sanjeev Srivastava, "Cola Row in India." BBC News Online). By using popular local icons in their advertisements, U.S. corporations successfully associate what is fashionable in local cultures with what is fashionable in America. America essentially samples the world's cultures, repackages them with the American trademark of materialism, and resells them to the world.

Critics of the theory of American cultural imperialism argue that foreign consumers don't passively absorb the images America bombards upon them. In fact, foreign consumers do play an active role in the reciprocal relationship between buyer and seller. For example, according to Naomi Klein's *No Logo*, American cultural imperialism has inspired a "slow food movement" in Italy and a demonstration involving the burning of chickens outside of the first Kentucky Fried Chicken outlet in India. Though there have been countless other conspicuous and inconspicuous acts of resistance, the intense, unrelenting barrage of American cultural influence continues ceaselessly.

Compounding the influence of commercial images are the media and information industries, which present both explicit and implicit messages about the very real military and economic hegemony of the United States. Ironically, the industry that claims to be the source for "fair and balanced" information plays a large role in the propagation of American influence around the world. The concentration of media ownership during the 1990s enabled both American and British media organizations to gain control of the majority of the world's news services. Satellites allow over 150 million households in approximately 212 countries and territories worldwide to subscribe to CNN, a member of Time Warner, the world's largest media conglomerate. In the words of British sociologist Jeremy Tunstall, "When a government allows news importation, it is in effect importing a piece of another country's politics—which is true of no other import." In addition to politics and commercials, networks like CNN also present foreign countries with unabashed accounts of the military and economic superiority of the United States.

The Internet acts as another vehicle for the worldwide propagation of American influence. Interestingly, some commentators cite the new "information economy" as proof that American cultural imperialism is in decline. They argue that the global accessibility of this decentralized medium has decreased the relevance of the "core and periphery" theory of global influence. This theory describes an inherent imbalance in the primarily outward flow of information and influence from the stronger, more powerful "core" nations such as the United States. Additionally, such critics argue, unlike consumers of other types of media, Internet users must actively seek out information; users can consciously choose to avoid all messages of American culture. While these arguments are valid, they ignore their converse: if one so desires, anyone can access a wealth of information about American culture possibly unavailable through previous channels. Thus, the Internet can dramatically increase exposure to American culture for those who desire it.

Fear of the cultural upheaval that could result from this exposure to new information has driven governments in communist China and Cuba to strictly monitor and regulate their citizens' access to websites (these protectionist policies aren't totally effective, however, because they are difficult to implement and maintain). Paradoxically, limiting access to the Internet nearly ensures that countries will remain largely the recipients, rather than the contributors, of information on the Internet.

Not all social critics see the Americanization of the world as a negative phenomenon. Proponents of cultural imperialism, such as David Rothkopf, a former senior official in Clinton's Department of Commerce, argue that American cultural imperialism is in the interest not only of the United States but also of the world at large. Rothkopf cites Samuel Huntington's theory from *The Clash of Civilizations and the Beginning of the World Order* that, the greater the cultural disparities in the world, the more likely it is that conflict will occur. Rothkopf argues that the removal of cultural barriers through U.S. cultural imperialism will promote a more stable world, one in which American culture reigns supreme as "the most just, the most tolerant, the most willing to constantly reassess and improve itself, and the best model for the future." Rothkopf is correct in one sense: Americans are on the way to establishing a global society with minimal cultural barriers. However, one must question whether this projected society is truly beneficial for all involved. Is it worth sacrificing countless indigenous cultures for the unlikely promise of a world without conflict?

Around the world, the answer is an overwhelming "No!" Disregarding the fact that a world of homogenized culture would not necessarily guarantee a world without conflict, the complex fabric of diverse cultures around the world is a fundamental and indispensable basis of humanity. Throughout the course of human existence, millions have died to preserve their indigenous culture. It is a fundamental right of humanity to be allowed to preserve the mental, physical, intellectual, and creative aspects of one's society. A single "global culture" would be nothing more than a shallow, artificial "culture" of materialism reliant on technology. Thankfully, it would be nearly impossible to create one bland culture in a world of over six billion people. And nor should we want to. Contrary to Rothkopf's (and George W. Bush's) belief that, "Good and evil, better and worse coexist in this world," there are no such absolutes in this world. The United States should not be able to relentlessly force other nations to accept its definition of what is "good" and "just" or even "modern."

Fortunately, many victims of American cultural imperialism aren't blind to the subversion of their cultures. Unfortunately, these nations are often too weak to fight the strength of the United States and subsequently to preserve their native cultures. Some countries—such as France, China, Cuba, Canada, and Iran—have attempted to quell America's cultural influence by limiting or prohibiting access to American cultural programming through satellites and the Internet. However, according to the UN Universal Declaration of Human Rights, it is a basic right of all people to "seek, receive, and impart information and ideas through any media and regardless of frontiers," Governments shouldn't have to restrict their citizens' access to information in order to preserve their native cultures. We as a world must find ways to defend local cultures in a manner that does not compromise the rights of indigenous people.

The prevalent proposed solutions to the problem of American cultural imperialism are a mix of defense and compromise measures on behalf of the endangered cultures. In *The Lexus and the Olive Tree*, Thomas Friedman advocates the use of protective legislation such as zoning laws and protected area laws, as well as the appointment of politicians with cultural integrity, such as those in agricultural, culturally pure Southern France. However, many other nations have no voice in the nomination of their leadership, so those countries need a middle-class and elite committed to social activism. If it is utterly impossible to maintain the cultural purity of a country through legislation, Friedman suggests the country attempt to "glocalize," that is:

> To absorb influences that naturally fit into and can enrich [a] culture, to resist those things that are truly alien and to compartmentalize those things that, while different, can nevertheless be enjoyed and celebrated as different.

These types of protective filters should help to maintain the integrity of a culture in the face of cultural imperialism. In *Jihad vs. McWorld*, Benjamin Barber calls for the resuscitation of nongovernmental, noncapitalist spaces—to the "civic spaces"—such as village greens, places of religious worship, or community schools. It is also equally important to focus on the education of youth in their native values and traditions. Teens especially need a counterbalance images of American consumerism they absorb from the media. Even if individuals or countries consciously choose to become "Americanized" or "modernized," their choice should be made freely and independently of the coercion and influence of American cultural imperialism.

The responsibility for preserving cultures shouldn't fall entirely on those at risk. The United States must also recognize that what is good for its economy isn't necessarily good for the world at large. We must learn to put people before profits. The corporate and political leaders of the United States would be well advised to heed these words of Gandhi:

> I do not want my house to be walled in on all sides and my windows to be stuffed. I want the culture of all lands to be blown about my house as freely as possible. But I refuse to be blown off my feet by any.

The United States must acknowledge that no one culture can or should reign supreme, for the death of diverse cultures can only further harm future generations.

NO ↵

Philippe Legrain

In Defense of Globalization

Fears that globalization is imposing a deadening cultural uniformity are as ubiquitous as Coca-Cola, McDonald's, and Mickey Mouse. Many people dread that local cultures and national identifies are dissolving into a crass all-American consumerism. That cultural imperialism is said to impose American values as well as products, promote the commercial at the expense of the authentic, and substitute shallow gratification for deeper satisfaction.

Thomas Friedman, columnist for the *New York Times* and author of *The Lexus and the Olive Tree,* believes that globalization is "globalizing American culture and American cultural icons." Naomi Klein, a Canadian journalist and author of *No Logo,* argues that "Despite the embrace of polyethnic imagery, market-driven globalization doesn't want diversity; quite the opposite. Its enemies are national habits, local brands, and distinctive regional tastes."

But it is a myth that globalization involves the imposition of Americanized uniformity, rather than an explosion of cultural exchange. And although—as with any change—it can have downsides, this cross-fertilization is overwhelmingly a force for good.

The beauty of globalization is that it can free people from the tyranny of geography. Just because someone was born in France does not mean they can only aspire to speak French, eat French food, read French books, and so on. That we are increasingly free to choose our cultural experiences enriches our lives immeasurably. We could not always enjoy the best the world has to offer.

Globalization not only increases individual freedom, but also revitalizes cultures and cultural artifacts through foreign influences, technologies, and markets. Many of the best things come from cultures mixing: Paul Gauguin painting in Polynesia, the African rhythms in rock 'n' roll, the great British curry. Admire the many-colored faces of France's World Cup-winning soccer team, the ferment of ideas that came from Eastern Europe's Jewish diaspora, and the cosmopolitan cities of London and New York.

Fears about an Americanized uniformity are overblown. For a start, many "American" products are not as all-American as they seem; MTV in Asia promotes Thai pop stars and plays rock music sung in Mandarin. Nor are American products all-conquering. Coke accounts for less than two of the 64 fluid ounces that the typical person drinks a day. France imported a mere $620 million in

From *The International Economy* by Philippe Legrain, vol. 17, no. 3, Summer 2003, pp. 62–65.

food from the United States in 2000, while exporting to America three times that. Worldwide, pizzas are more popular than burgers and Chinese restaurants sprout up everywhere.

In fashion, the ne plus ultra is Italian or French. Nike shoes are given a run for their money by Germany's Adidas, Britain's Reebok, and Italy's Fila. American pop stars do not have the stage to themselves. According to the IFPI, the record-industry bible, local acts accounted for 68 percent of music sales in 2000, up from 58 percent in 1991. And although nearly three-quarters of television drama exported worldwide comes from the United States, most countries' favorite shows are homegrown.

Nor are Americans the only players in the global media industry. Of the seven market leaders, one is German, one French, and one Japanese. What they distribute comes from all quarters: Germany's Bertelsmann publishes books by American writers; America's News Corporation broadcasts Asian news; Japan's Sony sells Brazilian music.

In some ways, America is an outlier, not a global leader. Baseball and American football have not traveled well; most prefer soccer. Most of the world has adopted the (French) metric system; America persists with antiquated British Imperial measurements. Most developed countries have become intensely secular, but many Americans burn with fundamentalist fervor—like Muslims in the Middle East.

Admittedly, Hollywood dominates the global movie market and swamps local products in most countries. American fare accounts for more than half the market in Japan and nearly two-thirds in Europe. Yet Hollywood is less American than it seems. Top actors and directors are often from outside America. Some studios are foreign-owned. To some extent, Hollywood is a global industry that just happens to be in America. Rather than exporting Americana, it serves up pap to appeal to a global audience.

Hollywood's dominance is in part due to economics: Movies cost a lot to make and so need a big audience to be profitable; Hollywood has used America's huge and relatively uniform domestic market as a platform to expand overseas. So there could be a case for stuffing subsidies into a rival European film industry, just as Airbus was created to challenge Boeing's near-monopoly. But France's sub-sidies have created a vicious circle whereby European film producers fail in global markets because they serve domestic demand and the wishes of politicians and cinematic bureaucrats.

Another American export is also conquering the globe: English. By 2050, it is reckoned, half the world will be more or less proficient in it. A common glo-bal language would certainly be a big plus—for businessmen, scientists, and tourists—but a single one seems far less desirable. Language is often at the heart of national culture, yet English may usurp other languages not because it is what people prefer to speak, but because, like Microsoft software, there are compelling advantages to using it if everyone else does.

But although many languages are becoming extinct, English is rarely to blame. People are learning English as well as—not instead of—their native tongue, and often many more languages besides. Where local languages are dying, it is typically national rivals that are stamping them out. So although, within the United States, English is displacing American Indian tongues, it is not doing away with Swahili or Norwegian.

Even though American consumer culture is widespread, its significance is often exaggerated. You can choose to drink Coke and eat at McDonald's without becoming American in any meaningful sense. One newspaper photo of Taliban fighters in Afghanistan showed them toting Kalashnikovs—as well as a sports bag with Nike's trademark swoosh. People's culture—in the sense of their shared ideas, beliefs, knowledge, inherited traditions, and art—may scarcely be eroded by mere commercial artifacts that, despite all the furious branding, embody at best flimsy values.

The really profound cultural changes have little to do with Coca-Cola. Western ideas about liberalism and science are taking root almost everywhere, while Europe and North America are becoming multicultural societies through immigration, mainly from developing countries. Technology is reshaping culture: Just think of the Internet. Individual choice is fragmenting the imposed uniformity of national cultures. New hybrid cultures are emerging, and regional ones re-emerging. National identity is not disappearing, but the bonds of nationality are loosening.

Cross-border cultural exchange increases diversity within societies—but at the expense of making them more alike. People everywhere have more choice, but they often choose similar things. That worries cultural pessimists, even though the right to choose to be the same is an essential part of freedom.

Cross-cultural exchange can spread greater diversity as well as greater similarity: more gourmet restaurants as well as more McDonald's outlets. And just as a big city can support a wider spread of restaurants than a small town, so a global market for cultural products allows a wider range of artists to thrive. If all the new customers are ignorant, a wider market may drive down the quality of cultural products: Think of tourist souvenirs. But as long as some customers are well informed (or have "good taste"), a general "dumbing down" is unlikely. Hobbyists, fans, artistic pride, and professional critics also help maintain (and raise) standards.

A bigger worry is that greater individual freedom may undermine national identity. The French fret that by individually choosing to watch Hollywood films they might unwittingly lose their collective Frenchness. Yet such fears are overdone. Natural cultures are much stronger than people seem to think. They can embrace some foreign influences and resist others. Foreign influences can rapidly become domesticated, changing national culture, but not destroying it. Clearly, though, there is a limit to how many foreign influences a culture can absorb before being swamped. Traditional cultures in the developing world that have until now evolved (or failed to evolve) in isolation may be particularly vulnerable.

In *The Silent Takeover*, Noreena Hertz describes the supposed spiritual Eden that was the isolated kingdom of Bhutan in the Himalayas as being defiled by such awful imports as basketball and Spice Girls T-shirts. But is that such a bad thing? It is odd, to put it mildly, that many on the left support multiculturalism in the West but advocate cultural purity in the developing world—an attitude they would tar as fascist if proposed for the United States. Hertz appears to want people outside the industrialized West preserved in unchanging but supposedly pure poverty. Yet the Westerners who want this supposed paradise preserved in aspic rarely feel like settling there. Nor do most people in developing countries want to lead an "authentic" unspoiled life of isolated poverty.

In truth, cultural pessimists are typically not attached to diversity per se but to designated manifestations of diversity, determined by their preferences. Cultural pessimists want to freeze things as they were. But if diversity at any point in time is desirable, why isn't diversity across time? Certainly, it is often a shame if ancient cultural traditions are lost. We should do our best to preserve them and keep them alive where possible. Foreigners can often help, by providing the new customers and technologies that have enabled reggae music, Haitian art, and Persian carpet making, for instance, to thrive and reach new markets. But people cannot be made to live in a museum. We in the West are forever casting off old customs when we feel they are no longer relevant. Nobody argues that Americans should ban nightclubs to force people back to line dancing. People in poor countries have a right to change, too.

Moreover, some losses of diversity are a good thing. Who laments that the world is now almost universally rid of slavery? More generally, Western ideas are reshaping the way people everywhere view themselves and the world. Like nationalism and socialism before it, liberalism is a European philosophy that has swept the world. Even people who resist liberal ideas, in the name of religion (Islamic and Christian fundamentalists), group identity (communitarians), authoritarianism (advocates of "Asian values") or tradition (cultural conservatives), now define themselves partly by their opposition to them.

Faith in science and technology is even more widespread. Even those who hate the West make use of its technologies. Osama bin Laden plots terrorism on a cellphone and crashes planes into skyscrapers. Antiglobalization protesters organize by e-mail and over the Internet. China no longer turns its nose up at Western technology: It tries to beat the West at its own game.

Yet globalization is not a one-way street. Although Europe's former colonial powers have left their stamp on much of the world, the recent flow of migration has been in the opposite direction. There are Algerian suburbs in Paris, but not French ones in Algiers. Whereas Muslims are a growing minority in Europe, Christians are a disappearing one in the Middle East.

Foreigners are changing America even as they adopt its ways. A million or so immigrants arrive each year, most of them Latino or Asian. Since 1990, the number of foreign-born American residents has risen by 6 million to just over 25 million, the biggest immigration wave since the turn of the 20th century. English may be all-conquering outside America, but in some parts of the United States, it is now second to Spanish.

The upshot is that national cultures are fragmenting into a kaleidoscope of different ones. New hybrid cultures are emerging. In "Amexica" people speak Spanglish. Regional cultures are reviving. The Scots and Welsh break with British monoculture. Estonia is reborn from the Soviet Union. Voices that were silent dare to speak again.

Individuals are forming new communities, linked by shared interests and passions, that cut across national borders. Friendships with foreigners met on holiday. Scientists sharing ideas over the Internet. Environmentalists campaigning together using e-mail. Greater individualism does not spell the end of community. The new communities are simply chosen rather than coerced, unlike the older ones that communitarians hark back to.

So is national identity dead? Hardly. People who speak the same language, were born and live near each other, face similar problems, have a common experience, and vote in the same elections still have plenty in common. For all our awareness of the world as a single place, we are not citizens of the world but citizens of a state. But if people now wear the bonds of nationality more loosely, is that such a bad thing? People may lament the passing of old ways. Indeed, many of the worries about globalization echo age-old fears about decline, a lost golden age, and so on. But by and large, people choose the new ways because they are more relevant to their current needs and offer new opportunities.

The truth is that we increasingly define ourselves rather than let others define us. Being British or American does not define who you are: It is part of who you are. You can like foreign things and still have strong bonds to your fellow citizens. As Mario Vargas Llosa, the Peruvian author, has written: "Seeking to impose a cultural identity on a people is equivalent to locking them in a prison and denying them the most precious of liberties—that of choosing what, how, and who they want to be."

POSTSCRIPT

Is the World a Victim of American Cultural Imperialism?

Globalization is a process of technological change and economic expansion under largely capitalist principles. The key actors driving the globalization process are multinational corporations like McDonald's, Coca-Cola, Nike, and Exxon Mobil. These companies are rooted in the American-Western cultural experience, and their premise is based on a materialistic world culture that is striving for greater and greater wealth. That value system is Western and American in origin and evolution. It is therefore logical to assume that as globalization goes, so goes American culture.

Evidence of "American" culture can be seen across the planet: kids in Djakarta or Lagos wearing Michael Jordan jerseys and Nike shoes, for example, and millions of young men and women from Cairo to Lima listening to Michael Jackson records. Symbols of American culture abound in almost every corner of the world, and most of that is associated with economics and the presence of multinational corporations.

As the youth of the world are seduced into an American cultural form and way of life, other cultures are often eclipsed. They lose traction and fade with generational change. Many would argue that this loss is unfortunate, but others would counter that it is part of the historical sweep of life. Social historians suggest that the cultures of Rome, Carthage, Phoenicia, and the Aztecs, while still influential, were eclipsed by a variety of forces that were dominant and historically rooted. While tragic, it was inevitable in the eyes of some social historians.

Regardless of whether this eclipse is positive or negative, the issue of cultural imperialism remains. Larger and more intrusive networks of communication, trade, and economic exchange bring values. In this world of value collision comes choices and change. Unfortunately, millions will find themselves drawn toward a lifestyle of materialism that carries with it a host of value choices. The losers in this clash are local cultures and traditions that, as so often is the case among the young, are easily jettisoned and discarded. It remains to be seen whether or not they will survive the onslaught.

Works on this subject include Benjamin Barber, "Democracy at Risk: American Culture in a Global Culture," *World Policy Journal* (Summer 1998); "Globalism's Discontents," *The American Prospect* (January 1, 2002); Seymour Martin Lipset, *American Exceptionalism: A Double-Edged Sword* (W. W. Norton, 1996); and Richard Barnet and John Cavanagh, *Global Dreams: Imperial Corporations and the New World Order* (Simon & Schuster, 1995).

ISSUE 14

Do International Financial Institutions and Multinational Corporations Exploit the Developing World?

YES: Joseph E. Stiglitz, from "Globalism's Discontents," *The American Prospect* (Winter 2002)

NO: Jagdish Bhagwati, from "Do Multinational Corporations Hurt Poor Countries?" *The American Enterprise* (June 2004)

ISSUE SUMMARY

YES: Nobel Prize winner Joseph Stiglitz argues that when international financial institutions (IFIs) such as the International Monetary Fund (IMF) and multinational corporations (MNCs) dictate terms of economic exchange for the rest of the world, exploitation occurs. He believes that states must negotiate tough terms that protect their national interests if they are to avoid such exploitation.

NO: Economist Jagdish Bhagwati contends that the evidence supports the contention that MNCs do more good than harm in the Third World and that critics are misguided and overly emotional in their condemnations.

The global economy is one in which capitalist principles of free trade, credit accessibility, free market, and comparative advantage reign supreme. States with alternative models, whether they are communist or socialist, still must operate within this sea of capitalist exchange.

Aside from states, the dominant actors in this world are IFIs and MNCs. These actors possess the bulk of the world's liquidity (i.e., cash, technology, transportations networks, and access to markets). Working with and among states, these entities are able to dominate trade routes, technology, access to markets, production capability, and access to funds necessary to jump-start trade, investment, and growth. Third World states must deal with these actors if they hope to produce market and sell products on the world stage.

This economic reality places Third World states in a precarious position. They must deal with actors who possess a clear bias for capitalist development and who also have vested economic interests in a variety of activities, including loaning funds, investing money, and removing barriers to trade and local economic policies. Is this the recipe for an exploitive relationship to one of mutual benefit?

The historical evidence suggests both situations have occurred in earnest. Third World countries have for centuries been victims of economic exploitation of natural resources, cheap labor, and location. The case of most states in Africa, Latin America, the Middle East, and Asia support this contention. However, other states that have dealt with these actors and built successful and diversified economies, such as South Korea and Malaysia, point to a different conclusion.

The following articles debate this issue and reach very different conclusions. Nobel Prize–winning economist Joseph E. Stiglitz argues that IFIs and by extension MNCs naturally attempt to exploit Third World states by negotiating deals of maximum return for themselves. They are less concerned with development in the Third World and more with immediate and substantial return for their stock holders. As such, Third World states must negotiate deals that are advantageous to them and thus manage globalization with a vigilant eye toward IFI exploitation. Economist Jagdish Bhagwati uses data analysis to argue that these businesses and lenders have done more positive things to improve the lot of Third World people, and myths regarding MNC/IFI exploitation must be debunked because they do not stand up to empirical scrutiny.

YES ↵

Globalism's Discontents

$\mathbf{F}$ew subjects have polarized people throughout the world as much as globalization. Some see it as the way of the future, bringing unprecedented prosperity to everyone, everywhere. Others, symbolized by the Seattle protestors of December 1999, fault globalization as the source of untold problems, from the destruction of native cultures to increasing poverty and immiseration. In this article, I want to sort out the different meanings of globalization. In many countries, globalization has brought huge benefits to a few with few benefits to the many. But in the case of a few countries, it has brought enormous benefit to the many. Why have there been these huge differences in experiences? The answer is that globalization has meant different things in different places.

The countries that have managed globalization on their own, such as those in East Asia, have, by and large, ensured that they reaped huge benefits and that those benefits were equitably shared; they were able substantially to control the terms on which they engaged with the global economy. By contrast, the countries that have, by and large, had globalization managed for them by the International Monetary Fund and other international economic institutions have not done so well. The problem is thus not with globalization but with how it has been managed.

The international financial institutions have pushed a particular ideology—market fundamentalism—that is both bad economics and bad politics; it is based on premises concerning how markets work that do not hold even for developed countries, much less for developing countries. The IMF has pushed these economics policies without a broader vision of society or the role of economics within society. And it has pushed these policies in ways that have undermined emerging democracies.

More generally, globalization itself has been governed in ways that are undemocratic and have been disadvantageous to developing countries, especially the poor within those countries. The Seattle protestors pointed to the absence of democracy and of transparency, the governance of the international economic institutions by and for special corporate and financial interests, and the absence of countervailing democratic checks to ensure that these informal and *public* institutions serve a general interest. In these complaints, there is more than a grain of truth.

Beneficial Globalization

Of the countries of the world, those in East Asia have grown the fastest and done most to reduce poverty. And they have done so, emphatically, via "globalization." Their growth has been based on exports—by taking advantage of the global market for exports and by closing the technology gap. It was not just gaps in capital and other resources that separated the developed from the less-developed countries, but differences in knowledge. East Asian countries took advantage of the "globalization of knowledge" to reduce these disparities. But while some of the countries in the region grew by opening themselves up to multinational companies, others, such as Korea and Taiwan, grew by creating their own enterprises. Here is the key distinction: Each of the most successful globalizing countries determined its own pace of change; each made sure as it grew that the benefits were shared equitably; each rejected the basic tenets of the "Washington Consensus," which argued for a minimalist role for government and rapid privatization and liberalization.

In East Asia, government took an active role in managing the economy. The steel industry that the Korean government created was among the most efficient in the world—performing far better than its private-sector rivals in the United States (which, though private, are constantly turning to the government for protection and for subsidies). Financial markets were highly regulated. My research shows that those regulations promoted growth. It was only when these countries stripped away the regulations, under pressure from the U.S. Treasury and the IMF, that they encountered problems.

During the 1960s, 1970s, and 1980s, the East Asian economies not only grew rapidly but were remarkably stable. Two of the countries most touched by the 1997–1998 economic crisis had had in the preceding three decades not a single year of negative growth; two had only one year—a better performance than the United States or the other wealthy nations that make up the Organization for Economic Cooperation and Development (OECD). The single most important factor leading to the troubles that several of the East Asian countries encountered in the late 1990s—the East Asian crisis—was the rapid liberalization of financial and capital markets. In short, the countries of East Asia benefited from globalization because they made globalization work for them; it was when they succumbed to the pressures from the outside that they ran into problems that were beyond their own capacity to manage well.

Globalization can yield immense benefits. Elsewhere in the developing world, globalization of knowledge has brought improved health, with life spans increasing at a rapid pace. How can one put a price on these benefits of globalization? Globalization has brought still other benefits: Today there is the beginning of a globalized civil society that has begun to succeed with such reforms as the Mine Ban Treaty and debt forgiveness for the poorest highly indebted countries (the Jubilee movement). The globalization protest movement itself would not have been possible without globalization.

The Darker Side of Globalization

How then could a trend with the power to have so many benefits have produced such opposition? Simply because it has not only failed to live up to its potential but frequently has had very adverse effects. But this forces us to ask, why has it had such adverse effects? The answer can be seen by looking at each of the economic elements of globalization as pursued by the international financial institutions and especially by the IMF.

The most adverse effects have arisen from the liberalization of financial and capital markets—which has posed risks to developing countries without commensurate rewards. The liberalization has left them prey to hot money pouring into the country, an influx that has fueled speculative real-estate booms; just as suddenly, as investor sentiment changes, the money is pulled out, leaving in its wake economic devastation. Early on, the IMF said that these countries were being rightly punished for pursuing bad economic policies. But as the crisis spread from country to country, even those that the IMF had given high marks found themselves ravaged.

The IMF often speaks about the importance of the discipline provided by capital markets. In doing so, it exhibits a certain paternalism, a new form of the old colonial mentality: "We in the establishment, we in the North who run our capital markets, know best. Do what we tell you to do, and you will prosper." The arrogance is offensive, but the objection is more than just to style. The position is highly undemocratic: There is an implied assumption that democracy by itself does not provide sufficient discipline. But if one is to have an external disciplinarian, one should choose a good disciplinarian who knows what is good for growth, who shares one's values. One doesn't want an arbitrary and capricious taskmaster who one moment praises you for your virtues and the next screams at you for being rotten to the core. But capital markets are just such a fickle taskmaster; even ardent advocates talk about their bouts of irrational exuberance followed by equally irrational pessimism.

Lessons of Crisis

Nowhere was the fickleness more evident than in the last global financial crisis. Historically, most of the disturbances in capital flows into and out of a country are not the result of factors inside the country. Major disturbances arise, rather, from influences outside the country. When Argentina suddenly faced high interest rates in 1998, it wasn't because of what Argentina did but because of what happened in Russia. Argentina cannot be blamed for Russia's crisis.

Small developing countries find it virtually impossible to withstand this volatility. I have described capital-market liberalization with a simple metaphor: Small countries are like small boats. Liberalizing capital markets is like setting them loose on a rough sea. Even if the boats are well captained, even if the boats are sound, they are likely to be hit broadside by a big wave and capsize. But the IMF pushed for the boats to set forth into the roughest parts of the sea before they were seaworthy, with untrained captains and crews, and without life vests. No wonder matters turned out so badly!

To see why it is important to choose a disciplinarian who shares one's values, consider a world in which there were free mobility of skilled labor. Skilled labor would then provide discipline. Today, a country that does not treat capital well will find capital quickly withdrawing; in a world of free labor mobility, if a country did not treat skilled labor well, it too would withdraw. Workers would worry about the quality of their children's education and their family's health care, the quality of their environment and of their own wages and working conditions. They would say to the government: If you fail to provide these essentials, we will move elsewhere. That is a far cry from the kind of discipline that free-flowing capital provides.

The liberalization of capital markets has not brought growth: How can one build factories or create jobs with money that can come in and out of a country overnight? And it gets worse: Prudential behavior requires countries to set aside reserves equal to the amount of short-term lending; so if a firm in a poor country borrows $100 million at, say, 20 percent interest rates short-term from a bank in the United States, the government must set aside a corresponding amount. The reserves are typically held in U.S. Treasury bills—a safe, liquid asset. In effect, the country is borrowing $100 million from the United States and lending $100 million to the United States. But when it borrows, it pays a high interest rate, 20 percent; when it lends, it receives a low interest rate, around 4 percent. This may be great for the United States, but it can hardly help the growth of the poor country. There is also a high *opportunity* cost of the reserves; the money could have been much better spent on building rural roads or constructing schools or health clinics. But instead, the country is, in effect, forced to lend money to the United States. . . .

The Costs of Volatility

Capital-market liberalization is inevitably accompanied by huge volatility, and this volatility impedes growth and increases poverty. It increases the risks of investing in the country, and thus investors demand a risk premium in the form of higher-than-normal profits. Not only is growth not enhanced but poverty is increased through several channels. The high volatility increases the likelihood of recessions—and the poor always bear the brunt of such downturns. Even in developed countries, safety nets are weak or nonexistent among the self-employed and in the rural sector. But these are the dominant sectors in developing countries. Without adequate safety nets, the recessions that follow from capital-market liberalization lead to impoverishment. In the name of imposing budget discipline and reassuring investors, the IMF invariably demands expenditure reductions, which almost inevitably result in cuts in outlays for safety nets that are already threadbare.

But matters are even worse—for under the doctrines of the discipline of the capital markets," if countries try to tax capital, capital flees. Thus, the IMF doctrines inevitably lead to an increase in tax burdens on the poor and the middle classes. Thus, while IMF bailouts enable the rich to take their money out of the country at more favorable terms (at the overvalued exchange rates), the burden of repaying the loans lies with the workers who remain behind.

The reason that I emphasize capital-market liberalization is that the case against it—and against the IMF's stance in pushing it—is so compelling. It illustrates what can go wrong with globalization. Even economists like Jagdish Bhagwati, strong advocates of free trade, see the folly in liberalizing capital markets. Belatedly, so too has the IMF—at least in its official rhetoric, though less so in its policy stances—but too late for all those countries that have suffered so much from following the IMF's prescriptions.

But while the case for trade liberalization—when properly done—is quite compelling, the way it has been pushed by the IMF has been far more problematic. The basic logic is simple: Trade liberalization is supposed to result in resources moving from inefficient protected sectors to more efficient export sectors. The problem is not only that job destruction comes before the job creation—so that unemployment and poverty result—but that the IMF's "structural adjustment programs" (designed in ways that allegedly would reassure global investors) make job creation almost impossible. For these programs are often accompanied by high interest rates that are often justified by a single-minded focus on inflation. Sometimes that concern is deserved; often, though, it is carried to an extreme. In the United States, we worry that small increases in the interest rate will discourage investment. The IMF has pushed for far higher interest rates in countries with a far less hospitable investment environment. The high interest rates mean that new jobs and enterprises are not created. What happens is that trade liberalization, rather than moving workers from low-productivity jobs to high-productivity ones, moves them from low-productivity jobs to unemployment. Rather than enhanced growth, the effect is increased poverty. To make matters even worse, the unfair trade-liberalization agenda forces poor countries to compete with highly subsidized American and European agriculture. . . .

Governance Through Ideology

Consider the contrast between how economic decisions are made inside the United States and how they are made in the international economic institutions. In this country, economic decisions within the administration are undertaken largely by the National Economic Council, which includes the secretary of labor, the secretary of commerce, the chairman of the Council of Economic Advisers, the treasury secretary, the assistant attorney general for antitrust, and the U.S. trade representative. The Treasury is only one vote and often gets voted down. All of these officials, of course, are part of an administration that must face Congress and the democratic electorate. But in the international arena, only the voices of the financial community are heard. The IMF reports to the ministers of finance and the governors of the central banks, and one of the important items on its agenda is to make these central banks more independent—and less democratically accountable. It might make little difference if the IMF dealt only with matters of concern to the financial community, such as the clearance of checks; but in fact, its policies affect every aspect of life. It forces countries to have tight monetary and fiscal policies: It evaluates the trade-off between inflation and unemployment, and in that trade-off it always puts far more weight on inflation than on jobs.

The problem with having the rules of the game dictated by the IMF—and thus by the financial community—is not just a question of values (though that is important) but also a question of ideology. The financial community's view of the world predominates—even when there is little evidence in its support.

Indeed, beliefs on key issues are held so strongly that theoretical and empirical support of the positions is viewed as hardly necessary.

Recall again the IMF's position on liberalizing capital markets. As noted, the IMF pushed a set of policies that exposed countries to serious risk. One might have thought, given the evidence of the costs, that the IMF could offer plenty of evidence that the policies also did some good. In fact, there was no such evidence; the evidence that was available suggested that there was little if any positive effect on growth. Ideology enabled IMF officials not only to ignore the absence of benefits but also to overlook the evidence of the huge costs imposed on countries. . . .

Globalization and September 11

September 11 brought home a still darker side of globalization—it provided a global arena for terrorists. But the ensuing events and discussions highlighted broader aspects of the globalization debate. It made clear how untenable American unilateralist positions were. President Bush, who had unilaterally rejected the international agreement to address one of the long-term global risks perceived by countries around the world—global warming, in which the United States is the largest culprit—called for a global alliance against terrorism. The administration realized that success would require concerted action by all.

One of the ways to fight terrorists, Washington soon discovered, was to cut off their sources of funding. Ever since the East Asian crisis, global attention had focused on the secretive offshore banking centers. Discussions following that crisis focused on the importance of good information—transparency, or openness—but this was intended for the developing countries. As international discussions turned to the lack of transparency shown by IMF and the offshore banking centers, the U.S. Treasury changed its tune. It is not because these secretive banking havens provide better services than those provided by banks in New York or London that billions have been put there; the secrecy serves a variety of nefarious purposes—including avoiding taxation and money laundering. These institutions could be shut down overnight—or forced to comply with international norms—if the United States and the other leading countries wanted. They continue to exist because they serve the interests of the financial community and the wealthy. Their continuing existence is no accident. Indeed, the OECD drafted an agreement to limit their scope—and before September 11, the Bush administration unilaterally walked away from this agreement too. How foolish this looks now in retrospect! Had it been embraced, we would have been further along the road to controlling the flow of money into the hands of the terrorists.

There is one more aspect to the aftermath of September 11 worth noting here. The United States was already in recession, but the attack made matters worse. It used to be said that when the United States sneezed, Mexico caught a cold. With globalization, when the United States sneezes, much of the rest of the world risks catching pneumonia. And the United States now has a bad case of the flu. With globalization, mismanaged macroeconomic policy in the United States—the failure to design an effective stimulus package—has global consequences. But around the world, anger at the traditional IMF policies is growing. The developing countries are saying to the industrialized nations: "When you face a slowdown, you follow the precepts that we are all taught in our economic courses: You adopt expansionary monetary and fiscal policies. But when we face a slowdown, you insist on contractionary policies. For you, deficits are okay; for us, they are impermissible—even if we can raise the funds through 'selling for ward,' say, some natural resources." A heightened sense of inequity prevails, partly because the consequences of maintaining contraction-ary policies are so great.

Global Social Justice

Today, in much of the developing world, globalization is being questioned. For instance, in Latin America, after a short burst of growth in the early 1990s, stagna-tion and recession have set in. The growth was not sustained—some might say, was not sustainable. Indeed, at this juncture, the growth record of the so-called post-reform era looks no better, and in some countries much worse, than in the widely criticized import-substitution period of the 1950s and 1960s when Latin countries tried to industrialize by discouraging imports. Indeed, reform critics point out that the burst of growth in the early 1990s was little more than a "catch-up" that did not even make up for the lost decade of the 1980s.

Throughout the region, people are asking: "Has reform failed or has global-ization failed?" The distinction is perhaps artificial, for globalization was at the center of the reforms. Even in those countries that have managed to grow, such as Mexico, the benefits have accrued largely to the upper 30 percent and have been even more concentrated in the top 10 percent. Those at the bottom have gained little; many are even worse off. The reforms have exposed countries to greater risk, and the risks have been borne disproportionately by those least able to cope with them. Just as in many countries where the pacing and sequencing of reforms has resulted in job destruction outmatching job creation, so too has the exposure to risk outmatched the ability to create institutions for coping with risk, including effective safety nets.

In this bleak landscape, there are some positive signs. Those in the North have become more aware of the inequities of the global economic architecture. The agreement at Doha to hold a new round of trade negotiations—the "Devel-opment Round"—promises to rectify some of the imbalances of the past. There has been a marked change in the rhetoric of the international economic institutions—at least they talk about poverty. At the World Bank, there have been some real reforms; there has been some progress in translating the rhetoric into reality—in ensuring that the voices of the poor are heard and the concerns of the developing countries are listened to. But elsewhere, there is often a gap between the rhetoric and the reality. Serious reforms in governance, in who makes decisions and how they are made, are not on the table. If one of the problems at the IMF has been that the ideology, interests, and perspectives of the financial community in the advanced industrialized countries have been given disproportionate weight (in matters whose effects go well beyond finance), then the prospects for success in the current discussions of reform, in which the same parties continue to predominate, are bleak. They are more likely to result in slight changes in the shape of the table, not changes in who is *at* the table or what is on the agenda.

September 11 has resulted in a global alliance against terrorism. What we now need is not just an alliance *against* evil, but an alliance *for* something positive—a global alliance for reducing poverty and for creating a better environment, an alliance for creating a global society with more social justice.

NO ⬅

Do Multinational Corporations Hurt Poor Countries?

There is a fierce debate today between those who consider globalization to be a malign influence on poor nations and those who find it a positive force. This debate focuses not just on trade, but also on multinational corporations. The hard evidence strongly suggests that the positive view is more realistic. There are many reasons to believe that multinationals in particular do good, not harm, in the developing world.

If any conviction strongly unites the critics of multinationals today, it is that they exploit workers in poor countries. Ire has been aroused by the assumption that rich, deep-pocketed corporations pay "unfair" or "inadequate" wages overseas. More generally, companies are condemned for violating "labor rights."

The typical critique asserts that if a Liz Claiborne jacket sells for $190 in New York, while the female worker abroad who sews it gets only 60 cents an hour, that is obviously exploitation. But there is no necessary relationship between the price of a specific product and the wage paid by a company. For starters, for every jacket that sells, there may be nine that do not. So the effective price of a jacket one must consider is a tenth of the sold jacket: $19, not $190. And distribution costs and tariff duties on apparel almost double the price of a jacket between the time it arrives at the dock or airport in New York and finds its way to a Lord & Taylor display.

It is often assumed that multinationals earn huge monopoly profits while paying their workers minimal wages, and that these firms should therefore share their "excess" profit with their workers. But nearly all multinationals such as Liz Claiborne and Nike operate in fiercely competitive environments. A recent study of the profits performance of 214 companies in the 1999 Fortune Global 500 list showed a rather sorry achievement—about 8.3 percent profit on foreign assets. Where are the huge spoils to be shared with workers?

Let's look at the facts on wage payments. Good empirical studies have been conducted in Bangladesh, Mexico, Shanghai, Indonesia, Vietnam, and elsewhere. And these studies find that multinationals actually pay what economists call a "wage premium," that is, an average wage that *exceeds* the going rate

From *The American Enterprise* by Jagdish Bhagwati, vol. 15, no. 4, June 2004, pp. 28–30. Copyright © 2004 by American Enterprise Institute. Reprinted by permission.

in the area where they are located. Affiliates of some U.S. multinationals pay a premium over local wages that ranges from 40 to 100 percent.

In one careful and convincing study, the economist Paul Glewwe, using Vietnamese household data for 1997–98, was able to isolate the incomes of workers employed in foreign-owned firms, joint ventures, and Vietnamese-owned enterprises. About half the Vietnamese workers in the study worked in the foreign textile or leather firms that are so often criticized. Contrary to the steady refrain from the critics, Glewwe found that workers in foreign-owned enterprises generally make almost twice the salary of the average worker employed at a Vietnamese company.

As Glewwe points out:

> The data also show that people who obtained employment in foreign-owned enterprises and joint ventures in Vietnam in the 1990s experienced increases in household income (as measured by per capita consumption expenditures) that exceeded the average increases for all Vietnamese households. This appears to contradict the claims that foreign-owned enterprises in poor countries such as Vietnam are "sweatshops." On the other hand, it is clear that the wages paid by these enterprises . . . are a fraction of wages paid in the U.S. and other wealthy countries. Yet Vietnam is so poor that it is better for a Vietnamese person to obtain this kind of employment than almost any other kind available in Vietnam.

But there remains the accusation that global corporations violate labor rights. Many damning charges are made, and anti-globalization activists are not beyond trumpeting the occasional lie, much like the corporations, politicians, and bureaucrats they excoriate. Only after IKEA was accused of exploitative child labor by its suppliers was it discovered that the German film documenting the abuse was simply faked by activists.

Another recent example, much-repeated by critics of multinationals and picked up by many sympathetic reporters, was the claim that the chocolate sold in rich countries relies on slave labor by children in the cocoa plantations of the Ivory Coast. Let me quote from Norimitsu Onishi's story in the *New York Times* uncovering the falsity of these charges.

> Many accounts in British and American news media last year spoke breathlessly of 15,000 child slaves . . . producing the chocolate you eat. The number first appeared in Malian newspapers, citing the UNICEF office in Mali. But UNICEF's Mali office had never researched the issue of forced child laborers in Ivory Coast. . . . Still, repeated often enough, the number was gladly accepted by some private organizations, globalization opponents seeking a fight with Nestle and Hershey, and some journalists. . . . This month, the results of the first extensive survey of child labor in cocoa plantations in Ivory Coast and three other African nations were released by the International Institute of Tropical Agriculture, a nonprofit organization. . . . The survey found that almost all children working in cocoa fields were children of the plantation owners, not forced laborers. . . . None reported being forced against their will.

Unfortunately, as Onishi observes, "politics is sometimes more influential than precision. . . . Since they were released early this month, the institute's findings have received little attention—perhaps only 1 percent of what the '15,000 slaves' figure received."

Sometimes when critics of multinationals attack, it is not egregious violations of local laws that are at issue. It is rather the claim that the companies do not meet the demands of "decency," or Western norms, or perhaps international law. This route to condemning multinationals is quite problematic, however.

For one thing, developing-world regulations may be less demanding than international ones (just as American standards are often below those of Europe and even Canada) for good reasons. Take the case of working hours, which can be quite long in some poor countries. As Nicholas Kristof and Sheryl WuDunn have pointed out in the *New York Times Magazine*, the young Third World workers who toil long hours at multinational factories generally do so voluntarily. Why? Because they want to make money as quickly as possible so they can return to their rural homes. Like many of us who work long hours, they are not being exploited, they drive themselves.

Kristof and WuDunn quote workers in a leather-stitching factory in the Chinese boomtown of Dongguan, who tell them they all regard it as a plus that the factory allows them to work long hours every day. Indeed, some had sought out this factory precisely because it offered them the chance to earn more. "It's actually pretty annoying how hard they want to work," said the factory manager, a Hong Kong man. "It means we have to worry about security and have a supervisor around almost constantly."

Not only are multinationals wrongly accused of exploitation in the developing countries, but economists have also noted a number of good effects they bring in their wake. Perhaps the chief good effect is what economists call spillover. This refers to the fact that domestic firms learn productivity-enhancing techniques from foreign corporations with better technology and management practices. Production workers often learn better techniques while employed by foreign firms. Managers may learn about better practices by observing, or by having previously worked at multinationals themselves. And increased competition pushes all companies in an area where multinationals are operating to become more productive.

In the movie *Manhattan*, Woody Allen's character talks about the hotel where the food is dreadful, and there was not enough of it, either! The critics of multinationals often make similar complaints. After arguing that multinationals must be condemned for exploiting workers and harming host countries wherever they go, critics sometimes inveigh against these same corporations for bypassing countries that need them, thereby widening the gap between the rich and the poor.

If multinationals avoid some poor countries, that is surely not surprising. They are businesses that must survive by making a profit—no corporation ever managed to do sustained good by continually posting losses. If a country wants to attract investment, it has to provide an attractive environment. That generally implies having political stability and economic advantages such as economical labor or useable natural resources. In the game of attracting investment, some countries are going to lose because they lack these attributes. The truly unfortunate countries are those experiencing acute problems of governance, as in the African countries ravaged by war.

It is unrealistic to expect multinationals to invest in these countries and "save" them. Instead, the international community has to help them put their paralyzing conflicts and inadequate governance behind them over the long haul—a truly heroic task. In the meantime, the answer to such nations' pressing humanitarian and developmental needs must be public aid, technical assistance, and altruism from corporations and civil society groups. The World Bank ought to concentrate more on these problem states and should correspondingly turn away from lending to countries such as India and China, which now have the ability to develop by themselves. But, of course, the World Bank leadership seeks to maximize influence by distributing largesse to all; even altruistic institutions will occasionally be run by men whose private ambitions, rather than the social good, are the primary determinants of their policies.

POSTSCRIPT

Do International Financial Institutions and Multinational Corporations Exploit the Developing World?

Determining whether IFIs and MNCs are exploitive often centers on the impact of their actions combined with the degree to which Third World states are complicit or resist such unequal terms of exchange. Evidence regarding such exploitation was more pronounced during the colonial period when states literally dominated and extracted resources and cheap labor from the third world. Today, in the current globalizing economic system, the evidence is more mixed.

From recent history, we can make certain determinations that may help frame this debate for future discussion. First, MNCs are motivated by profit and not altruence. Thus, if unchecked, they may and often do negotiate terms beneficial to themselves and not necessarily beneficial to Third World states and people. Second, IFIs possess a particular economic philosophy that works in some states and regions and apparently does not in others. As a result, some people are adversely effected by structural adjustment reforms and policies enforced by IFIs. Third, Third World governments have an obligation to protect the interests of their people in such dealings, and the extent that exploitation does occur may be a result of their diligence in such exchanges. Fourth, the nature of whether these entities are exploitive will be determined by the long-term impact of trade and investment. Whatever your position, the final arbiters of this question will be the millions of people in the Third World whose poverty will either improve or be exacerbated by the current globalizing capitalist system and its main actors, IFIs and MNCs.

Some key literature on this subject includes a debate between Thomas Friedman and Robert Kaplan, "States of Discord, *Foreign Policy* (March/April 2002), Thomas Friedman, *The Lexus and the Olive Tree* (Farrar, Straus and Giroux, 1999), David C. Korten, *When Corporations Rule the World* (1997), Donald Marsh, "Free Trade and Their Critics: The Need for Education" Washington Council on International Trade, (2000) and Caroline Thomas and Melvyn Reader, "Development and Inequality," in *Issues in World Politics*, edited by Brian White, Richard Little, and Michael Smith (2001).

ISSUE 15

Is the Transnational Media Hurting Global Society?

YES: Herbert I. Schiller, from "Transnational Media: Creating Consumers Worldwide," *Journal of International Affairs* (Summer 1993)

NO: Benjamin Compaine, from "Global Media," *Foreign Policy* (November/December 2002)

ISSUE SUMMARY

YES: Herbet Schiller argues that the globalization of media places image-building and information in the hands of private entities whose interests are not educative or informational but rather designed to create a worldwide consumer society.

NO: Benjamin Compaine contends that the image of a global media colossus is overblown and that large private media conglomerates may be a good thing for the public interest as a whole.

Satellite technology has allowed us to expand communication networks far beyond our conception even a few years ago. With this expediential growth in technology had come the expansion of multimedia conglomerates like Viacom, Disney, Gannett, Fox, and others to truly global status as purveyors of information and twenty-four-hour-a-day programming on all forms of communication. The corporatization of global media is happening as deregulation increases in countries like the United States and new smaller networks are merged with larger sources of funding to provide outlets with regional or global reach. Even the tiny island nation of Fiji recently sold its airwaves to a media conglomerate to generate funding for other more pressing needs.

What is the impact of this globalization of transnational media? Does it mean more information, knowledge, and access for more people, or does it spell a network of automatons and consumers hostage to the programming and messages that a few Western executives put forth?

If the answer is the former, then what positive signs do we see of increased knowledge and awareness for millions? Are people more tuned in to what is happening in the world, and are the information and messages that they are receiving accurate, balanced, and representative of global society and its many subparts?

If the answer is the latter, then what is the impact of a corporate media that generates consistent and tailored images for global society and reinforces them with advertising by like multinationals? Can people escape this dumbing down from these entities, or are people losing their ability to gain multiple perspectives from independent sources?

In the first section, Herbert Schiller explores the transnational media and argues that this image is generating a homogenous consumer culture modeled on a Western construct and as such their interests are not compatible with individual societies and their need to educate and inform their populations.

In the next section, Benjamin Compaine contends that this argument of transnational media control is overblown and spurious. He states that many media outlets are owned by non-U.S. firms and that the privatization of media is a good thing since it allows these companies the freedom to explore multiple forms of programming and imaging that will ultimately serve people's interests.

Herbert I. Schiller    **YES**

Transnational Media: Creating Consumers Worldwide

> By the end of the 1980s, "globalization" had become the term for accelerating interdependence.... The primary agent of globalization is the transnational corporation. The primary driving force is the revolution in information and communication technologies.[1]

Contrary to the conventional wisdom that recent advances in communications have led to the emergence of the "global village," I do not believe that globalization of the media industries sector has resulted in the formation of an international civil society as such.[2] Rather, this process has resulted in an international order organized by transnational economic interests that are largely unaccountable to the nation-states in which they operate. This transnational corporate system is the product of a rationalized and commercialized communications infrastructure, which transmits massive flows of information and has extended its marketing reach to every corner of every hemisphere. While the U.S. role in the creation and reproduction of this worldwide consumer society has lessened, the supporting institutions and the content of the information still bear a heavy American imprint.

The Hegemony of International Media Industries

The reality of American global information mastery was strikingly on display throughout the war in the Persian Gulf. During the actual hostilities, one account—that of the transnational U.S.-based Cable News Network (CNN)—dominated television screens around the world.[3] Though press interpretations of the war may have varied from country to country, the broadcast images of high technology combat were identical worldwide. However remarkable a demonstration of the American information monopoly—now challenged by an expanded British Broadcasting Corporation (BBC) World Service Television and France's newly created Euronews programming—even this barely suggests the vast capabilities of American broadcasters and U.S.-based cultural industries to define reality.

From *Journal of International Affairs* by Herbert I. Schiller, vol. 47, no. 1, Summer 1993, pp. 47–55, 57–58. Copyright © 1993 by Journal of International Affairs. Reprinted by permission.

CNN's broadcasts are but one kind of image, sound and symbol production. Such output also comes to us in the familiar forms of films, television programs, video cassettes, compact discs, books, magazines, on-line data and computer software. The transmission of this production is neatly explained by Walter Wriston, former chief executive officer of Citicorp:

> The single most powerful development in global communities has been the satellite, born a mere thirty-one years ago. . . . Satellites now bind the world for better or worse, in an electronic infrastructure that carries news, money, and data anywhere on the planet at the speed of light. Satellites have made borders utterly porous to information.[4]

Wriston properly makes no distinction between news, money and data: ". . . [H]undreds of millions of people around the world are plugged into what has become essentially a single network . . . of popular communication."[5] Those global corporations and media-cultural conglomerates that have the capability to use the global satellite systems are indifferent to formal communication boundaries; digitized electronic communication transforms all messages and images into a uniform information stream.

This globalization of communication since the 1960s can be best understood as the phenomenal growth of such transnational media-information corporations as Time Warner, Disney, Reuters, SONY, Murdoch and Bertelsman—based mostly in the developed economies—in achieving a worldwide market share.[6] While state, non-governmental and non-corporate organizations have made good use of these new electronic networks, their use is dwarfed by that of the transnational companies. The capability of the private, resource-rich conglomerates to shift capital, currency, production and data—almost at will—constitutes the true levers of contemporary power.

Edward Herman describes the integration of broadcasting into a global market in recent decades, achieved largely through "cross-border acquisition of interests in and control of program production and rights, cable and broadcasting facilities and the sale and rental of program stocks, technology and equipment."[7] This international economic expansion in broadcasting "[has] tended to increase the strength of commercial broadcasting and reduce that of public systems."[8] Herman concludes that

> the strength and momentum of the forces of the market in the last decade of the twentieth century are formidable. It therefore seems likely that the U.S. patterns of commercial hegemony over broadcasting will be gradually extended over the entire globe.[9]

Herman's predictions have been validated with astonishing rapidity and singular effect. While the American cultural product—film, television, fashions and tapes—still dominates screens, homes and shops throughout the world, local outputs are also increasing. Yet, invariably, they are fashioned on the American model and serve as the same kind of bait with which to snare the potential consumer. French television dramas, for example, repeat worn U.S. formulae; Brazil's powerful television-production industry is at the beck and call of the same transnational advertisers who dominate North American television screens.[10] The American pop cultural product has obvious hegemonic properties, which can be attributed to a century of marketing experience and the rapid utilization of state-of-the-art technologies to achieve compelling special effects.[11]

As Wriston enthusiastically makes clear, efforts by individual states to protect and insulate their societies from these stimuli have been futile. Global notions of what constitutes freedom, individual choice, a good life and a desirable future come largely from their output. Because of market imperatives, institutional infrastructures in country after country have been recast to facilitate the transmission of the American informational and cultural product.

Clearly, the media industries' unexceptional quest for profitability has had a direct—albeit immeasurable—impact on human consciousness. While the ultimate effect of their cultural packages on the human senses is impossible to assess concretely, the existence of the effect cannot be ignored. The worldwide output of America's cultural industries probably has as great an impact as any other form of American power. Already it has actively assisted in the transformation of broadcasting and telecommunications systems around the world. People everywhere are consumers of American images, sounds, ideas, products and services.

The Silencing of Public Debate

In the United States, despite a seemingly thick network of organizations and social groups comprising a rich civil society, the voice of the corporate speaker has succeeded in dominating the national discourse. Although the corporate perspective has held a privileged place in American society for generations, it was balanced in earlier times by the opposing voices of farmers' movements, organized labor and civil rights organizations on the national stage. Since the end of the Second World War, however, structural economic change and the evolution of media industries have contributed to a decline of opposing voices—such as the American labor movement—and an eclipse of a comprehensive adversarial view.[12]

Corporate control over the means of communication thus has immediate political implications: The rising price of television air time has caused the cost of political candidacy to spiral. In 1992, the *New York Times* reported that

> spending for House seats by 427 Democrats, 416 Republicans and 294 candidates not affiliated with either party totalled $313.7 million, compared with about $220 million two years ago. . . . The combined spending for House and Senate seats increased to $504 million in 1992, $113 more than in the same period two years ago.[13]

This means that in order to wage a successful campaign, a political candidate must either be independently wealthy or be able to convince those who have resources to offer support. In either case, the electoral process is transformed into a mechanism for representing the advantaged.

More broadly, corporations enjoy the protection of law. More than a century ago, the Supreme Court ruled that corporations had the constitutional rights of individual citizens.[14] With such protection, it has been exceedingly difficult to monitor and control corporate activities and behavior. In the late 1970s, the corporation was once again the beneficiary of a Supreme Court decision, which stated that corporations had First Amendment rights, and that their speech—with some limitations—was as protected as individual expression.[15]

This and related rulings codified the pre-eminent role of corporate expression in the contemporary American cultural landscape. Corporate expression literally has no serious competition. Public television, which was supposed to be a non-commercial alternative to advertiser-supported television, has been co-opted by sponsorship. Cable television, although receiving most of its revenues from subscriptions, is steadily drawing more support from advertisers. Given the overwhelming reliance of American radio and television on commercial advertising, the domestic informational system has become, in effect, a marketing and ideological apparatus of corporate influence. Robert McChesney finds that the media are the national and ultimate interpreters of reality;[16] it is a reality fashioned according to their own corporate advantage.

Media and cultural power, already awesome, is further enhanced by its capability to define and present its own role to the public. This self-constructed picture never fails to emphasize the objectivity, dedication to the public interest and fragility of the cultural industries' activities. Its hegemonic effect is evident: Corporate ascendancy, untouched by social accountability or federal oversight, has gone almost unchallenged and largely unremarked in the for a of public opinion.

As could be expected, the realm of permissible debate has narrowed appreciably in recent decades. For all the talk shows, personal witness programs and endless hours of sports spectaculars and crime dramas, the national discourse is astonishingly bland—except insofar as personal accounts of behavioral excesses are concerned—and almost totally reticent about the structural determinants of American existence. Programming that might shed some light on the country's deepening social crisis does not seem to impress the program decision makers as worthy of much attention. Only after South Central Los Angeles burned did the cameras turn—and only briefly—to the American urban condition.

And so it goes. While single-issue constituencies sometimes receive prominence and some public issues generate a modicum of excitement, consensus on the essential features of the social order prevails within the media industries. The main business of corporate America—marketing—proceeds without interruption. Fundamental institutions have been reshaped to accommodate the dominant presence of the corporation in American life, thereby offering seeming confirmation that their hegemony must be the outcome of inescapable natural forces.

The rich fabric of American history, a story of unceasing struggles against plutocratic privilege and continual efforts to achieve social dignity and equality for working people—including women and African-Americans—is rarely visible to a national audience; the little that does get noted is generally either decontextualized or fragmented. This thin and largely expurgated presentation of the national experience is the underside of the daily retailing of corporate images and messages, and the endless affirmations of commercial culture. In recent years, these highly selective accounts of society and history are no longer confined within national boundaries; they have become globalized through the massive export of American television programs and films.[17]

Corporate Strategems

There is nothing unanticipated about the increasing authority of corporate media actors. These powerful private economic conglomerates are moved by common impulses: the search for markets, cheap and non-union labor, low taxes, compliant governments and secure property rights. Corporate enterprise insists on concessions in these areas wherever it undertakes operations. In the period following the Second World War, corporations demanded "deregulation," which was essentially the removal of limitations from the unrestrained pursuit of profit. The achievement of this freedom has been a successful and indispensable achievement of the international corporate system; the American contribution to this trend has been substantial and decisive.

In particular, the transnational corporate order places the highest priority on deregulation in the broadcasting and telecommunications spheres. Telecommunications provide the means of linking and coordinating globally dispersed operations, a crucial requirement for transnational corporate operations. Broadcasting, when deregulated, enables the super-companies and their advertising agencies unrestricted access to national television screens. Utilizing this access, they can transmit in ever-increasing volume their advertising messages and general programming, the latter of which is no less a carrier of the sales message.

Consequently, there exists today a corporate-induced and -administered global environment of consumer capitalism that follows identical prescriptions and uses a uniform rhetoric.[18] This includes the espousal and protection of corporate speech and the justification of whatever programming is produced and transmitted as the proof of consumer choice and sovereignty. International efforts to combat or counter the now-pervasive condition of corporate dominance have been defeated by the counterattack of the transnational corporate order and its national surrogates.

The Decline of Opposing International Voices

In the 1960s and 1970s, a group of post-colonial Third World states made mostly rhetorical efforts to create a New World Information and Communication Order (NWICO) that challenged the Western—mostly American—domination of world news, and information and cultural flows. The NWICO proponent's views have been summed up by Zimbabwean prime minister Robert Mugabe:

> In the information and communication field, the Non-Aligned Nations and other developing countries are adversely affected by the monopoly which the developed nations hold over the world's communications systems. . . . The old order has ensured the continued dependence of our information and communication infrastructures and systems on those of the developed nations. Such dependence constitutes a serious threat to the preservation of our respective cultures and indigenous life-styles.[19]

Third World efforts on behalf of the NWICO agenda crested in 1978; the concept, however, was overwhelmingly rejected by the United States and its few developed allies. Further, the unity of NWICO advocates was shattered by a U.S. offer of limited assistance for a development program in communication technologies, calculated to win some Third World support. This was complemented by a frontal assault—concentrated in the Western mass media—on the U.N. Educational, Scientific and Cultural Organization (UNESCO), which was an important locus for NWICO advocates.[20] This campaign culminated in the U.S. withdrawal from UNESCO in 1984, and was part of the Reagan Administration's agenda to browbeat the international community into accepting U.S. global information policy.

Global corporate actors have sought to cripple other international agencies and state structures that might have served as shields against unlimited transnational corporate power. For example, in Europe there has been unrelenting pressure to eliminate or marginalize the Post, Telephone and Telecommunications entities (PTTs). These governmental bureaucracies, for all their faults, at least represented in part national public communication interests. Branded by their transnational corporate adversaries as "monopolies," however, their authority has been eroded by liberalization and privatization initiatives—advanced by the transnational corporate sector and its allies. Their capability to monitor and prescribe the behavior of the communication companies operating in their national space has been largely lost and their survival is threatened. . . .

The extreme sensitivity of the corporate order to the global information climate was further demonstrated by the successful effort to expunge the subject of transborder data flows (TDF) from the language and the agendas of international economic meetings. For a brief period in the 1970s, TDF—the term for mostly electronic data crossing national frontiers—was a subject of great debate. Yet its implications for the examination and possible oversight of the data flows of the global companies came too close to the nerve centers of the transnational business system. For this reason, the term itself was neatly shelved and subsumed under the opaque and innocuous trade-in-services category of the GATT.[21]

Having neutralized international and state opposition, the media supercompanies can carry on their worldwide operations, almost completely outside of any scrutiny; their activities are completely ignored in the general discussions of American economic public policy. International organizations like the United Nations, the ITU, UNESCO and the U.N. Centre for Transnational Corporations have either been bypassed, restructured, weakened or neutered.[22]

Conclusion

Publicly unaccountable media-cultural power today constitutes the ultimate "Catch-22" situation. The public interest demands information that is, however, dependent on private image providers whose own interests are often incompatible.[23] To begin to confront this condition is the one of the greatest challenges of the next century.

Notes

1. Sylvia Ostry, "The Domestic Domain: The New International Policy Area," *Transnational Corporations* 1, no. 1 (February 1992) p. 7.
2. For an opposing perspective, see Mike Featherstone, ed., *Global Culture* (Newbury Park, CA: Sage, 1990).

3.   Hamid Mowlana, George Gerbner and Herbert I. Schiller, *Triumph of the Image: The Media's War in the Persian Gulf: A Global Perspective* (Boulder, CO: Westview Press, 1992).

4.   Walter B. Wriston, *The Twilight of Sovereignty* (New York: Charles Scribner's Sons, 1992) p. 12.

5.   *ibid.*, p. 130.

6.   "America's Most Valuable Companies," *Business Week,* 1993 Special Bonus Issue, *passim.*

7.   Edward S. Herman, "The Externalities Effects of Commercial and Public Broadcasting," in K. Nordenstreng & H.I. Schiller, eds., *Beyond National Sovereignty: International Communications in the 1990s* (Norwood, NJ: Ablex Publishing Corp., 1993) pp. 108–9.

8.   *ibid.*, p. 108.

9.   *ibid.*

10. O.S. Oliveira, "Brazilian Soaps Outshine Hollywood: Is Cultural Imperialism Fading Out?" Paper presented at the meetings of the Deutsche Gesellschaft für Semiotik (German Society for Semiotics), Internationaler Kongress, Universität Passau, 8–10 October 1990.

11. These and other factors are described in more detail in Herbert Schiller, "La Culture Americaine au service des marchands," *Le Monde Diplomatique,* October 1992, p. 28.

12. Herbert I. Schiller, *Culture Inc., The Corporate Takeover of Public Expression* (New York: Oxford University Press, 1989).

13. "Spending on Races for U.S. House Soars to a Record \$313.7 Million," *New York Times,* 2 January 1993, p. 12.

14. Santa Clara County v. Southern Pacific Railroad, 118 U.S. 394 (1886).

15. First National Bank of Boston et al. v. Bellotti, Attorney General of Massachusetts et al., 435 U.S. 765 (1978).

16. Robert W. McChesney, "Off Limits: An Inquiry Into the Lack of Debate over the Ownership, Structure and Control of the Mass Media in U.S. Political Life," *Communication* 13 (1992) pp. 1–19.

17. Herbert I. Schiller, *Mass Communications and American Empire* (New York: A. Kelley, 1969; 2nd ed., Boulder, CO: Westview Press, 1992).

18. Leslie Sklair, *Sociology of the Global System* (Baltimore, MD: John Hopkins University Press, 1991).

19. Speech delivered at the official opening of the Second Conference of Ministers of Information of Non-Aligned Countries, Harare, Zimbabwe, 10 June 1987. A good summary of NWICO argumentation and positions can be found in "Many Voices, One World," International Commission for the Study of Communication Problems (New York: Unipub, 1980).

20. William Preston, Jr., Edward Herman and Herbert I. Schiller, *Hope and Folly: The United States and UNESCO, 1945–1985* (Minneapolis, MN: University of Minnesota Press, 1989).

21. William Drake, "Territoriality and Intangibility: Transborder Data Flow and National Sovereignty," in Nordenstreng and Schiller, pp. 259–313.

22. The U.N. Centre for Transnational Corporations was reorganized and put into the Transnational Corporations and Management Division of the United Nations Department of Economic and Social Development in March 1992.

23. C. Edwin Baker, "Advertising and a Democratic Press," *University of Pennsylvania Law Review* 140, no. 6 (June 1992) pp. 2097–243.

NO ↵

Global Media

Big media barons are routinely accused of dominating markets, dumbing down the news to plump up the bottom line, and forcing U.S. content on world audiences. But these companies are not as big, bad, dominant, or American as critics claim. And company size is largely irrelevant to many of the problems facing today's Fourth Estate.

"A Few Big Companies Are Taking Over the World's Media"

No. Much of the debate on media structure is too black-and-white. A merger of Time Inc. with Warner Communications and then with America Online dominates headlines, but the incremental growth of smaller companies from the bottom does not. Breakups and divestitures do not generally receive front-page treatment, nor do the arrival and rapid growth of new players or the shrinkage of once influential players.

In the United States, today's top 50 largest media companies account for little more of total media revenue than did the companies that made up the top 50 in 1986. CBS Inc., for example, was then the largest media company in the United States. In the 1990s, it sold off its magazines, divested its book publishing, and was not even among the 10 largest U.S. media companies by the time it agreed to be acquired by Viacom, which was a second tier player in 1986. Conversely, Bertelsmann, though a major player in Germany in 1986, was barely visible in the United States. By 1997, it was the third largest player in the United States, where it owns book publisher Random House. Companies such as Amazon.com, Books-A-Million, Comcast, and C-Net were nowhere to be found on a list of the largest media companies in 1980. Others, such as Allied Artists, Macmillan, and Playboy Enterprises, either folded or grew so slowly as to fall out of the top ranks.

Indeed, media merger activity is more like rearranging the furniture: In the past 15 years, MCA with its Universal Pictures was sold by its U.S. owners to Matsushita (Japan), who sold to Seagram's (Canada), who sold to Vivendi (France). Vivendi has already announced that it will divest some major media assets, including textbook publisher Houghton-Mifflin. Bertelsmann also has

From *Foreign Policy* by Benjamin Compaine, November/December 2002, pp. 20–28. Copyright © 2002. Permission conveyed through Copyright Clearance Center. Reprinted by permission. www.foreignpolicy.com

had difficulty maintaining all the parts of its global enterprise: It recently fired its top executive and is planning to shed its online bookstore. There is an ebb as well as a flow among even the largest media companies.

The notion of the rise of a handful of all-powerful transnational media giants is also vastly overstated. Some media companies own properties internationally or provide some content across borders (for example, Vivendi's Canal+ distributes movies internationally), but no large media conglomerate owns newspapers, book publishers, radio stations, cable companies, or television licenses in all the major world markets. News Corp. comes closest to being a global media enterprise in both content and distribution, but on a global scale it is still a minor presence—that is, minor as a percentage of global media revenue, global audience, and in the number of markets it covers.

Media companies have indeed grown over the past 15 years, but this growth should be understood in context. Developed economies have grown, so expanding enterprises are often simply standing still in relative terms. Or their growth looks less weighty. For example, measured by revenue, Gannett was the largest U.S. newspaper publisher in 1986, its sales accounting for 3.4 percent of all media revenue that year. In 1997, it accounted for less than 2 percent of total media revenue. Helped by major acquisitions, Gannett's revenue had actually increased by 69 percent, but the U.S. economy had grown 86 percent. The media industry itself had grown 188 percent, making a "bigger" Gannett smaller in relative terms. Similar examples abound.

"U.S. Companies Dominate the Media"

No.   Long before liberalization of ownership in television in the 1980s, critics around the world were obsessed by the reach of U.S. programming, which cultural elites often considered too mass market and too infused with American cultural values. However, in most of the world, decisions of what programming to buy traditionally lay in the hands of managers who worked for government-owned or government-controlled broadcasters. Then, as now, no nation's media companies could require a programmer to buy their offerings or force consumers to watch them. As the market becomes more competitive, with content providers such as Canal+ and the BBC marketing their products globally, it is even more important that media enterprises offer programming that people want to watch.

While Viacom, Disney, and AOL Time Warner are U.S. owned, many non-U.S.-owned companies dominate the roster of the largest media groups: News Corp. (Australia), Bertelsmann (Germany), Reed-Elsevier (Britain/Netherlands), Vivendi and Lagadere-Hachette (France), and Sony Corp. (Japan).

The pervasiveness of a handful of media companies looks even less relevant when one looks at media ownership across countries. The United Nations' "Human Development Report 2002" examined ownership of the five largest newspaper and broadcast enterprises in 97 countries. It found that 29 percent of the world's largest newspapers are state owned and another 57 percent are family owned. Only 8 percent are owned by employees or the public. For radio stations, 72 percent are state owned and 24 percent family owned. For television stations, 60 percent are state owned, 34 percent family owned. These data suggest there is little foreign direct investment in the media sectors of most countries.

News media can tap wire services from around the globe such as Reuters, Agence France-Presse, the Associated Press, Kyodo News, Xinhua News Agency, and Itar-Tass. TV news editors can use video feeds from sources as diverse as U.S.-based CNN to the Qatar-based Al Jazeera. The variety and ownership of TV content in general has substantially increased—a reality media critics ignore. From two state-owned channels in many European countries and from three U.S. networks plus the Public Broadcasting Service, there are now dozens, often hundreds, of video options via terrestrial, cable, and satellite transmission, not to mention the offline variety of videocassettes and DVDs and the online availability of music and movies. In addition, book and magazine publishing continues to be robust worldwide. Encouraged by relatively low start-up costs, new publishers are popping up constantly.

"Corporate Ownership Is Killing Hard-Hitting Journalism"

A bright red herring. When exactly was this golden age of hard-hitting journalism? One might call to mind brief periods: the muckrakers in the early 20th century or Watergate reporting in the 1970s. But across countries and centuries, journalism typically has not been "hard-hitting." With more news outlets and competition today, there is a greater range of journalism than was typical in the past. Further, a 2000 comparison of 186 countries by Freedom House, a nonprofit devoted to promoting democracy, suggests that press independence, including journalists' freedom from economic influence, remained high in all but two members (Mexico and Turkey) of the Organisation for Economic Co-operation and Development, where global media's markets are concentrated.

Also underlying the complaint that news has been "dumbed down" is an assumption that the media ought to be providing a big dose of policy-relevant content. Japan's dominant public broadcaster, NHK, does so, yet is Japan a more vibrant democracy as a result? More to the point, with so many media outlets today, readers and viewers can get more and better news from more diverse perspectives, if that is what they want. Or they can avoid it altogether. The alternative is to limit the number of outlets and impose content requirements on those remaining.

The third problem with this notion of corporations killing journalism is that it assumes ownership matters. In the old days of media moguls it may have: William Randolph Hearst, William Loeb, and Robert McCormick were attracted to the media because they each had political agendas, which permeated their newspapers. Nearly a century before Italian media owner Silvio Berlusconi rose to the top of Italian politics, Hearst, whose newspapers dominated in the United States, was elected to the U.S. Congress and harbored presidential aspirations. But Hearst's dual roles did not affect U.S. politics or democracy in any lasting way. The jury is still out on the effect of Berlusconi's dual roles.

Corporate-owned newspapers may actually provide better products than those that are family owned: Research suggests that large, chain-owned newspapers devote more space to editorial material than papers owned by small firms. In many parts of South America, where regulation has restricted or prevented corporate ownership, family-run enterprises have often been closely identified with ideological biases or even with using political influence to benefit other businesses. Brazilian media enterprise Globo, owned by the politically involved Marinho family, encompasses a TV network, radio, cable, and magazines. Yet Globo no longer opposes recent moves to liberalize Brazilian media ownership because then it could gain access to desirable foreign investment. As Latin American media shift from family-owned, partisan media to corporations, observes Latin American media scholar Silvio Waisbord, the media become less the "public avenues for the many ambitions of their owners," and their coverage of government corruption "is more likely to be informed by marketing calculations and the professional aspirations of reporters." This trade-off may not be bad.

Global media will not necessarily introduce aggressive journalism in places where press freedom has traditionally been constricted. For instance, News Corp. was criticized for dropping BBC news programming from Star TV presumably to mollify Chinese leaders in the mid-1990s. Yet satellite broadcaster Phoenix TV (in which News Corp.'s Star TV maintains a 37.6 percent stake, alongside that of the local Chinese owners) sometimes pushes the envelope in China, as when it reported on the election of Chen Shui-bian as president in Taiwan.

"Global Media Drown Out Local Content"

Absolutely not.   Most media—like politics—are inherently local. Global firms peddle wholly homogeneous content across markets at their peril. Thus, MTV in Brazil plays a mix of music videos and other programming determined by local producers, even though it shares a recognizable format with MTV stations elsewhere. News Corp.'s newspapers in the United Kingdom look and read differently from those in the United States. When Star TV, an Asian subsidiary of News Corp., began broadcasting satellite television into India, few tuned in to *Dallas* and *The Bold and the Beautiful* dubbed in Hindi. The network only succeeded in India once it hired an executive with experience in Indian programming to create Indian soap operas and when an Indian production house took over news and current affairs programming.

Often viewed as a negative, consolidation may have considerable social benefits. It took the deep pockets of News Corp. to create and sustain a long-awaited fourth broadcast network in the United States. And the 1990 merger in the United Kingdom of Sky Channel and BSB created a viable television competitor from two money-losing satellite services.

"The Internet Has Leveled the Playing Field"

Yes.   Or more accurately, it's helping to level the terrain because it is a relatively low-cost conduit for all content providers. As the old adage goes, "Freedom of the press is guaranteed only to those who own one." Make no mistake: an activist with a dial-up Internet connection and 10 megabytes of Web server space cannot easily challenge Disney for audiences. But an individual or a small group can reach the whole world and, with a little work and less money, can actually find an audience.

Worldwide, an estimated 581 million people were online by 2002, more than one third of whom lived outside North America and Europe. Yet the Internet is in its infancy. The number of users is still growing and will continue to expand to the literate population as access costs decrease.

Once online, Internet users have access to thousands of information providers. Some are the same old players—Disney with its stable of cartoon icons, Infinity with its familiar music and talk-radio broadcasting, and old government-run stations still operating in much of the world. But these coexist with newer, Internet-only options such as those found at Realguide.com, which links to 2,500 real-time audio streams from around the world, or NetRadio, which outdraws many traditional stations. These Internet-only "broadcasters" have not had to invest in government-sanctioned licenses and generally have no limits on their speech.

In countries where governments strictly control print and broadcast media, governments also can try to restrict Internet access, as China does. But some may choose not to do so: In Malaysia, the government pledged not to censor the Internet to promote its version of Silicon Valley to foreign investors. As a consequence, Malaysian cyberspace media are free of the restrictions their print and broadcast brethren face [see "Mahathir's Paradox," page 100].

"Proliferating Media Outlets Balkanize Public Opinion"

No. The flip side of concerns that media concentration has limited available information is the concern that technology has made it possible to access so many voices that people in democratic societies can and will seek only information that supports their prejudices. A fragmented public, tuning in only to select cable channels or specific Web sites, could thus wall itself off from healthy public debate.

Recent U.S. studies show that as users gain experience with the Internet, they use it not to replace other sources of information but for more practical applications. They perform work-related tasks, make purchases and other financial transactions, write e-mail messages, and seek information that is important to their everyday lives.

Although news is low on the list of its uses, the Internet functions in much the same way as older news media: offering opportunities for both those who directly seek news sites and those who chance upon news links serendipitously. The Pew Internet and American Life Project reports that 42 percent of those who read news on the Web typically find news while they are doing other things online. This picture is not consistent with the notion that Web audiences routinely tune out information with which they disagree.

"Media Coverage Drives Foreign Policy"

Probably not often. Analyzing media coverage is often a chicken-and-egg dilemma: What stimulated the media to cover an event or issue? And if public policy responds to an event the news media cover, does that mean the media (or those who run the media) set the agenda?

The idea that media coverage of international crises can spark a response from politicians is termed the "CNN effect." The classic case is the coverage of starving children in Somalia in the early 1990s, which was followed by U.S. military involvement in humanitarian relief efforts. But even in the case of Somalia, some administration officials actually used the media to get the attention of other officials, and the majority of the coverage in Somalia followed rather than preceded official action.

In many places, governments are even more likely to be driving media coverage rather than the other way around, although it may suit governments to appear as if they have bowed to public opinion. The Chinese government delayed release of the crew of the U.S. EP-3 spy plane that made an emergency landing on Hainan Island in 2001, claiming that an embittered Chinese public demanded it. Angry Web comments did precede and were then reflected in media coverage of the incident. But at the same time, the government had been fanning the flames, cultivating nationalistic sentiment through the selection and treatment of stories in the news. At other times, the Chinese government both censors Web comments and withholds information from the media when it needs to preserve its foreign policy options.

"Stricter Regulation of Media Is in the Public Interest"

Just the opposite.   Beware when someone claims to be speaking for the "public interest." In most cases, those who invoke the term really mean "interested publics." For example, advertisers' sense of which policies on media ownership are in their interest may differ from that of regular newspaper readers or that of satellite TV subscribers.

Fostering competition has long been a central tenet of U.S. media regulation. What if preventing two newspapers from merging results in both having to trim news budgets or pages, neither having the resources to engage in investigative reporting, or worse yet, one closing shop? Media concentration may be in the public interest if it provides a publisher with greater profit margins and the wherewithal to spend some of that on editorial content, and research in fact shows this is the case.

Licensing acts as an entry barrier to new players, and antitrust laws often lag behind reality. In the market for video program distribution, for instance, terrestrial broadcast licensees compete with cable operators and networks, who in turn compete with satellite providers. Regulation and policy limits will always be necessary, but having different regulatory frameworks for each media segment makes less sense today.

Governments that give can also take. Japanese law makes public broadcaster NHK one of the world's most autonomous public broadcasters, yet the ruling Liberal Democratic Party (LDP) strongly influences the agencies that control media licenses and that select NHK's governing board. Not coincidentally, NHK provides neutral, policy-relevant news but avoids controversial topics and investigative reporting. Where Japanese commercial television has tried to fill this gap, LDP politicians have reacted: in one case, asking an advertiser to withdraw sponsorship and in another, seeking the withdrawal of a broadcasting license.

Paradoxically, relaxing broadcast regulation may expand competition. When News Corp. put together a fourth network in the United States in 1986, the timing was not random. It followed two regulatory decisions: the Federal Communications Commission raised the limit on local licenses that a single firm could own from seven to twelve and waived a rule that kept TV networks from owning their programming. The first change allowed News Corp. to assemble a core of stations in larger markets that gave it a viable base audience, and the second sanctioned News Corp.'s purchase of 20th Century Fox, with its television production studio. Fox was thus able to launch the first successful alternative to the Big Three in 30 years. Its success also paved the way for three other large media players to initiate networks.

<hr />

Want to Know More?

Shanthi Kalathil's "Chinese Media and the Information Revolution" (Washington: Carnegie Endowment for International Peace, 2002) explains why the Internet is not a force for press freedom in China. Other fascinating country-specific glimpses are Laurie Anne Freeman's *Closing the Shop: Information Cartels and Japan's Mass Media* (Princeton: Princeton University Press, 2000) and Anil Dharker's look at the Star Network's Indian debut in his review of Edward S. Herman and Robert W. McChesney's *The Global Media: The New Missionaries of Corporate Capitalism* (London: Cassell, 1997) in the Summer 1998 issue of FOREIGN POLICY.

On the OpenDemocracy Web site, see the spirited debate between Compaine and McChesney on the impact of corporate media ownership on democracy. The site also includes several articles on media in such places as Italy, Latin America, and Japan.

POSTSCRIPT

Is the Transnational Media Hurting Global Society?

There is a communication revolution happening in the world today. Cell phones, computers, digital television, satellite radio, and a host of other devices provide access for people, ideas, values, and things to be transmitted around the globe. "Being wired" is no longer a term rooted in the drug culture but rather a communication term for the mainstream of many societies.

Yet, inexorably linked with the globalization of the media is the fact that it is driven and controlled by a cartel of private transnational corporations whose aim is to profit from this revolution. As technology becomes more sophisticated, the number of actors developing, patenting, and disseminating it also constricts. States and people are increasingly dependent on several media giants whose control in an integrated set of communication enterprises gives them enormous power to shape ideas, values, and events.

While the jury is still out on the positive or negative impact of this control, one fact appears certain. Hundreds of millions of people are becoming dependent upon media giants for their access to the global superhighway of information. The extent to which that relationship is positive or corrosive will be determined by the needs of people and the desires of media giants to meet or shape those needs.

Some important literature on this includes Ella Shohat and Robert Stam, editors, *Multiculturalism, Postcoloniality and Transnational Media* (2003), William A. Hatchen and James F. Scotton, *The World News Prism: Global Media in the Age of Terrorism* (2003), S.O. Siochru, B. Girard, and A. Mahan, *Global Media Governance* (2003) and Jabbar Al-Obaidi, "Communication and the Culture of Democracy: Global Media and Promotion of Democracy in the Middle East," *International Journal of Instructional Media* (2003).

Nuclear Terrorism: How to Prevent It

This site of the Nuclear Control Institute discusses nuclear terrorism and how best to prevent it. Topics include terrorists' ability to build nuclear weapons, the threat of "dirty bombs," and whether or not nuclear reactors are adequately protected against attack. This site features numerous links to key nuclear terrorism documents and Web sites as well as to recent developments and related news items.

http://www.nci.org/nuketerror.htm

CDI Terrorism Project

The Center for Defense Information's (CDI) Terrorism Project is designed to provide insights, in-depth analysis, and facts on the military, security, and foreign policy challenges of terrorism. The project looks at all aspects of fighting terrorism, from near-term issues of response and defense to long-term questions about how the United States should shape its future international security strategy.

http://www.cdi.org/terrorism/

Exploring Global Conflict: An Internet Guide to the Study of Conflict

Exploring Global Conflict: An Internet Guide to the Study of Conflict is an Internet resource designed to provide understanding of global conflict. Information related to specific conflicts in areas such as Northern Ireland, the Middle East, the Great Lakes region in Africa, and the former Yugoslavia is included on this site. Current news and educational resource sites are listed as well.

http://www.uwm.edu/Dept/CIS/conflict/congeneral.html

Central Intelligence Agency

The U.S. government agency with major responsibility for the war on terrorism provides substantial information on its Web site.

http://wwwcia.gov/terrorism/

The Cato Institute

This U.S. public organization conducts research on a wide range of public policy issues. It subscribes to what it terms "basic American principles." One important research issue is civil liberties concerns relating to the war on terrorism.

http://cato.org/current/civil-liberties/

Center for Strategic and International Studies (CSIS)

CSIS now provides substantial information on terrorism in the aftermath of 9/11. The URL noted here leads directly to its thinking on the issue of homeland defense.

http://www.csis.org/homeland

PART 4

The New Global
Security Dilemma

*W*ith the end of the Cold War, the concept of security was freed from
its bipolar constraints of great power calculations. And as a consequence
of 9/11, the definition of security and how to achieve it were once again
redefined to encompass new kinds of threats from a new group of perpetrators. In short, our concept of security in a post-modern age has broad-
ened considerably.

As a result, an entirely new set of issues has come to occupy the
attention of policymakers and citizens alike. These include a clash of civilizations based on culture and ethnicity, the likelihood of a future nuclear
threat, strategies such as preemption for dealing with new threats, and the
fear of compromising civil liberties in the process of addressing these
emerging problems.

This section examines some of the key issues shaping the security
dilemma of the twenty-first century.

- Are We Headed Toward a Nuclear 9/11?

- Is Preemption a Viable Policy Option in Today's Global Security
  Environment?

- Are Civil Liberties Being Compromised in the War Against Terrorism?

- Are Cultural and Ethnic Rivalries the Defining Dimensions of Twenty-
  First Century Conflict?

ISSUE 16

Are We Headed Toward a Nuclear 9/11?

YES: Graham Allison and Andrei Kokoshin, from "The New Containment: An Alliance Against Nuclear Terrorism," *The National Interest* (Fall 2002)

NO: Jessica Stern, from "A Rational Response to Dirty Bombs," *Financial Times* (June 11, 2002)

ISSUE SUMMARY

YES: Graham Allison and Andrei Kokoshin contend that a U.S.-Russian alliance against terrorism is needed before terrorists acquire nuclear weapons and "launch." They surmise that nuclear terrorism will occur in just a matter of time if these two nuclear powers do not act quickly.

NO: Jessica Stern, a lecturer in public policy, argues that Americans are in danger of overestimating terrorist capabilities and thus creating a graver threat than actually exists. She warns that the United States must not overreact in its policy response and that prudent security measures will greatly reduce such threats now and in the future.

Since the terrorist attacks of September 11, 2001, much has been written about the specter of nuclear terrorism and the releasing of a dirty bomb (one loaded with radioactive material) in an urban/civilian setting. The events of September 11 have all but ensured the world's preoccupation with such an event for the foreseeable future. Indeed, the arrest of a U.S. man with dirty bomb materials indicates that such plans may indeed be in the works between Al Qaeda and other terrorist cells. When this horror is combined with the availability of elements of nuclear-related material in places like the states of the former Soviet Union, Pakistan, India, Iraq, Iran, North Korea, and many other states, one can envision a variety of sobering scenarios.

Hollywood feeds these views with such films as *The Sum of All Fears* and *The Peacemaker,* in which nuclear terrorism is portrayed as all too easy to carry out and likely to occur. It is difficult in such environments to separate fact from fiction and to ascertain objectively the probabilities of such events. So many factors go into a successful initiative in this area. One needs to find a committed cadre of terrorists, sufficient financial backing, technological know-how, intense security and secrecy, the means of delivery, and many other variables, including luck. In truth, such acts may have already been advanced and thwarted by governments, security services, or terrorist mistakes and incompetence. We do not know, and we may never know.

Regional and ethnic conflicts of a particularly savage nature in places like Chechnya, Kashmir, Colombia, and Afghanistan help to fuel fears that adequately financed zealots will see in nuclear weapons a swift and catastrophic answer to their demands and angers. Osama bin Laden's contribution to worldwide terrorism has been the success of money over security and the realization that particularly destructive acts with high levels of coordination can be "successful." This will undoubtedly encourage others with similar ambitions against real or perceived enemies.

Conversely, many argue that fear of the terrorist threat has left us imagining that which is not likely. They point to a myriad of roadblocks to terrorist groups' obtaining all of the elements necessary for a nuclear or dirty bomb. They cite technological impediments, monetary issues, lack of sophistication, and inability to deliver. They also cite governments' universal desire to prevent such actions. Even critics of Iraqi leader Saddam Hussein have argued that were he to develop such weapons, he would not deliver them to terrorist groups nor would he use them except in the most dire of circumstances, such as his own regime's survival. They argue that the threat is overblown and, in some cases, merely used to justify increased security and the restriction of civil liberties.

The following selections reflect this dichotomy of views. Graham Allison and Andrei Kokoshin make the argument that nuclear terrorism is just over the horizon, real, likely, and worthy of direct action. They contend that the United States and Russia should form an alliance to combat this threat as the two largest nuclear powers. They contend that a mere "half dozen nuclear explosions across the United States or Russia would shift the course of history." Thus, action to counter this real threat is imperative.

In the second selection, Jessica Stern argues that although a threat exists, the United States should not overplay it. She does suggest that the probability of a nuclear terrorist attack can be greatly reduced by using preventive security measures and education. Essentially, Stern contends that overreaction on America's part is an element of the terrorist arsenal; we must not overestimate the dangers.

Graham Allison and
Andrei Kokoshin

➡ YES

The New Containment: An Alliance Against Nuclear Terrorism

. . . In the aftermath of Osama bin Laden's September 11 assault, which awakened the world to the reality of global terrorism, it is incumbent upon serious national security analysts to think again about the unthinkable. Could a nuclear terrorist attack happen today? Our considered answer is: yes, unquestionably, without any doubt. It is not only a possibility, but in fact the most urgent unaddressed national security threat to both the United States and Russia.

Consider this hypothetical: A crude nuclear weapon constructed from stolen materials explodes in Red Square in Moscow. A 15-kiloton blast would instantaneously destroy the Kremlin, Saint Basil's Cathedral, the ministries of foreign affairs and defense, the Tretyakov Gallery, and tens of thousands of individual lives. In Washington, an equivalent explosion near the White House would completely destroy that building, the Old Executive Office Building and everything within a one-mile radius, including the Departments of State, Treasury, the Federal Reserve and all of their occupants—as well as damaging the Potomac-facing side of the Pentagon.

Psychologically, such a hypothetical is as difficult to internalize as are the plot lines of a writer like Tom Clancy (whose novel Debt of Honor ends with terrorists crashing a jumbo jet into the U.S. Capitol on Inauguration Day, and whose The Sum of All Fears contemplates the very scenario we discuss—the detonation of a nuclear device in a major American metropolis by terrorists). That these kinds of scenarios are physically possible, however, is an undeniable, brute fact.

After the first nuclear terrorist attack, the Duma, Congress—or what little is left of them—and the press will investigate: Who knew what, when? They will ask what could have been done to prevent the attack. Most officials will no doubt seek cover behind the claim that "no one could have imagined" this happening. But that defense should ring hollow. We have unambiguous strategic warning today that a nuclear terrorist attack could occur at any moment. Responsible leaders should be asking hard questions now. Nothing prevents the governments of Russia, America and other countries from taking effective action immediately—nothing, that is, but a lack of determination.

The argument made here can be summarized in two propositions: first, nuclear terrorism poses a clear and present danger to the United States, Russia and other nations; second, nuclear terrorism is a largely preventable disaster. Preventing nuclear terrorism is a large, complex, but ultimately finite challenge that can be met by a bold, determined, but nonetheless finite response. The current mismatch between the seriousness of the threat on the one hand, and the actions governments are now taking to meet it on the other, is unacceptable. Below we assess the threat and outline a solution that begins with a U.S.-Russian led Alliance Against Nuclear Terrorism.

Assessing the Threat

A comprehensive threat assessment must consider both the likelihood of an event and the magnitude of its anticipated consequences. As described above, the impact of even a crude nuclear explosion in a city would produce devastation in a class by itself. A half dozen nuclear explosions across the United States or Russia would shift the course of history. The question is: how likely is such an event?

Security studies offer no well-developed methodology for estimating the probabilities of unprecedented events. Contemplating the possibility of a criminal act, Sherlock Holmes investigated three factors: motive, means and opportunity. That framework can be useful for analyzing the question at hand. If no actor simultaneously has motive, means and opportunity, no nuclear terrorist act will occur. Where these three factors are abundant and widespread, the likelihood of a nuclear terrorist attack increases. The questions become: Is anyone motivated to instigate a nuclear attack? Could terrorist groups acquire the means to attack the United States or Russia with nuclear weapons? Could these groups find or create an opportunity to act?

I. Motive

There is no doubt that Osama bin Laden and his associates have serious nuclear ambitions. For almost a decade they have been actively seeking nuclear weapons, and, as President Bush has noted, they would use such weapons against the United States or its allies "in a heartbeat." In 2000, the CIA intercepted a message in which a member of Al-Qaeda boasted of plans for a "Hiroshima" against America. According to the Justice Department indictment for the 1998 bombings of the American embassies in Kenya and Tanzania, "At various times from at least as early as 1993, Osama bin Laden and others, known and unknown, made efforts to obtain the components of nuclear weapons." Additional evidence from a former Al-Qaeda member describes attempts to buy uranium of South African origin, repeated travels to three Central Asian states to try to buy a complete warhead or weapons-usable material, and discussions with Chechen criminal groups in which money and drugs were offered for nuclear weapons.

Bin Laden himself has declared that acquiring nuclear weapons is a religious duty. "If I have indeed acquired [nuclear] weapons," he once said, "then I thank God for enabling me to do so." When forging an alliance of terrorist organizations in 1998, he issued a statement entitled "The Nuclear Bomb of

Islam." Characterized by Bernard Lewis as "a magnificent piece of eloquent, at times even poetic Arabic prose," it states: "It is the duty of Muslims to prepare as much force as possible to terrorize the enemies of God." If anything, the ongoing American-led war on global terrorism is heightening out adversary's incentive to obtain and use a nuclear weapon. Al-Qaeda has discovered that it can no longer attack the United States with impunity. Faced with an assertive, determined opponent now doing everything it can to destroy this terrorist network, Al-Qaeda has every incentive to take its best shot.

Russia also faces adversaries whose objectives could be advanced by using nuclear weapons. Chechen terrorist groups, for example, have demonstrated little if any restraint on their willingness to kill civilians and may be tempted to strike a definitive blow to assert independence from Russia. They have already issued, in effect, a radioactive warning by planting a package containing cesium-137 at Izmailovsky Park in Moscow and then tipping off a Russian reporter. Particularly as the remaining Chechen terrorists have been marginalized over the course of the second Chechen war, they could well imagine that by destroying one Russian city and credibly threatening Moscow, they could persuade Russia to halt its campaign against them. . . .

II. Means

To the best of our knowledge, no terrorist group can now detonate a nuclear weapon. But as Secretary of Defense Donald Rumsfeld has stated, "the absence of evidence is not evidence of absence." Are the means beyond terrorists' reach, even that of relatively sophisticated groups like Al-Qaeda?

Over four decades of Cold War competition, the superpowers spent trillions of dollars assembling mass arsenals, stockpiles, nuclear complexes and enterprises that engaged hundreds of thousands of accomplished scientists and engineers. Technical know-how cannot be UN-invented. Reducing arsenals that include some 40,000 nuclear weapons and the equivalents of more than 100,000 nuclear weapons in the form of highly enriched uranium (HEU) and plutonium to manageable levels is a gargantuan challenge.

Terrorists could seek to buy an assembled nuclear weapon from insiders or criminals. Nuclear weapons are known to exist in eight states: the United States, Russia, Great Britain, France, China, Israel, India and Pakistan. Security measures, such as "permissive action links" designed to prevent unauthorized use, are most reliable in the United States, Russia, France and the United Kingdom. These safeguards, as well as command-and-control systems, are much less reliable in the two newest nuclear states—India and Pakistan. But even where good systems are in place, maintaining high levels of security requires constant attention from high-level government officials.

Alternatively, terrorists could try to build a weapon. The only component that is especially difficult to obtain is the nuclear fissile material—HEU or plutonium. Although the largest stockpiles of weapons-grade material are predominantly found in the nuclear weapons programs of the United States and Russia, fissile material in sufficient quantifies to make a crude nuclear weapon can also be found in many civilian settings around the globe. Some 345 research reactors in 58 states together contain twenty metric tons of HEU, many in quantities sufficient to build a bomb. Other civilian reactors produce enough weapons-grade nuclear material to pose a proliferation threat; several European states, Japan, Russia and India reprocess spent fuel to separate out plutonium for use as new fuel. The United States has actually facilitated the spread of fissile material in the past-over three decades of the Atoms for Peace program, the United States exported 749 kg of plutonium and 26.6 metric tons of HEU to 39 countries.

Terrorist groups could obtain these materials by theft, illicit purchase or voluntary transfer from state control. There is ample evidence that attempts to steal or sell nuclear weapons or weapons-usable material are not hypothetical, but a recurring fact. Just last fall, the chief of the directorate of the Russian Defense Ministry responsible for nuclear weapons reported two recent incidents in which terrorist groups attempted to perform reconnaissance at Russian nuclear storage sites. The past decade has seen repeated incidents in which individuals and groups have successfully stolen weapons material from sites in Russia and sought to export them—but were caught trying to do so. In one highly publicized case, a group of insiders at a Russian nuclear weapons facility in Chelyabinsk plotted to steal 18.5 kg (40.7 lbs.) of HEU, which would have been enough to construct a bomb, but were thwarted by Russian Federal Security Service agents.

In the mid-1990s, material sufficient to allow terrorists to build more than twenty nuclear weapons—more than 1,000 pounds of highly enriched uranium-sat unprotected in Kazakhstan. Iranian and possibly Al-Qaeda operatives with nuclear ambitions were widely reported to be in Kazakhstan. Recognizing the danger, the American government itself purchased the material and removed it to Oak Ridge, Tennessee. In February 2002, the U.S. National Intelligence Council reported to Congress that "undetected smuggling [of weapons-usable nuclear materials from Russia] has occurred, although we do not know the extent of such thefts." Each assertion invariably provokes blanket denials from Russian officials. Russian Atomic Energy Minister Aleksandr Rumyantsev has claimed categorically: "Fissile materials have not disappeared." President Putin has stated that he is "absolutely confident" that terrorists in Afghanistan do not have weapons of mass destruction of Soviet or Russian origin.

For perspective on claims of the inviolable security of nuclear weapons or material, it is worth considering the issue of "lost nukes." Is it possible that the United States or Soviet Union lost assembled nuclear weapons? At least on the American side the evidence is clear. In 1981, the U.S. Department of Defense published a list of 32 accidents involving nuclear weapons, many of which resulted in lost bombs. One involved a submarine that sank along with two nuclear torpedoes. In other cases, nuclear bombs were lost from aircraft. Though on the Soviet/Russian side there is no official information, we do know that four Soviet submarines carrying nuclear weapons have sunk since 1968, resulting in an estimated 43 lost nuclear warheads. These accidents suggest the complexity of controlling and accounting for vast nuclear arsenals and stockpiles.

Nuclear materials have also been stolen from stockpiles housed at research reactors. In 1999, Italian police seized a bar of enriched uranium from an organized crime group trying to sell it to an agent posing as a Middle Eastern businessman with presumed ties to terrorists. On investigation, the Italians found that the uranium originated from a U.S.-supplied research reactor in the former Zaire, where it presumably had been stolen or purchased sub rosa.

Finally, as President Bush has stressed, terrorists could obtain nuclear weapons or material from states hostile to the United States. In his now-infamous phrase, Bush called hostile regimes developing WMD and their terrorist allies an "axis of evil." He argued that states such as Iraq, Iran and North Korea, if allowed to realize their nuclear ambitions, "could provide these arms to terrorists, giving them the means to match their hatred." The fear that a hostile regime might transfer a nuclear weapon to terrorists has contributed to the Bush Administration's development of a new doctrine of preemption against such regimes, with Iraq as the likeliest test case. It also adds to American concerns about Russian transfer of nuclear technologies to Iran. While Washington and Moscow continue to disagree over whether any safeguarded civilian nuclear cooperation with Iran is justified, both agree on the dangers a nuclear-armed Iran would pose. Russia is more than willing to agree that there should be no transfers of technology that could help Iran make nuclear weapons.

III. Opportunity

Security analysts have long focused on ballistic missiles as the preferred means by which nuclear weapons would be delivered. But today this is actually the least likely vehicle by which a nuclear weapon will be delivered against Russia or the United States. Ballistic weapons are hard to produce, costly and difficult to hide. A nuclear weapon delivered by a missile also leaves an unambiguous return address, inviting devastating retaliation. As Robert Walpole, a National Intelligence Officer, told a Senate subcommittee in March, "Nonmissile delivery means are less costly, easier to acquire, and more reliable and accurate." Despite this assessment, the U.S. government continues to invest much more heavily in developing and deploying missile defenses than in addressing more likely trajectories by which weapons could arrive.

Terrorists would not find it very difficult to sneak a nuclear device or nuclear fissile material into the United States via shipping containers, trucks, ships or aircraft. Recall that the nuclear material required is smaller than a football. Even an assembled device, like a suitcase nuclear weapon, could be shipped in a container, in the hull of a ship or in a trunk carried by an aircraft. After this past September 11, the number of containers that are x-rayed has increased, to about 500 of the 5,000 containers currently arriving daily at the port of New York/New Jersey—approximately 10 percent. But as the chief executive of CSX Lines, one of the foremost container-shipping companies, put it: "If you can smuggle heroin in containers, you may be able to smuggle in a nuclear bomb."

Effectively countering missile attacks will require technological breakthroughs well beyond current systems. Success in countering covert delivery of weapons will require not just technical advances but a conceptual breakthrough. Recent efforts to bolster border security are laudable, but they only begin to scratch the surface. More than 500 million people, 11 million trucks and 2 million rail cars cross into the United States each year, while 7,500 foreign-flag ships make 51,000 calls in U.S. ports. That's not counting the tens of thousands of people, hundreds of aircraft and numerous boats that enter illegally and uncounted. Given this volume and the lengthy land and sea borders of the United States, even a radically renovated and reorganized system cannot aspire to be airtight.

The opportunities for terrorists to smuggle a nuclear weapon into Russia or another state are even greater. Russia's land borders are nearly twice as long as America's, connecting it to more than a dozen other states. In many places, in part because borders between republics were less significant in the time of the Soviet Union, these borders are not closely monitored. Corruption has been a major problem among border patrols. Visa-free travel between Russia and several of its neighbors creates additional opportunities for weapons smugglers and terrorists. The "homeland security" challenge for Russia is truly monumental.

In sum: even a conservative estimate must conclude that dozens of terrorist groups have sufficient motive to use a nuclear weapon, several could potentially obtain nuclear means, and hundreds of opportunities exist for a group with means and motive to make the United States or Russia a victim of nuclear terrorism. The mystery before us is not how a nuclear terrorist attack could possibly occur, but rather why no terrorist group has yet combined motive, means and opportunity to commit a nuclear attack. We have been lucky so far, but who among us trusts luck to protect us in the future?

Chto Delat?

The good news about nuclear terrorism can be summarized in one line: no highly enriched uranium or plutonium, no nuclear explosion, no nuclear terrorism. Though the world's stockpiles of nuclear weapons and weapons-usable materials are vast, they are finite. The prerequisites for manufacturing fissile material are many and require the resources of a modern state. Technologies for locking up super-dangerous or valuable items—from gold in Fort Knox to treasures in the Kremlin Armory—are well developed and tested. While challenging, a specific program of actions to keep nuclear materials out of the hands of the most dangerous groups is not beyond reach, if leaders give this objective highest priority and hold subordinates accountable for achieving this result.

The starting points for such a program are already in place. In his major foreign policy campaign address at the Ronald Reagan Library, then—presidential candidate George W. Bush called for "Congress to increase substantially our assistance to dismantle as many Russian weapons as possible, as quickly as possible." In his September 2000 address to the United Nations Millennium Summit, Russian President Putin proposed to "find ways to block the spread of nuclear weapons by excluding use of enriched uranium and plutonium in global atomic energy production." The Joint Declaration on the New Strategic Relationship between the United States and Russia, signed by the two presidents at the May 2002 summit, stated that the two partners would combat the "closely linked threats of international terrorism and the proliferation of weapons of mass destruction." Another important result yielded by the summit was the upgrading of the Armitage/Trubnikov-led U.S.-Russia Working Group on Afghanistan to the U.S.-Russia Working Group on Counterterrorism, whose agenda is to thwart nuclear, biological and chemical terrorism.

Operationally, however, priority is measured not by words, but by deeds. A decade of Nunn-Lugar Cooperative Threat Reduction Programs has accomplished much in safeguarding nuclear materials. Unfortunately, the job of upgrading security to minimum basic standards is mostly unfinished: according to Department of Energy reports, two-thirds of the nuclear material in Russia remains to be adequately secured. Bureaucratic inertia, bolstered by mistrust and misperception on both sides, leaves these joint programs bogged down on timetables that extend to 2008. Unless implementation improves significantly, they will probably fail to meet even this unacceptably distant target. What is required on both sides is personal, presidential priority measured in commensurate energy, specific orders, funding and accountability. This should be embodied in a new U.S.-Russian led Alliance Against Nuclear Terrorism.

Five Pillars of Wisdom

When it comes to the threat of nuclear terrorism, many Americans judge Russia to be part of the problem, not the solution. But if Russia is welcomed and supported as a fully responsible non-proliferation partner, the United States stands to accomplish far more toward minimizing the risk of nuclear terrorism than if it treats Russia as an unreconstructed pariah. As the first step in establishing this alliance, the two presidents should pledge to each other that his government will do everything technically possible to prevent criminals or terrorists from stealing nuclear weapons or weapons-usable material, and to do so on the fastest possible timetable. Each should make clear that he will personally hold accountable the entire chain of command within his own government to assure this result. Understanding that each country bears responsibility for the security of its own nuclear materials, the United States should nonetheless offer Russia any assistance required to make this happen. Each nation-and each leader-should provide the other sufficient transparency to monitor performance.

To ensure that this is done on an expedited schedule, both governments should name specific individuals, directly answerable to their respective presidents, to co-chair a group tasked with developing a joint Russian-American strategy within one month. In developing a joint strategy and program of action, the nuclear superpowers would establish a new world-class "international security standard" based on President Putin's Millennium proposal for new technologies that allow production of electricity with low-enriched, non-weapons-usable nuclear fuel.

A second pillar of this alliance would reach out to all other nuclear weapons states—beginning with Pakistan. Each should be invited to join the alliance and offered assistance, if necessary, in assuring that all weapons and weapons-usable material are secured to the new established international standard in a manner sufficiently transparent to reassure all others. Invitations should be diplomatic in tone but nonetheless clear that this is an offer that cannot be refused. China should become an early ally in this effort, one that could help Pakistan understand the advantages of willing compliance.

A third pillar of this alliance calls for global outreach along the lines proposed by Senator Richard Lugar in what has been called the Lugar Doctrine. All states that possess weapons-usable nuclear materials—even those without nuclear weapons capabilities—must enlist in an international effort to guarantee the security of such materials from theft by terrorists or criminals groups. In effect, each would be required to meet the new international security standard and to do so in a transparent fashion. Pakistan is particularly important given its location and relationship with Al-Qaeda, but beyond nuclear weapons states, several dozen additional countries hosting research reactors—such as Serbia, Libya and Ghana—should be persuaded to surrender such material (almost all of it either American or Soviet in origin), or have the material secured to acceptable international standards.

A fourth pillar of this effort should include Russian-American led cooperation in preventing any further spread of nuclear weapons to additional states, focusing sharply on North Korea, Iraq and Iran. The historical record demonstrates that when the United States and Russia have cooperated intensely, nuclear wannabes have been largely stymied. It was only during periods of competition or distraction, for example in the mid-1990s, that new nuclear weapons states realized their ambitions. India and Pakistan provide two vivid case studies. Recent Russian-American-Chinese cooperation in nudging India and Pakistan back from the nuclear brink suggests a good course of action. The failure and subsequent freeze of North Korean nuclear programs offers complementary lessons about the consequences of competition and distraction. The new alliance should reinvent a robust non-proliferation regime of controls on the sale and export of weapons of mass destruction, nuclear material and missile technologies, recognizing the threat to each of the major states that would be posed by a nuclear-armed Iran, North Korea or Iraq.

Finally, adapting lessons learned in U.S.-Russian cooperation in the campaign against bin Laden and the Taliban, this new alliance should be heavy on intelligence sharing and affirmative counter-proliferation, including disruption and pre-emption to prevent acquisition of materials and know-how by nuclear wannabes. Beyond joint intelligence sharing, joint training for pre-emptive actions against terrorists, criminal groups or rogue states attempting to acquire weapons of mass destruction would provide a fitting enforcement mechanism for alliance commitments.

As former Senator Sam Nunn has noted: "At the dawn of a new century, we find ourselves in a new arms race. Terrorists are racing to get weapons of mass destruction; we ought to be racing to stop them." Preventing nuclear terrorism will require no less imagination, energy and persistence than did avoiding nuclear war between the superpowers over four decades of Cold War. But absent deep, sustained cooperation between the United States, Russia and other nuclear states, such an effort is doomed to failure. In the context of the qualitatively new relationship Presidents Putin and Bush have established in the aftermath of last September 11, success in such a bold effort is within the reach of determined Russian-American leadership. Succeed we must.

Jessica Stern   ↩ **NO**

A Rational Response
to Dirty Bombs

The capture of Abdullah al-Muhajir, the US citizen accused of plotting a
"dirty bomb" attack on Washington DC, is another demonstration that the
world is now vulnerable to unconventional terrorism.

We have known for some time that al-Qaeda has been seeking weapons of
mass destruction. Operatives described their attempts to acquire uranium for
use in nuclear weapons in a New York City federal court over a year ago. Plans
and materials for designing unconventional weapons were found in Afghanistan.
And Osama bin Laden has repeatedly boasted about his success in hiring scientists
to assist him in producing nuclear and biological weapons.

The arrest of Mr al-Muhajir, a former Chicago gang member who changed
his name from Jose Padilla after converting to Islam, reminds us that radical
Islamists aim to turn rage, humiliation and perceived deprivation into a weapon
of war. They do not recruit exclusively in Muslim countries; they seek out dis-
affected youth all around the world.

The UK has become a popular gathering place for radical Islamists from
around the world, some of whom are recruited to join jihads from mosques. A
former FBI counter-terrorism official estimates that between 1,000 and 2,000
young men left the US to become Mujahideen in purported holy wars during
the 1990s. Two New York mosques sent 40 to 50 recruits per year overseas in
the mid-1990s, he says.

The jihadis were never tracked as a group. Immigration officials do not
keep records of US citizens travelling abroad, and a combination of legal controls and self-restraint stopped the FBI and CIA from monitoring their activities.

Nor should it be a surprise that al-Qaeda successfully recruited a former
gang member who had served time in a US prison. Terrorists have long used prisons as a source of operatives, and young men join gangs for some of the same
reasons they join terrorist groups. Hopelessness, the feeling of being left behind
by rapidly changing societies and fear are the fuels of terror, whether practised by
inner-city gangs or by global terrorist criminals.

Still, it is important to keep the threat in perspective. Dirty bombs are
far more frightening than lethal. Numerous studies by government and nongovernment scientists have shown that a dirty bomb would kill only people
in the immediate vicinity of the explosion. While people living downwind of
the blast would be exposed to additional radiation, there would be very few
additional cancer deaths, probably undetectable in the statistical noise.

The US government considered developing radiological weapons during the second world war, but abandoned the project as impracticable. Unlike chemical and biological agents, radioactive poisons act slowly. They are difficult to disseminate in concentrations sufficient to cause death, radiation sickness or cancer. In contrast, chemical agents can be stored for a long time, and are easier to transport. That makes them more attractive to terrorists than radiation devices if the main objective is to kill many people.

But radioactive weapons can be effective instruments of terror because of their psychological impact—the human fear of radiation makes them inherently terrifying. For more than a quarter of a century, psychologists and risk analysts have sought to identify the attributes of risks that are especially feared. Studies of perceived risk show that fear is disproportionately evoked by certain characteristics of radiation: it is mysterious, unfamiliar, indiscriminate, uncontrollable, inequitable and invisible. Exposure is involuntary, and the effects are delayed.

The media also tend to highlight terrorist incidents, heightening dread and panic still further. Because Belfast is still considered a terrorist city, many people consider it to be more dangerous than Washington DC, although there are far more murders per head of population in Washington than in Belfast.

We feel a gut-level fear of terrorism, and are prone to trying to eradicate the risk entirely, with little regard to the cost. In contrast, when risky activities are perceived as voluntary and familiar, danger is likely to be underestimated. On average, more than 100 US citizens a day die in car accidents. Yet people expose themselves to the risk because it is a voluntary act and drivers feel the illusion of control.

What can be done about the problem? First, we need to realise this is a new kind of war. Our enemies deliberately target civilians. But uncertainty, dread and disruption are their most important weapons. Our most important response, then, is an informed public that understands not only the risks we face, but also the role of fear.

But public education is only the first step. Many policy measures can reduce the likelihood and impact of such threats. Nuclear power plants must be secured. Evacuation and clean-up plans should be readied and hospitals should be prepared. Radiation detectors should be deployed at ports and borders. Tracking systems for radioactive isotopes must be improved. Despite the relatively low casualty rate for radiological attacks, the psychological impact will be far more devastating if governments are perceived to be unprepared.

Unconventional weapons, used in a total war, require an unconventional response. New agencies and organisations will have to be involved. Businesses will play an increasingly important role. The food industry needs to be aware that the enemy in this war will not be dressed as a soldier and may not carry a gun. Instead, he may be an insider working at a food processing plant aiming to steal radioactive sources or contaminate food products.

Terrorism is a form of psychological warfare, requiring a psychologically informed response. Our hardest challenge is not to overreact—the terrorists' fondest hope—and not to give in to fears. We will need to find the right balance between civil liberties and public safety. The news about Abdullah al-Muhajir makes us want to scour our cities in search of would-be killers. It would be better to act in a measured and deliberate way.

POSTSCRIPT

Are We Headed Toward a Nuclear 9/11?

There are many arguments to support the contention that nuclear and dirty bombs are hard to obtain, difficult to move and assemble, and even harder to deliver. There is also ample evidence to suggest that most, if not all, of the U.S. government's work is in one way or another designed to thwart such actions because of the enormous consequences were such acts to be carried out. These facts should make Americans rest easier and allay fears if only for the reasons of probability.

However, Allison's contention that failure to assume the worst may prevent the thwarting of such terrorist designs is persuasive. Since September 11 it is clear that the world has entered a new phase of terrorist action and a new level of funding, sophistication, and motivation. The attitude that because something is difficult it is unlikely to take place may be too dangerous to possess. The collapse of the USSR has unleashed a variety of forces, some positive and some more sinister and secretive. The enormous prices that radioactive material and nuclear devices can command on the black market make the likelihood of temptation strong and possibly irresistible.

If states are to err, perhaps they should err on the side of caution and preventive action rather than on reliance on the statistical probability that nuclear terrorism is unlikely. We may never see a nuclear terrorist act in this century, but it is statistically likely that the reason for this will not be for lack of effort on the part of motivated terrorist groups.

Some important research and commentary on nuclear terrorism can be found in Elaine Landau, *Osama bin Laden: A War Against the West* (Twenty-First Century Books, 2002); Jan Lodal, *The Price of Dominance: The New Weapons of Mass Destruction and Their Challenge to American Leadership* (Council on Foreign Relations Press, 2001); and Jessica Stern, *The Ultimate Terrorists* (Harvard University Press, 1999).

ISSUE 17

Is Preemption a Viable Policy Option in Today's Global Security Environment?

YES: Endy Zemenides, from "The Doctrine of Preemption: Precedents and Problems," *The Officer* (April 2004)

NO: Neta C. Crawford, from "The Slippery Slope to Preventive War," *Ethics and International Affairs* (2003)

ISSUE SUMMARY

YES: Endy Zemenides, a member of the National Strategy Forum Review Editorial Board, writes that international laws against preemption are too strict and ignore the nature of self-defense in today's security world. He contends that such a policy, while tricky, must be part of the optional arsenal of U.S. foreign policy.

NO: Neta Crawford, a professor of international affairs at Brown University, contends that preemption is a faulty policy because it creates conflict where none existed and presupposes a level of certainty that no government can adhere to.

In 2002, in the wake of the 9/11 attacks, the Bush administration articulated a foreign policy strategy revolutionary in verbalized U.S. foreign policy history. The president stated that the United States would use preemptive war against America's enemies to prevent subsequent 9/11-like attacks or worse. It also articulated a policy that targeted supporters of terrorism in the same category as the terrorists themselves. This policy of preemption is not new. The state of Israel has made this official policy for decades, and other states have informally practiced this policy when it has served their interests. What is new is that the lone superpower and most powerful military state in the world has articulated such a clear stance, signaling to the world its intent and following through in Iraq.

The ramifications of such a policy and the questions that it engenders are substantial and far-ranging. For example, if preemption is based on imminent threat, then how is that threat calculated and by whom? What does international law say about the policy of preemption, and doesn't this policy violate the United Nations charter? How can a society expect to be immune from such a policy directed against them?

The proponents of preemption argue that the world of terrorism is such that identifying, locating, and acting before terrorists strike is imperative given the enormous human cost of not acting. The argument that it is better to be safe than sorry is strong among these proponents. They further contend that global technology networks and freedom of movement mean that governments must act when they have credible evidence that an attack is about to occur or that other states are supporting and nurturing such an attack.

The debate is an emotional one because it involves the application of military force around the globe and the potential impact on countless civilians amidst such conflicts.

In the following selections, Endy Zemenides argues that to ignore preemption as a tool of foreign policy is to ignore the nature of international relations today. Terrorism of various forms along with rogue states and ethnic conflicts demand that not preemption but international law be reexamined to allow for such actions by legitimate states aimed at self-defense. Neta Crawford contends that no state can be certain of its information or its motives despite its own self-image. Further, she argues that preemption provides "a great temptation" to act for other less-than-immediate interests, and consequently regime self-interest and not national self-defense becomes the primary motivation.

YES

Endy Zemenides

The Doctrine of Preemption: Precedents and Problems

In the aftermath of the September 11 attacks, the Bush administration released a new national security strategy. No aspect of that strategy, or of the Bush administration's practice of foreign policy, has been more widely criticized as the concept of preemptive war declared in the strategy. This concept has been called "unprecedented," "illegal" and "dangerous" by critics at home and abroad. The critiques grew sharper in the wake of the war in Iraq, which many have identified as the first exercise of the Bush administration's doctrine of preemptive war. Unfortunately, the controversy surrounding the war in Iraq and the Bush administration's foreign policy have not resulted in a substantive debate over how to reconcile the historic practice of preemptive war/self-defense and the threat of weapons of mass destruction (WMD) and terrorists.

U.S. Diplomatic History and Preemption

U.S. foreign policy has long contemplated the legality of preemptive self-defense. On December 29, 1837, British troops crossed the Niagara River to the U.S. side and attacked the *Caroline,* a ship owned by U.S. nationals that was allegedly providing assistance to anti-British rebels in Canada. Despite the British claim of self-defense, heated diplomatic exchanges—which ultimately led to a British apology—ensued. Communications between British diplomats and Secretary of State Daniel Webster recognized a right to preemptive self-defense. Webster, however, argued that the *Caroline* incident was not a case of permissible self-defense because of the absence of key criteria.

In a letter to Lord Ashburton, special British representative to Washington, Webster argued that a state properly exercising a right to preemptive self-defense would first have to demonstrate that the "necessity of that self-defense is instant, overwhelming, and leaving no choice of means, and no moment of deliberation." After establishing such necessity, the state acting preemptively could only respond in a manner proportionate to the threat. Thus, the leading nations in the coalition that went to war with Iraq in 2003—the United States

From National Strategy Forum's Spring 2004 Review by Endy Zemendies, pp. 31–32. Copyright © by National Strategy Forum. Reprinted by permission.

and the United Kingdom—established the criteria of necessity and proportionality as the standard for permissible preemptive action.

Preemption and the U.N. Charter

The experience of two World Wars led the international community to establish an order for the use of force. This order took form in the U.N. Charter, which prohibited states from the threat or use of force. The U.N. Charter allowed two exceptions to this prohibition: force authorized by the Security Council under Article 42 of the Charter or the exercise of "the inherent right of individual of collective self-defense if an armed attack occurs against a Member of the United Nations" under Article 51 of the Charter. No provision of the Charter permits, or even acknowledges, the right to preemptive self-defense. The U.N. Charter, however, did not bring about the end of preemptive war in international relations, nor did it necessarily result in the censure of states that conducted such action.

The United States faced the issue of the legality of preemptive action during Security Council debate over the Cuban missile crisis in 1962. A military preemption action to prevent the Soviet Union from introducing nuclear-capable, intermediate-range ballistic missiles into Cuba was considered by the Kennedy administration. President Kennedy ultimately ordered a military "quarantine" of Cuba, and a diplomatic solution was reached.

Although war was ultimately averted, the Cuban missile crisis serves as an important precedent for the right to preemptive self-defense. Not only did the Security Council fail to repudiate a right to preemptive self-defense, the debate in the Security Council included consideration of the necessity criteria established in the *Caroline* case.

Five years after the Cuban missile crisis, Israel launched a preemptive military action against Egypt, Syria, and Jordan. This action, which resulted in a decisive Israeli victory and is known as the Six-Day War, is a classic example of preemptive self-defense. On June 5, 1967, after several weeks of frantic diplomacy while hostile troops were massing in the Sinai, the Golan Heights, and the West Bank, Israel launched its operation. The Security Council took the crisis under advisement, but refused (as did the General Assembly) to condemn the Israeli military action.

The value of the Six-Day War as a precedent for preemptive self-defense is based on the nature of the threat faced by Israel. Hostile nations were undertaking a massive mobilization on Israel's border, threatening the very existence of the state of Israel, and diplomatic recourse had been all but exhausted. The imminence of the threat was undeniable, and Israel's only other choice of means was to withstand an attack.

Self-Defense and Post 9-11 World

Despite the restrictions of the U.N. Charter, there is no explicit prohibition of preemptive self-defense in international law, and customary international law appears to recognize a right to such self-defense when certain criteria are met.

The U.N. Charter and customary international law were formed in a world where states were the main actors to be regulated by law. Moreover, the most dangerous weapons in the world today are WMD, which can be mobilized and undetected, unlike the Egyptian tanks, planes and troops that massed on Israel's borders in 1967.

In such a world, is it reasonable to require states to engage in preemptive self-defense only when an overwhelming threat is *imminent*? Given the difficulty of detecting either mobilized WMD or terrorists, do not states essentially forego a right to effective self-defense if preemptive self-defense is not an option? On the other hand, if states follow the lead of the Bush Administration and do not require a threat to be imminent before acting against it, would preemptive actions remain defensive in nature? If the *Caroline* criteria are relaxed too far, what is to prevent India from acting against Pakistan, Russia acting against former Soviet Republics, and Israel acting against Iran, all in the name of pre-empting attacks by terrorists or WMD attacks?

Preemptive self-defense has always had, and will always have, a place in the foreign policy toolbox of U.S. presidents. The challenge for the international community is to reconsider the established customary international law criteria for preemptive self-defense. The new dimension is the threat of terrorism and WMD.

Neta C. Crawford  **NO**

The Slippery Slope
to Preventive War

The Bush administration's arguments in favor of a preemptive doctrine rest on the view that warfare has been transformed. As Colin Powell argues, "It's a different world . . . it's a new kind of threat."[1] And in several important respects, war has changed along the lines the administration suggests, although that transformation has been under way for at least the last ten to fifteen years. Unconventional adversaries prepared to wage unconventional war can conceal their movements, weapons, and immediate intentions and conduct devastating surprise attacks.[2] Nuclear, chemical, and biological weapons, though not widely dispersed, are more readily available than they were in the recent past. And the everyday infrastructure of the United States can be turned against it as were the planes the terrorists hijacked on September 11, 2001. Further, the administration argues that we face enemies who "reject basic human values and hate the United States and everything for which it stands."[3] Although vulnerability could certainly be reduced in many ways, it is impossible to achieve complete invulnerability.

Such vulnerability and fear, the argument goes, means the United States must take the offensive. Indeed, soon after the September 11, 2001, attacks, members of the Bush administration began equating self-defense with preemption:

> There is no question but that the United States of America has every right, as every country does, of self-defense, and the problem with terrorism is that there is no way to defend against the terrorists at every place and every time against every conceivable technique. Therefore, the only way to deal with the terrorist network is to take the battle to them. That is in fact what we're doing. That is in effect self-defense of a preemptive nature.[4]

The character of potential threats becomes extremely important in evaluating the legitimacy of the new preemption doctrine, and thus the assertion that the United States faces rogue enemies who oppose everything about the United States must be carefully evaluated. There is certainly robust evidence to believe that al-Qaeda members desire to harm the United States and American citizens.

From *Ethics & International Affairs* by Neta C. Crawford, vol. 17, no. 1, 2003, pp. 30–36. Copyright © 2003 by Carnegie Council on Ethics & International Affairs. Reprinted by permission.

The National Security Strategy makes a questionable leap, however, when it assumes that "rogue states" also desire to harm the United States and pose an imminent military threat. Further, the administration blurs the distinction between "rogue states" and terrorists, essentially erasing the difference between terrorists and those states in which they reside: "We make no distinction between terrorists and those who knowingly harbor or provide aid to them."[5] But these distinctions do indeed make a difference.

Legitimate preemption could occur if four necessary conditions were met. First, the party contemplating preemption would have a narrow conception of the "self" to be defended in circumstances of self-defense. Preemption is not justified to protect imperial interests or assets taken in a war of aggression. Second, there would have to be strong evidence that war was inevitable and likely in the immediate future. Immediate threats are those which can be made manifest within days or weeks unless action is taken to thwart them. This requires clear intelligence showing that a potential aggressor has both the capability and the intention to do harm in the near future. Capability alone is not a justification. Third, preemption should be likely to succeed in reducing the threat. Specifically, there should be a high likelihood that the source of the military threat can be found and the damage that it was about to do can be greatly reduced or eliminated by a preemptive attack. If preemption is likely to fail, it should not be undertaken. Fourth, military force must be necessary; no other measures can have time to work or be likely to work.

A Defensible Self

On the face of it, the self-defense criteria seem clear. When our lives are threatened, we must be able to defend ourselves using force if necessary. But self-defense may have another meaning, that in which our "self" is expressed not only by mere existence, but also by a free and prosperous life. For example, even if a tyrant would allow us to live, but not under institutions of our own choosing, we may justly fight to free ourselves from political oppression. But how far do the rights of the self extend? If someone threatens our access to food, or fuel, or shelter, can we legitimately use force? Or if they allow us access to the material goods necessary for our existence, but charge such a high price that we must make a terrible choice between food and health care, or between mere existence and growth, are we justified in using force to secure access to a good that would enhance the self? When economic interests and vulnerabilities are understood to be global, and when the moral and political community of democracy and human rights are defined more broadly than ever before, the self-conception of great powers tends to enlarge. But a broad conception of self is not necessarily legitimate and neither are the values to be defended completely obvious.

For example, the U.S. definition of the self to be defended has become very broad. The administration, in its most recent Quadrennial Defense Review, defines "enduring national interests" as including "contributing to economic well-being," which entails maintaining "vitality and productivity of the global economy" and "access to key markets and strategic resources." Further, the goal of U.S. strategy, according to this document, is to maintain "preeminence."[6] The

National Security Strategy also fuses ambitious political and economic goals with security: "The U.S. national security strategy will be based on a distinctly American internationalism that reflects the fusion of our values and our national interests. The aim of this strategy is to help make the world not just safer but better." And "today the distinction between domestic and foreign affairs is diminishing."[7]

If the self is defined so broadly and threats to this greater "self" are met with military force, at what point does self-defense begin to look like aggression? As Richard Betts has argued, "When security is defined in terms broader than protecting the near-term integrity of national sovereignty and borders, the distinction between offense and defense blurs hopelessly. . . . Security can be as insatiable an appetite as acquisitiveness—there may never be enough buffers."[8] The large self-conception of the United States could lead to a tendency to intervene everywhere that this greater self might conceivably be at risk of, for example, losing access to markets. Thus, a conception of the self that justifies legitimate preemption in self-defense must be narrowly confined to immediate risks to life and health within borders or to the life and health of citizens abroad.

Threshold and Conduct of Justified Preemption

The Bush administration is correct to emphasize the United States' vulnerability to terrorist attack. The administration also argues that the United States cannot wait for a smoking gun if it comes in the form of a mushroom cloud. There may be little or no evidence in advance of a terrorist attack using nuclear, chemical, or biological weapons. Yet, under this view, the requirement for evidence is reduced to a fear that the other has, or might someday acquire, the means for an assault. But the bar for preemption seems to be set too low in the Bush administration's National Security Strategy. How much and what kind of evidence is necessary to justify preemption? What is a credible fear that justifies preemption?

As Michael Walzer has argued persuasively in *Just and Unjust Wars,* simple fear cannot be the only criterion. Fear is omnipresent in the context of a terrorist campaign. And if fear was once clearly justified, when and how will we know that a threat has been significantly reduced or eliminated? The nature of fear may be that once a group has suffered a terrible surprise attack, a government and people will, justifiably, be vigilant. Indeed they may, out of fear, be aware of threats to the point of hypervigilance—seeing small threats as large, and squashing all potential threats with enormous brutality.

The threshold for credible fear is necessarily lower in the context of contemporary counterterrorism war, but the consequences of lowering the threshold may be increased instability and the premature use of force. If this is the case, if fear justifies assault, then the occasions for attack will potentially be limitless since, according to the Bush administration's own arguments, we cannot always know with certainty what the other side has, where it might be located, or when it might be used. If one attacks on the basis of fear, or suspicion that a potential adversary may someday have the intention and capacity to harm you, then the line between preemptive and preventive war has been crossed. Again, the problem is knowing the capabilities and intentions of potential adversaries.

There is thus a fine balance to be struck. The threshold of evidence and warning cannot be too low, where simple apprehension that a potential adversary might be out there somewhere and may be acquiring the means to do the United States harm triggers the offensive use of force. This is not preemption, but paranoid aggression. We must, as stressful as this is psychologically, accept some vulnerability and uncertainty. We must also avoid the tendency to exaggerate the threat and inadvertently to heighten our own fear. For example, although nuclear weapons are more widely available than in the past, as are delivery vehicles of medium and long range, these forces are not yet in the hands of dozens of terrorists. A policy that assumes such a dangerous world is, at this historical juncture, paranoid. We must, rather than assume this is the present case or will be in the future, work to make this outcome less likely.

On the other hand, the threshold of evidence and warning for justified fear cannot be so high that those who might be about to do harm get so advanced in their preparations that they cannot he stopped or the damage limited. What is required, assuming a substantial investment in intelligence gathering, assessment, and understanding of potential advisories, is a policy that both maximizes our understanding of the capabilities and intentions of potential adversaries and minimizes our physical vulnerability. While uncertainty about intentions, capabilities, and risk can never be eliminated, it can be reduced.

Fear of possible future attack is not enough to justify preemption. Rather, aggressive intent, coupled with a capacity and plans to do immediate harm, is the threshold that may trigger justified preemptive attacks. We may judge aggressive intent if the answer to these two questions is yes: First, have potential aggressors said they want to harm us in the near future or have they harmed us in the recent past? Second, are potential adversaries moving their forces into a position to do significant harm? . . .

The conduct of preemptive actions must be limited in purpose to reducing or eliminating the immediate threat. Preemptive strikes that go beyond this purpose will, reasonably, be considered aggression by the targets of such strikes. Those conducting preemptive strikes should also obey the *jus in bello* limits of just war theory, specifically avoiding injury to noncombatants and avoiding disproportionate damage. For example, in the case of the plans for the September 11, 2001, attacks, on these criteria—and assuming intelligence warning of preparations and clear evidence of aggressive intent—a justifiable preemptive action would have been the arrest of the hijackers of the four aircraft that were to be used as weapons. But, prior to the attacks, taking the war to Afghanistan to attack al-Qaeda camps or the Taliban could not have been justified preemption.

The Risks of Preventive War

Foreign policies must not only be judged on grounds of legality and morality, but also on grounds of prudence. Preemption is only prudent if it is limited to clear and immediate dangers and if there are limits to its conduct—proportionality, discrimination, and limited aims. If preemption becomes a regular practice or if it becomes the cover for a preventive offensive war doctrine, the strategy then may become self-defeating as it increases instability and insecurity.

Specifically, a legitimate preemptive war requires that states identify that potential aggressors have both the capability and the intention of doing great harm to you in the immediate future. However, while capability may not be in dispute, the motives and intentions of a potential adversary may be misinterpreted. Specifically, states may mobilize in what appear to be aggressive ways because they are fearful or because they are aggressive. A preemptive doctrine which has, because of great fear and a desire to control the international environment, become a preventive war doctrine of eliminating potential threats that may materialize at some point in the future is likely to create more of both fearful and aggressive states. Some states may defensively arm because they are afraid of the preemptive-preventive state; others may arm offensively because they resent the preventive war aggressor who may have killed many innocents in its quest for total security.

In either case, whether states and groups armed because they were afraid or because they have aggressive intentions, instability is likely to grow as a preventive war doctrine creates the mutual fear of surprise attack. In the case of the U.S. preemptive-preventive war doctrine, instability is likely to increase because the doctrine is coupled with the U.S. goal of maintaining global preeminence and a military force "beyond challenge."[9]

Further, a preventive offensive war doctrine undermines international law and diplomacy, both of which can be useful, even to hegemonic powers. Preventive war short-circuits nonmilitary means of solving problems. If all states reacted to potential adversaries as if they faced a clear and present danger of imminent attack, security would be destabilized as tensions escalated along already tense borders and regions. Article 51 of the UN Charter would lose much of its force. In sum, a preemptive-preventive doctrine moves us closer to a state of nature than a state of international law. Moreover, while preventive war doctrines assume that today's potential rival will become tomorrow's adversary, diplomacy or some other factor could work to change the relationship from antagonism to accommodation. As Otto von Bismarck said to Wilhelm I in 1875, "I would . . . never advise Your Majesty to declare war forthwith, simply because it appeared that our opponent would begin hostilities in the near future. One can never anticipate the ways of divine providence securely enough for that."[10]

One can understand why any administration would favor preemption and why some would be attracted to preventive wars if they think a preventive war could guarantee security from future attack. But the psychological reassurance promised by a preventive offensive war doctrine is at best illusory, and at worst, preventive war is a recipe for conflict. Preventive wars are imprudent because they bring wars that might not happen and increase resentment. They are also unjust because they assume perfect knowledge of an adversary's ill intentions when such a presumption of guilt may be premature or unwarranted. Preemption can be justified, on the other hand, if it is undertaken due to an immediate threat, where there is no time for diplomacy to be attempted, and where the action is limited to reducing that threat. There is a great temptation, however, to step over the line from preemptive to preventive war, because that line is vague and because the stress of living under the threat of war is great. But that temptation should be avoided, and the stress of living in fear should be assuaged by true prevention—arms control, disarmament, negotiations, confidence-building measures, and the development of international law.

Notes

1. Colin Powell, "Perspectives: Powell Defends a First Strike as Iraq Option," interview, *New York Times,* September 8, 2002, sec. 1, p. 18.

2. For more on the nature of this transformation, see Neta C. Crawford, "Just War Theory and the U.S. Counterterror War," *Perspectives on Politics* 1 (March 2003), forthcoming.

3. "The National Security Strategy of the United States of America September 2002," p. 14; available at www.whitehouse.gov/nsc/nss.pdf.

4. Donald H. Rumsfeld, "Remarks at Stakeout Outside ABC TV Studio," October 28, 2001; available at www.defenselink.mil/news/Oct2001/t10292001_t1028sd3.html.

5. "National Security Strategy," p. 5.

6. Department of Defense, "Quadrennial Defense Review" (Washington, D.C.: U.S. Government Printing Office, September 30, 2001), pp. 2, 30, 62.

7. "National Security Strategy," pp. 1, 31.

8. Richard K. Betts, *Surprise Attack: Lessons for Defense Planning* (Washington, D.C.: Brookings Institution, 1982), pp. 14–43.

9. Department of Defense, "Quadrennial Defense Review," pp. 30, 62; and "Remarks by President George W. Bush at 2002 Graduation Exercise of the United States Military Academy, West Point, New York," June 1, 2002; available at www.whitehouse.gov/news/releases/2002/06/20020601-3.html.

10. Quoted in Gordon A. Craig, *The Politics of the Prussian Army, 1640–1945* (Oxford: Oxford University Press, 1955), p. 255.

POSTSCRIPT

Is Preemption a Viable Policy Option in Today's Global Security Environment?

The United States has entered a new era of foreign policy. The end of the Cold War, the fall of the USSR, and the events of 9/11 have changed America's position in the world and arguably how we view the international arena. We are the most powerful country on earth, unchallenged in our military power and reach, and also vulnerable as we have not been for over half a century. We have the means to destroy traditional enemies (states) but find ourselves in a conflict with a shadowy array of forces, difficult to see and thwart.

From out of this complex and new position, the Bush administration has crafted a new foreign policy predicated on preemption as a primary tool to combat enemies. The success or failure of such a policy will be determined by its application over time. The Iraqi example has thus far proven to be a mixed bag of tactical success and strategic failure.

How does the rest of the world view such a policy? Views here are shaded by fear, apprehension, and a bit of skepticism regarding the United States rue motives and ability to "get it right." The only state in recent times that has practiced a strongly overt form of preemption is Israel. Clearly, its internal and external security issues have not been solved with such a policy, although as a military entity, she is unchallenged in the Middle East.

The wisdom and righteousness of preemption will largely be determined in the halls of popular opinion across the United States and the world. Can the United States sustain such a policy and the concomitant increasing applications of U.S. force around the world, and will the global community view the United States as benevolent protectors of democratic values, or imperial bullies bent on changing regimes that they don't like? The jury on these questions is still out.

Some valuable literature here includes: Max Boot, "NeoCons," *Foreign Policy* (January/February 2004), Bruce Nussbaum, "Building a New Multilateral World," *Business Week* (April 2003), Alan W. Dowd, "In Search of Monster's to Destroy: The Causes and Costs of the Bush Doctrine," *The World and I* (January 2003), and Arthur Schlesinger, "The Immorality of Preemptive War," *New Perspectives Quarterly* (Fall 2002).

Are Civil Liberties Being Compromised in the War Against Terrorism?

YES: David Cole, from "An Ounce of Detention," *The American Prospect* (September 2003)

NO: Kim R. Holmes and Edwin Meese III, from "The Administration's Anti-Terrorism Package: Balancing Security and Liberty," *The Heritage Foundation Backgrounder* (October 3, 2001)

ISSUE SUMMARY

YES: David Cole, writing for *The American Prospect,* argues that the Bush administration has restricted civil liberties and begun a process of detention and isolation all in the name of security. Cole contends that these policies are an anathema to the rule of law and hurt our conflict with terrorism, not aid it.

NO: Kim R. Holmes, of the Heritage Foundation, and Edwin Meese III, the Ronald Reagan Distinguished Fellow in Public Policy at the Heritage Foundation, argue that the Bush administration's approach to fighting terrorism, including the Patriot Act, strikes a proper balance between liberty and security.

Two thousand years ago, the Roman statesman Marcus Tullius Cicero said, "Inter arma silent leges," which translates into "In times of war the law is silent." Cicero's point was that governments must act differently during wartime and that one of its likely casualties is the rule of law. Given Western democracies' commitment to the latter principle, particularly to the precept that individuals possess a long list of rights because they are members of the human race, Cicero's words represent a chilling reminder of the inherent tension between security and liberty.

It is not surprising that an overwhelming number of Americans were quick to criticize the George W. Bush administration's approach to winning the war against terrorism—the USA Patriot Act, which gives the executive branch greater power and more tools with which to fight terrorism. Specifically, citizens are concerned about the potential for the curtailment of civil rights.

Efforts to address potential civil rights violations brought about by the war on terrorism have been varied. The MacArthur Foundation awarded over $1 million to 10 groups to investigate the presumed assault on civil liberties by new antiterrorism laws. Three hundred law professors signed a letter of protest against the creation of military tribunals. A Web site has been established for the specific purpose of demanding that the president maintain civil liberties; six months after the terrorist attack on the World Trade Center, over 200,000 e-mails with signatures had been forwarded to the White House. And editorial after editorial echoes the fear that the war on terrorism's first victim would be freedom.

The average American citizen seems far less concerned about the impact of the USA Patriot Act on individual liberties. For example, the NPR/Kaiser/Kennedy School National Survey on Civil Liberties, released in November 2001, concluded that despite holding strong opinions on civil liberties in principle, a "vast majority of Americans are willing to forgo some civil liberties to fight terrorism." A *Newsweek* poll taken in November 2001 found that 72 percent of those polled thought that the level of restrictions on civil liberties imposed by the White House is "about right." And an article in the December 20, 2001, issue of *The Christian Science Monitor* reports that most Americans are happy to have Big Brother watching.

At the same time, scholars have reexamined history to ascertain how the tension between security and liberty was resolved in earlier times of trouble. Not surprisingly, critics on opposite sides of the issue reach different conclusions about what history teaches us. Those who oppose the Bush administration's policies point to the Alien and Sedition Acts (which, among other things, made it a crime to criticize the government), a variety of questionable steps taken by President Abraham Lincoln during the Civil War, legislation passed during World War I curtailing criticism of America's war efforts, McCarthyism of the 1950s, and President Richard Nixon's vigorous pursuit of anti–Vietnam War protesters as prime examples of how the government has used external threats to ignore the norms of civil liberties. Others suggest that America has underestimated past threats and has typically resisted tilting the balance away from liberty and toward higher security until almost too late.

In the following selections, David Cole critiques the Bush administration on the grounds that restricting civil liberties and detaining thousands of people run counter to justice, fairness, and the rule of law. He believes that only through a strict adherence to the law will we prevail over terrorism and its perpetrators. Kim R. Holmes and Edwin Meese III, while agreeing that civil liberties in American political life are of paramount importance, argue that critics of the administration have ignored two important points: that one must distinguish between constitutional liberties on the one hand and mere privileges and conveniences on the other, and that liberty depends on security. They conclude that the administration's antiterrorism measures draw these distinctions properly.

YES ←

David Cole

An Ounce of Detention

In Steven Spielberg's *Minority Report,* set in the not-all-that-distant future, police in Washington, D.C., have hit upon a way—through the enslavement of psychic visionaries—to predict and prevent future crimes. Would-be criminals are apprehended before they actually break the law and are punished for their intent to do so. But as one might expect, things go awry when one officer learns that the psychics' visions can be manipulated, and an innocent man is implicated in a future murder he does not intend to commit.

Neither President George W. Bush nor Attorney General John Ashcroft has discovered any psychic visionaries—with the possible exception of Karl Rove, and his field of vision is limited—but in fighting the war on terrorism, they have nonetheless adopted sweeping new "preventive" strategies that depend on the ability to predict the future. At home, the Department of Justice's goal is no longer simply to prosecute criminals after the fact but to keep violent acts from occurring in the first place—in Ashcroft's terms, "a paradigm of prevention." Abroad, the Bush administration's national-security strategy has redefined self-defense to encompass preventive war—the initiation of hostilities to forestall not only imminent threats but also dangers that might develop at some point down the road. These strategies are rarely considered together, but they are in fact two sides of the same coin. They share not only a common origin and justification but a common philosophy—one that ultimately depends upon double standards and secrecy, disdains the rule of law for the rule of force and is very likely to render us less, not more, secure.

The impetus to strike first is understandable. All other things being equal, preventing a terrorist act is certainly preferable to responding after the fact, all the more so when the threats include weapons of mass destruction and our adversaries are difficult to detect, undeterrable and seemingly unconstrained by considerations of law, morality or human dignity.

But all other things are not equal. Detention and killing, whether through the justice system or waging war, are the two most extreme acts a state can take, and both carry substantial risks of abuse. For these reasons, both the criminal law and the law of war strongly disfavor locking up human beings or launching a war for preventive purposes. As long as the future remains unpredictable,

From *American Prospect* by David Cole, September 1, 2003. Copyright © 2003 by American Prospect. Reprinted by permission.

preventive strategies are bound to harm innocents and to substitute subjective will for the ideal of objective justice.

We've seen this kind of approach before. The federal government justified the excesses of the McCarthy era and the Japanese internment of World War II in preventive terms. In 1951, the Supreme Court adopted that reasoning to uphold the conviction of several American Communist Party leaders for subversive speech. In *Dennis v. United States,* the Court reasoned that in assessing whether speech posed a "clear and present danger," courts "must ask whether the gravity of the evil, discounted by its improbability, justifies such invasion of free speech as is necessary to avoid the danger." Because the threat posed by a communist overthrow of the United States was so great, it did not matter that there was no evidence that it was likely to come to pass.

Similarly, the Court in *Korematsu v. United States* upheld the internment of 110,000 Japanese Americans in the absence of any actual evidence that they posed a danger, deferring instead to the military's unsupported assertions of national-security concerns. Today, *Dennis* is widely seen as a low point in the Court's protection of speech and its standard has been abandoned, while *Korematsu* is universally repudiated. Yet the Bush administration has invoked the same failed reasoning to defend both the domestic and foreign sides of its war on terrorism.

The administration's domestic and foreign-policy preventive strategies share three common features. First, they rely on double standards. Most of the government's domestic incursions on fundamental rights have been targeted at foreign nationals, including the detention of more than 5,000 noncitizens in an initial roundup immediately following September 11 and two subsequent initiatives directed at registering and deporting Arab and Muslim foreign nationals. Only one of these detainees has been convicted of any terrorist crime; nonetheless, the most sweeping campaign of ethnic profiling the country has undertaken since the Japanese internment continues. By contrast, when the government has proposed measures that would affect citizens more directly, the political process has often imposed constraints—as when Congress last year killed post-9-11 proposals to create a national identity card and to set up "Operation TIPS," a network of 11 million citizen spies. We want prevention, it appears, only when the costs are borne by someone else.

The national-security strategy abroad is also predicated on double standards. We would not tolerate a world in which every nation that was concerned about another nation's potential threat could use that as a justification for unilateral offensive military action—a world in which Pakistan could attack India, India attack Pakistan, Iran attack Israel and so on. And as at home, our preventive strategy abroad targets the most vulnerable. As Jonathan Schell has convincingly argued in *The Nation,* we attacked Iraq rather than Pakistan, North Korea, Russia or Iran, all of which pose much more serious dangers, "not because it [was] the worst proliferator, but because it [was] the weakest."

Second, the administration's strategies seek to circumvent processes designed to forestall precipitate action by requiring objective proof of wrongdoing. As the government's treatment of Zacarias Moussaoui and Jose Padilla has demonstrated, guarantees associated with the criminal process—such as the right to a lawyer, to call witnesses and to confront the government's evidence—are obstacles to a preventive strategy. By using noncriminal proceedings—including immigration hearings, detention of people as material witnesses and military custody of "enemy combatants"—the administration has denied most of its targets these basic rights. Indeed, the inspector general's recent report on the treatment of immigration detainees labeled "of interest" to the September 11 investigation makes clear that the administration's modus operandi has been to use immigration law for precisely this end.

Where the administration has resorted to the criminal process, it has generally relied on a sweeping statute that allows it to obtain convictions without proof of individual wrongdoing. Virtually every post-9-11 terrorism prosecution has included a charge under a 1996 statute making it a crime to provide "material support" to designated "terrorist organizations." Under this statute, which was hardly ever used before September 11, an individual can be convicted for providing anything of value—from a book to his own time—to any group designated "terrorist" in a secret administrative process. It is no defense that one's support furthered only lawful, nonviolent activity. This statute essentially resurrects "guilt by association," a tempting tool for preventive law enforcement. [See *"Immaterial and Unsupportable,"* Alexander Gourevitch.]

The preventive national-security strategy similarly relies on shortcuts, as Bush's decision to bypass the United Nations Security Council when it would not do his bidding illustrated. And the administration's revision of the standard for going to war virtually paraphrases the Supreme Court in *Dennis*: "The greater the threat, the greater is the risk of inaction—and the more compelling the case for taking anticipatory action to defend ourselves, even if uncertainty remains as to the time and place of the enemy's attack."

Finally, prevention at home and abroad depends on secrecy. The government has refused to identify most of the approximately 1,200 people it arrested in the first seven weeks after September 11, and to disclose even the number of people detained since then. It tried several hundred foreign nationals in secret. And it has refused to reveal the most basic facts concerning its use of broad new surveillance powers granted it in the wake of September 11.

Secrecy has also ruled the day in foreign affairs. Throughout the run-up to the Iraq War, the administration claimed it knew the Iraqis had weapons of mass destruction and links to al-Qaeda, but simultaneously suggested that it could not reveal the evidence because that would expose critical sources of intelligence. But as months go by without finding evidence of either one, it appears that the administration exploited claims of secrecy to conceal the fact that it simply lacked solid evidence.

The reason that international and domestic laws have long rejected preventive detention and preventive war is that these three defining characteristics— double standards, avoidance of procedural safeguards and secrecy—are anathema to the rule of law, which depends upon consistency, procedural regularity and transparency.

Throwing off the constraints of law does not make us more secure. It undermines the legitimacy of our efforts to quell terrorism and makes it less likely that Arab and Muslim communities, the targets of our double standards, will work cooperatively with us to root out al-Qaeda enemies. And it fuels today's unprecedented anti-Americanism, which in turn supports recruitment by the other side.

Our long-term security in the world rests neither on locking up thousands of suspected terrorists who turn out to have no connection to terrorism nor on attacking countries that have not threatened to attack us. On the contrary, it lies in a commitment to fairness, justice and the rule of law. That is the only true strategy of prevention.

NO

**Kim R. Holmes and
Edwin Meese III**

The Administration's Anti-Terrorism Package: Balancing Security and Liberty

The Bush Administration's anti-terrorist proposal has been the only issue since the terrorist attacks on September 11 to spark serious disagreement in Congress. Some fear that adopting the Administration's proposed new law enforcement measures to increase security would endanger civil liberties.

This is a necessary debate; however, it is also one that could easily be misunderstood. Before the debate endangers the unity needed to fight the war on terrorism, it is appropriate for policymakers to stop and think more carefully about the relationship between liberty and security. Above all, care must be taken to avoid artificially polarizing the discussion into two hostile camps—one favoring security and the other favoring civil liberties.

This can be accomplished in two ways. First, policymakers must distinguish between constitutional liberties on the one hand, and mere privileges and conveniences on the other. Second, they must understand that liberty depends on security and that freedom in the long run depends on eliminating the threat of terrorism as soon as possible.

Indeed, policymakers must do everything in their power to preserve the basic liberties protected by the U.S. Constitution, such as the due process of law (including the need to show probable cause and judicial review for issuing warrants and the right to a hearing); the right to be free of unreasonable search and seizures; the right of free speech and religion; and the right to assembly. While it may be permissible to suspend some rights temporarily in a state of emergency—as in a formal declaration of war by Congress—so far this has not been done.

However committed Americans must be to civil liberties, they do *not* have a constitutional right to complete privacy if it endangers the lives of others. Investigators should not be denied access to potentially critical information gained overseas by foreign intelligence sources that could save lives, merely because the methods by which it was obtained do not conform to the U.S. Constitution. Nor should sensitive intelligence information on terrorists be compromised by disclosure in open court proceedings. There must be a reasonable balance between privacy and security.

From Kim R. Holmes and Edwin Meese III, "The Administration's Anti-Terrorism Package: Balancing Security and Liberty," *The Heritage Foundation Backgrounder,* no. 1484 (October 3, 2001). Copyright © 2001 by The Heritage Foundation. Reprinted by permission.

The Administration's proposed anti-terrorism package for the most part draws these distinctions properly. Specifically, it would:

- *Update* wiretapping laws to keep up with changing technologies concerning cell phones, voice mail, and e-mail surveillance;
- *Permit* the sharing of information between law enforcement agencies and intelligence services;
- *Give* courts the authority to review terrorism cases without compromising classified information; and
- *Enable* the government to detain alien individuals who are found to pose a threat to national security until they are actually removed or until the Attorney General determines the person no longer poses a threat. These people would be detained while charges (such as violation of immigration laws) are pending against them. Once those charges were resolved in favor of the individual, he or she would go free.

Reasonable people may disagree with some of these measures and seek improvements. For example, it seems that some time limitations could be established on the detention of foreigners suspected of terrorist ties. Moreover, some provisions could be modified to ensure that information collected on U.S. citizens is not leaked or shared with someone in or out of government who has no need or right to see it. Finally, every effort should be made to ensure that changes made in criminal law are focused as much as possible on the threat of terrorism and could not be abused or used to broaden the investigative powers of the government for non-terrorist–related cases. The new legal tools, developed as emergency measures to fight terrorism, should not necessarily apply to routine criminal investigations. Information collected on people in terrorist investigations should not be used in criminal cases unrelated to terrorism—in tax evasion cases, for example.

Stop Terrorism to Protect Civil Liberties

The attacks of September 11 inevitably will effect changes in some aspects of criminal law. Foreign groups are attacking Americans within the United States, which cannot help but blur the previously bright lines of distinction between national security and criminal law. Compromises will have to be made. People flying on airplanes will have to be subjected to more intrusive questioning, and while any information an airline gathers should not be used for purposes other than security, people must realize that flying is a voluntary act. Some amount of privacy will have to be sacrificed for the sake of the common safety of the crew and passengers.

If the Administration's anti-terrorism proposal is adopted by Congress and enacted into law, there will be cases in which notice of a warrant on suspected terrorists may have to be delayed in order to avoid tipping off a suspect. And there will be times when aliens suspected of terrorism will be detained by rules and legal procedures that are more restrictive. All of these new provisions are reasonable and necessary. None of them represents an infringement on the constitutional rights of American citizens.

The nation must realize that the nature of the terrorist threat has changed and that some laws must change to deal with this new threat. The distinction between domestic and international terrorism—indeed, even between domestic and foreign threats—is now artificial. The foreign threat is here at home, and the laws that normally protect citizens from criminal activity are not sufficient to deal with the complexities of a threat that is both foreign and domestic. Facing a threat as serious as this, civil liberties are in greater danger from a "business as usual" attitude than they are from the minor changes proposed in the Administration's anti-terrorist package.

The Need to Act

However moderate the Administration's proposal may be, the future could be very different if terrorism is not stopped. The longer Americans remain insecure and vulnerable to terrorist attacks, the greater the likelihood that their constitutional liberties will be eroded in the long run. Surely, the nation needs a balanced approach to fighting terrorism at home, but the government must be able to take whatever measures are necessary and consistent with constitutional liberties to quickly re-establish an environment of security.

Until the federal government takes measures that re-establish Americans' confidence in air travel, the economy will be hampered. Until the government makes Americans feel safe and secure, Muslim Americans may be wrongly attacked. Until the government demonstrates that terrorist cells have been eradicated inside the United States, many Americans will be demanding the kind of excessive law enforcement measures that civil libertarians decry. And until the government takes decisive action against terrorist networks and states abroad, Americans will not feel confident that the threat has been removed from their shores.

Imagine what would happen if the war against terrorism fails. Repeated attacks would create panic, and a terrible backlash against civil liberties would ensue. As the casualty toll grew, the calls for draconian measures would make the rather modest provisions in the Administration's anti-terrorist package pale by comparison. A long twilight struggle against terrorism that proves ineffective would chip away at the Constitution in ways Americans can scarcely imagine. Over time, fear and loathing—particularly if America were victim to an attack from weapons of mass destruction that killed millions—would create a tremendous demand from the American people to restrict liberties in the name of security.

That is why the United States must act quickly and decisively now to destroy terrorism at home and abroad. While the President is correct in saying that the anti-terrorist campaign may take many years, the war on terrorism should not become a permanent feature of American life. The goal should be to destroy terrorism as quickly as possible. The United States needs to be patient, to be sure, but not too patient, especially as far as foreign operations against terrorism are concerned. It must be forceful and determined in the short run in order to protect Americans' constitutional liberties in the long run.

This means that the foreign campaign must be as broad as necessary to ensure that the threat of terrorism is not merely contained, but defeated. It means going after the regimes that harbor terrorists. If the international effort merely strikes at one man, one group, or even one network of terrorists without changing or fundamentally altering the pro-terrorist policies of the regimes that protect them, another terrorist leader, group, or network will arise in their place. The more forceful and effective the foreign anti-terrorist campaign, the less pressure there will be against civil liberties at home.

Americans should not forget that there is no more basic civil liberty than the right not to be blown to bits. Civil governments are formed not only to protect liberties, but to protect the lives of their citizens—from each other and from foreign attacks. The Preamble to the U.S. Constitution states that among the purposes of the federal government are to "insure domestic Tranquility, provide for the common defense, [and] promote the general Welfare" of the American people. There would be no "Blessings of Liberty"—another constitutional goal—if it were not for the order and security provided by the federal government.

Conclusion

The changes in law contained in the Administration's anti-terrorist proposal would be a small price to pay to enhance the nation's capabilities to apprehend terrorists. Whatever limited sacrifice in privacy and privileges there may be in these proposed measures is small in comparison to the long-term risks posed to civil liberties by terrorism.

John Adams said in 1765 that

> Liberty must at all hazards be supported. We have a right to it, derived from our Maker. But if we had not, our fathers have earned and bought it for us, at the expense of their ease, their estates, their pleasure, and their blood.

While American troops may be asked to pay for liberty in blood, most Americans will be asked merely to give up a few privileges and conveniences. Surely, this is a sacrifice they can afford to make.

Americans will never be free so long as terrorists are threatening their homeland. It would be ironic indeed if an inordinate fear of losing some rights were sufficient to deny the nation the tools it needs to stop the very thing that would doom the Constitution—the scourge of terrorism. Americans cannot be free unless they are secure any more than they, in the long run, can be secure unless they are free. The United States must stop terrorism in America if it is to preserve freedom.

POSTSCRIPT

Are Civil Liberties Being Compromised in the War Against Terrorism?

Cole's article is representative of many that appeared in the months following September 11, 2001. In his opinion, the Bush administration was too eager to have its antiterrorism agenda enacted into law. Whereas the administration would respond that the sense of urgency required shortcuts in the legislative process, its detractors saw only an executive branch that believed itself suddenly freed from the system of checks and balances. Moreover, in arguing that because of past transgressions the government cannot be trusted with any additional power, Cole ignores the dramatically different nature of the current situation.

Holmes and Meese assert that critics confuse liberties on the one hand and privileges and conveniences on the other. For Holmes and Meese, the bottom line is that the administration made the proper distinction. Their second point—that liberty depends on security, which in turn depends on the elimination of terrorism as soon as possible—is less defensible in the eyes of neutral observers. Nowhere in history can a definitive case be made that the abdication of civil liberties was a prerequisite for security and victory in wartime.

While much of the literature takes the Bush administration to task for presumably failing to emphasize the protection of civil liberties in its quest, there are a number of sources that argue either that freedoms are not in jeopardy at all or that history shows that any transgressions tend to be temporary and less evasive than previously assumed. A good place to begin additional reading is the Senate Judiciary Committee hearing of November 6, 2001 (http://judiciary.senate.gov/hearing.cfm?id=121), in which Attorney General John Ashcroft makes a strong case in support of the government's intention to uphold civil liberties in its pursuit of terrorists. A particularly thoughtful essay is James V. DeLong's "Liberty and Order: A Delicate but Clear Balance," National Review Online, http://www.nationalreview.com/comment/comment-delong100201.shtml (October 2, 2001). The article "Liberty v Security," *The Economist* (September 29, 2001) addresses the question of where the balancing point between security and liberty must be set. An excellent discussion of the negative implications for civil liberties can be found in the December 4, 2001, testimony of Nadine Strossen, president of the American Civil Liberties Union, before the Senate Committee on the Judiciary (http://judiciary.senate.gov/hearing.cfm?id=128). Finally, a global report addressing the liberty/security issue can be found in the Amnesty International report *Rights at Risk* (January 18, 2002).

ISSUE 19

Are Cultural and Ethnic Rivalries the Defining Dimensions of Twenty-First Century Conflict?

YES: Samuel P. Huntington, from "The Clash of Civilizations?" *Foreign Affairs* (Summer 1993)

NO: Amartya Sen, from "A World Not Neatly Divided," *The New York Times* (November 23, 2001)

ISSUE SUMMARY

YES: Political scientist Samuel P. Huntington argues that the emerging conflicts of the twenty-first century will be cultural and not ideological. He identifies the key fault lines of conflict and discusses how these conflicts will reshape global policy.

NO: Amartya Sen, a Nobel Prize–winning economist, responds that categorizing people into neat cultural containers is artificial and not indicative of the reality of the human experience. He contends that such divisions do not mirror the real world but rather promote division and conflicts, not explain them.

Ethnic conflicts seem to be flaring up around the world with greater and greater frequency. The last few years have witnessed ethnic fighting in places such as Northern Ireland, the Middle East, Southeast Asia, and southern Africa. Ethnic clashes have also broken out in places like Bosnia, Kosovo, Rwanda, East Timor, and Chechnya. Certainly, such ethnic conflicts have flared up throughout the centuries in various places. Yet is it possible that ethnic conflict and clashes between cultures are on the rise and will dominate understanding of conflict in the twenty-first century?

For most of the twentieth century, ideological battles between nations took center stage. From the growth of communism and fascism in the 1920s and 1930s came ideological battles centered on notions of governance, economics, race, and the role of people in relation to the state. In that major battle communism and capitalist democracy won out. In the subsequent years the ideological bipolar conflict dominated, as the United States, the USSR, and their surrogates fought on battlefields from Angola to El Salvador and from Vietnam to Afghanistan. At least in part, the battle centered over which system and method of governance would achieve preeminence. In the end the United States won, and capitalist democracy triumphed over communism.

Today scholars and policymakers grapple with the new dynamics of the global system and search for unifying elements that will help to explain why and how groups engage in conflict. Since ideology has lost its zest, and since no apparent philosophies stand ready to directly challenge capitalism, other rallying cries are being uttered and adopted. This period has witnessed the rise or reinvigoration of such philosophies as Islamic fundamentalism, environmentalism, national self-determination (omnipresent in global politics), and ethnic identity as movements and rallying cries for groups to challenge perceived or real oppressors.

Combined with this development is a technological revolution that has brought people in various parts of the world closer than ever before. Ethnic groups traditionally divided by closed borders or reduced contact are now increasingly thrust together by political, economic, and social factors. While this development is benign enough on the surface, many who feel that their cultures are threatened, their identities challenged, and their rights usurped have reacted with disdain and distrust for "the other." As a result, ethnic conflicts have flared in the Balkans, the Middle East, the former Soviet Union, the Great Lakes region of Africa, South America, and the Indian subcontinent.

With increased tensions, greater amounts of weaponry, and less restraint offered by a bipolar world, these conflicts have raged with devastating human consequences. Issues of ethnic cleansing, genocide, land mines, nuclear proliferation, and narcoterrorism have all sprung from or been fueled by these conflicts.

In the following selection, Samuel P. Huntington argues that cultural rivalries are the wave of the coming age. He contends that fault lines between civilizations (where dominant cultures meet) will be "the flash points for crisis and bloodshed" in the coming decades. He contends that cultures will predominate in the battle for hearts and minds and that groups will engage in conflict to defend against challenges to their cultures as they perceive them.

In the second selection, Nobel Prize–winning economist Amartya Sen makes the case that the entire clash-of-civilizations debate is spurious because it creates false divisions between cultures and peoples. He argues that human identity is far more interwoven and complex and that Huntington's analysis merely simplifies and accentuates differences and as such acerbates the very conflict it purports to explain.

Samuel P. Huntington

YES

The Clash of Civilizations?

The Next Pattern of Conflict

World politics is entering a new phase, and intellectuals have not hesitated to proliferate visions of what it will be—the end of history; the return of traditional rivalries between nation states, and the decline of the nation state from the conflicting pulls of tribalism and globalism, among others. Each of these visions catches aspects of the emerging reality. Yet they all miss a crucial, indeed a central, aspect of what global politics is likely to be in the coming years.

It is my hypothesis that the fundamental source of conflict in this new world will not be primarily ideological or primarily economic. The great divisions among humankind and the dominating source of conflict will be cultural. Nation states will remain the most powerful actors in world affairs, but the principal conflicts of global politics will occur between nations and groups of different civilizations. The clash of civilizations will dominate global politics. The fault lines between civilizations will be the battle lines of the future.

Conflict between civilizations will be the latest phase in the evolution of conflict in the modern world. For a century and a half after the emergence of the modern international system with the Peace of Westphalia [1648], the conflicts of the Western world were largely among princes—emperors, absolute monarchs and constitutional monarchs attempting to expand their bureaucracies, their armies, their mercantilist economic strength and, most important, the territory they ruled. In the process they created nation states, and beginning with the French Revolution the principal lines of conflict were between nations rather than princes. In 1973, as R. R. Palmer put it, "The wars of kings were over; the wars of peoples had begun." This nineteenth-century pattern lasted until the end of World War I. Then, as a result of the Russian Revolution and the reaction against it, the conflict of nations yielded to the conflict of ideologies, first among communism, fascism-Nazism and liberal democracy, and then between communism and liberal democracy. During the Cold War, this latter conflict became embodied in the struggle between the two superpowers, neither of which was a nation state in the classical European sense and each of which defined its identity in terms of its ideology.

From Samuel P. Huntington, "The Clash of Civilizations?" *Foreign Affairs*, vol. 72, no. 3 (Summer 1993). Copyright © 1993 by The Council on Foreign Relations, Inc. Reprinted by permission of *Foreign Affairs*.

These conflicts between princes, nation states and ideologies were primarily conflicts within Western civilization, "Western civil wars," as William Lind has labeled them. This was as true of the Cold War as it was of the world wars and the earlier wars of the seventeenth, eighteenth and nineteenth centuries. With the end of the Cold War, international politics moves out of its Western phase, and its centerpiece becomes the interaction between the West and non-Western civilizations and among non-Western civilizations. In the politics of civilizations, the peoples and governments of non-Western civilizations no longer remain the objects of history as targets of Western colonialism but join the West as movers and shapers of history.

The Nature of Civilizations

During the Cold War the world was divided into the First, Second and Third Worlds. Those divisions are no longer relevant. It is far more meaningful now to group countries not in terms of their political or economic systems or in terms of their level of economic development but rather in terms of their culture and civilization.

What do we mean when we talk of a civilization? A civilization is a cultural entity. Villages, regions, ethnic groups, nationalities, religious groups, all have distinct cultures at different levels of cultural heterogeneity. The culture of a village in southern Italy may be different from that of a village in northern Italy, but both will share in a common Italian culture that distinguishes them from German villages. European communities, in turn, will share cultural features that distinguish them from Arab or Chinese communities. Arabs, Chinese and Westerners, however, are not part of any broader cultural entity. They constitute civilizations. A civilization is thus the highest cultural grouping of people and the broadest level of cultural identity people have short of that which distinguishes humans from other species. It is defined both by common objective elements, such as language, history, religion, customs, institutions, and by the subjective self-identification of people. People have levels of identity: a resident of Rome may define himself with varying degrees of intensity as a Roman, an Italian, a Catholic, a Christian, a European, a Westerner. The civilization to which he belongs is the broadest level of identification with which he intensely identifies. People can and do redefine their identities and, as a result, the composition and boundaries of civilizations change.

Civilizations may involve a large number of people, as with China, ("a civilization pretending to be a state," as Lucian Pye put it), or a very small number of people, such as the Anglophone Caribbean. A civilization may include several nation states, as is the case with Western, Latin American and Arab civilizations, or only one, as is the case with Japanese civilization. Civilizations obviously blend and overlap, and may include subcivilizations. Western civilization has two major variants, European and North American, and Islam has its Arab, Turkic and Malay subdivisions. Civilizations are nonetheless meaningful entities, and while the lines between them are seldom sharp, they are real. Civilizations are dynamic; they rise and fall; they divide and merge. And, as any student of history knows, civilizations disappear and are buried in the sands of time.

Westerners tend to think of nation states as the principal actors in global affairs. They have been that, however, for only a few centuries. The broader reaches of human history have been the history of civilizations. In *A Study of History,* Arnold Toynbee identified 21 major civilizations; only six of them exist in the contemporary world.

Why Civilizations Will Clash

Civilization identity will be increasingly important in the future, and the world will be shaped in large measure by the interactions among seven or eight major civilizations. These include Western, Confucian, Japanese, Islamic, Hindu, Slavic-Orthodox, Latin American and possibly African civilization. The most important conflicts of the future will occur along the cultural fault lines separating these civilizations from one another.

Why will this be the case?

First, differences among civilizations are not only real; they are basic. Civilizations are differentiated from each other by history, language, culture, tradition and, most important, religion. The people of different civilizations have different views on the relations between God and man, the individual and the group, the citizen and the state, parents and children, husband and wife, as well as differing views of the relative importance of rights and responsibilities, liberty and authority, equality and hierarchy. These differences are the product of centuries. They will not soon disappear. They are far more fundamental than differences among political ideologies and political regimes. Differences do not necessarily mean conflict, and conflict does not necessarily mean violence. Over the centuries, however, differences among civilizations have generated the most prolonged and the most violent conflicts.

Second, the world is becoming a smaller place. The interactions between peoples of different civilizations are increasing; these increasing interactions intensify civilization consciousness and awareness of differences between civilizations and commonalities within civilizations. North African immigration to France generates hostility among Frenchmen and at the same time increased receptivity to immigration by "good" European Catholic Poles. Americans react far more negatively to Japanese investment than to larger investments from Canada and European countries. Similarly, as Donald Horowitz has pointed out, "An Ibo may be . . . an Owerri Ibo or an Onitsha Ibo in what was the Eastern region of Nigeria. In Lagos, he is simply an Ibo. In London, he is a Nigerian. In New York, he is an African." The interactions among peoples of different civilizations enhance the civilization-consciousness of people that, in turn, invigorates differences and animosities stretching or thought to stretch back deep into history.

Third, the processes of economic modernization and social change throughout the world are separating people from longstanding local identities. They also weaken the nation state as a source of identity. In much of the world religion has moved in to fill this gap, often in the form of movements that are labeled "fundamentalist." Such movements are found in Western Christianity, Judaism, Buddhism and Hinduism, as well as in Islam. In most countries and most religions the people active in fundamentalist movements are young, college-educated, middle-class technicians, professionals and business persons. The "unsecularization of the world," George Weigel has remarked, is one of the dominant social facts of life in the late twentieth century." The revival of religion, "la revanche de Dieu," as Gilles Kepel labeled it, provides a basis for identity and commitment that transcends national boundaries and unites civilizations.

Fourth, the growth of civilization-consciousness is enhanced by the dual role of the West. On the one hand, the West is at a peak of power. At the same time, however, and perhaps as a result, a return to the roots phenomenon is occurring among non-Western civilizations. Increasingly one hears references to trends toward a turning inward and "Asianization" in Japan, the end of the Nehru legacy and the "Hinduization" of India, the failure of Western ideas of socialism and nationalism and hence "re-Islamization" of the Middle East, and now a debate over Westernization versus Russianization in Boris Yeltsin's country. A West at the peak of its power confronts non-Wests that increasingly have the desire, the will and the resources to shape the world in non-Western ways.

In the past, the elites of non-Western societies were usually the people who were most involved with the West, had been educated at Oxford, the Sorbonne or Sandhurst, and had absorbed Western attitudes and values. At the same time, the populace in non-Western countries often remained deeply imbued with the indigenous culture. Now, however, these relationships are being reversed. A de-Westernization and indigenization of elites is occurring in many non-Western countries at the same time that Western, usually American, cultures, styles and habits become more popular among the mass of the people.

Fifth, cultural characteristics and differences are less mutable and hence less easily compromised and resolved than political and economic ones. In the former Soviet Union, communists can become democrats, the rich can become poor and the poor rich, but Russians cannot become Estonians and Azeris cannot become Armenians. In class and ideological conflicts, the key question was "Which side are you on?" and people could and did choose sides and change sides. In conflicts between civilizations, the question is "What are you?" That is a given that cannot be changed. And as we know, from Bosnia to the Caucasus to the Sudan, the wrong answer to that question can mean a bullet in the head. Even more than ethnicity, religion discriminates sharply and exclusively among people. A person can be half-French and half-Arab and simultaneously even a citizen of two countries. It is more difficult to be half-Catholic and half-Muslim.

Finally, economic regionalism is increasing. The proportions of total trade that were intraregional rose between 1980 and 1989 from 51 percent to 59 percent in Europe, 33 percent to 37 percent in East Asia, and 32 percent to 36 percent in North America. The importance of regional economic blocs is likely to continue to increase in the future. On the one hand, successful economic regionalism will reinforce civilization-consciousness. On the other hand, economic regionalism may succeed only when it is rooted in a common civilization. The European Community rests on the shared foundation of European culture and Western Christianity. The success of the North American Free Trade Area depends on the convergence now underway of Mexican, Canadian and American cultures. Japan, in contrast, faces difficulties in creating a comparable economic entity in East Asia because Japan is a society and civilization unique to itself. However strong the trade and investment links Japan may develop with other East Asian countries, its cultural differences with those countries inhibit and perhaps preclude its promoting regional economic integration like that in Europe and North America.

Common culture, in contrast, is clearly facilitating the rapid expansion of the economic relations between the People's Republic of China and Hong Kong, Taiwan, Singapore and the overseas Chinese communities in other Asian countries. With the Cold War over, cultural commanalities increasingly overcome ideological differences, and mainland China and Taiwan move closer together. If cultural commonality is a prerequisite for economic integration, the principal East Asian economic bloc of the future is likely to be centered on China. This bloc is, in fact, already coming into existence. As Murray Weidenbaum has observed,

> Despite the current Japanese dominance of the region, the Chinese-based economy of Asia is rapidly emerging as a new epicenter for industry, commerce and finance. This strategic area contains substantial amounts of technology and manufacturing capability (Taiwan), outstanding entrepreneurial, marketing and services acumen (Hong Kong), a fine communications network (Singapore), a tremendous pool of financial capital (all three), and very large endowments of land, resources and labor (mainland China). . . . From Guangzhou to Singapore, from Kuala Lumpur to Manila, this influential network—often based on extensions of the tranditional clans—has been described as the beckbone of the East Asian economy.[1]

Culture and religion also form the basis of the Economic Cooperation Organization, which brings together ten non-Arab Muslim countries: Iran, Pakistan, Turkey, Azerbaijan, Kazakhstan, Kyrgyzstan, Turkmenistan, Tadjikistan, Uzbekistan and Afghanistan. One impetus to the revival and expansion of this organization, founded originally in the 1960s by Turkey, Pakistan and Iran, is the realization by the leaders of several of these countries that they had no chance of admission to the European Community. Similarly, Caricom, the Central American Common Market and Mercosur rest on common cultural foundations. Efforts to build a broader Caribbean-Central American economic entity bridging the Anglo-Latin divide, however, have to date failed.

As people define their identity in ethnic and religious terms, they are likely to see an "us" versus "them" relation existing between themselves and people of different ethnicity or religion. The end of ideologically defined states in Eastern Europe and the former Soviet Union permits traditional ethnic identities and animosities to come to the fore. Differences in culture and religion create differences over policy issues, ranging from human rights to immigration to trade and commerce to the environment. Geographical propinquity gives rise to conflicting territorial claims from Bosnia to Mindanao. Most important, the efforts of the West to promote its values of democracy and liberalism as universal values, to maintain its military predominance and to advance its economic interests engender countering responses from other civilizations. Decreasingly able to mobilize support and form coalitions on the basis of ideology, governments and groups will increasingly attempt to mobilize support by appealing to common religion and civilization identity.

The clash of civilizations thus occurs at two levels. At the micro-level, adjacent groups along the fault lines between civilizations struggle, often violently, over the control of territory and each other. At the macro-level, states from different civilizations compete for relative military and economic power, struggle over the control of international institutions and third parties, and competitively promote their particular political and religious values.

The Fault Lines Between Civilizations

The fault lines between civilizations are replacing the political and ideological boundaries of the Cold War as the flash points for crisis and bloodshed. The Cold War began when the Iron Curtain divided Europe politically and ideologically. The Cold War ended with the end of the Iron Curtain. As the ideological division of Europe has disappeared, the cultural division of Europe between Western Christianity, on the one hand, and Orthodox Christianity and Islam, on the other, has reemerged. The most significant dividing line in Europe, as William Wallace has suggested, may well be the eastern boundary of Western Christianity in the year 1500. This line runs along what are now the boundaries between Finland and Russia and between the Baltic states and Russia, cuts through Belarus and Ukraine separating the more Catholic western Ukraine from Orthodox eastern Ukraine, swings westward separating Transylvania from the rest of Romania, and then goes through Yugoslavia almost exactly along the line now separating Croatia and Slovenia from the rest of Yugoslavia. In the Balkans this line, of course, coincides with the historic boundary between the Hapsburg and Ottoman empires. The peoples to the north and west of this line are Protestant or Catholic; they shared the common experiences of European history—feudalism, the Renaissance, the Reformation, the Enlightenment, the French Revolution, the Industrial Revolution; they are generally economically better off than the peoples to the east; and they may now look forward to increasing involvement in a common European economy and to the consolidation of democratic political systems. The peoples to the east and south of this line are Orthodox or Muslim; they historically belonged to the Ottoman or Tsarist empires and were only lightly touched by the shaping events in the rest of Europe; they are generally less advanced economically; they seem much less likely to develop stable democratic political systems. The Velvet Curtain of culture has replaced the Iron Curtain of ideology as the most significant dividing line in Europe. As the events in Yugoslavia show, it is not only a line of difference; it is also at times a line of bloody conflict.

Conflict along the fault line between Western and Islamic civilizations has been going on for 1,300 years. After the founding of Islam, the Arab and Moorish surge west and north only ended at Tours in 732. From the eleventh to the thirteenth century the Crusaders attempted with temporary success to bring Christianity and Christian rule to the Holy Land. From the fourteenth to the seventeenth century, the Ottoman Turks reversed the balance, extended their sway over the Middle East and the Balkans, captured Constantinople, and twice laid siege to Vienna. In the nineteenth and early twentieth centuries as Ottoman power declined Britain, France, and Italy established Western control over most of North Africa and the Middle East.

After World War II, the West, in turn, began to retreat; the colonial empires disappeared; first Arab nationalism and then Islamic fundamentalism manifested themselves; the West became heavily dependent on the Persian Gulf countries for its energy; the oil-rich Muslim countries became money-rich and, when they wished to, weapons-rich. Several wars occurred between Arabs and Israel (created by the West). France fought a bloody and ruthless war in Algeria for most of the 1950s; British and French forces invaded Egypt in 1956; American forces went into Lebanon in 1958; subsequently American forces returned to Lebanon, attacked Libya, and engaged in various military encounters with Iran; Arab and Islamic terrorists, supported by at least three Middle Eastern governments, employed the weapon of the weak and bombed Western planes and installations and seized Western hostages. This warfare between Arabs and the West culminated in 1990, when the United States sent a massive army to the Persian Gulf to defend some Arab countries against aggression by another. In its aftermath NATO planning is increasingly directed to potential threats and instability along its "southern tier."

This centuries-old military interaction between the West and Islam is unlikely to decline. It could become more virulent. The Gulf War left some Arabs feeling proud that Saddam Hussein had attacked Israel and stood up to the West. It also left many feeling humiliated and resentful of the West's military presence in the Persian Gulf, the West's overwhelming military dominance, and their apparent inability to shape their own destiny. Many Arab countries, in addition to the oil exporters, are reaching levels of economic and social development where autocratic forms of government become inappropriate and efforts to introduce democracy become stronger. Some openings in Arab political systems have already occurred. The principal beneficiaries of these openings have been Islamist movements. In the Arab world, in short, Western democracy strengthens anti-Western political forces. This may be a passing phenomenon, but it surely complicates relations between Islamic countries and the West.

Those relations are also complicated by demography. The spectacular population growth in Arab countries, particularly in North Africa, has led to increased migration to Western Europe. The movement within Western Europe toward minimizing internal boundaries has sharpened political sensitivities with respect to this development. In Italy, France and Germany, racism is increasingly open, and political reactions and violence against Arab and Turkish migrants have become more intense and more widespread since 1990.

On both sides the interaction between Islam and the West is seen as a clash of civilizations. The West's "next confrontation," observes M. J. Akbar, an Indian Muslim author, "is definitely going to come from the Muslim world. It is in the sweep of the Islamic nations from the Maghreb to Pakistan that the struggle for a new world order will begin." Bernard Lewis comes to a similar conclusion:

> We are facing a mood and a movement far transcending the level of issues and policies and the governments that pursue them. This is no less than a

clash of civilizations—the perhaps irrational but surely historic reaction of an ancient rival against our Judeo-Christian heritage, our secular present, and the worldwide expansion of both.[2]

Historically, the other great antagonistic interaction of Arab Islamic civilization has been with the pagan, animist, and now increasingly Christian black peoples to the south. In the past, this antagonism was epitomized in the image of Arab slave dealers and black slaves. It has been reflected in the on-going civil war in the Sudan between Arabs and blacks, the fighting in Chad between Libyan-supported insurgents and the government, the tensions between Orthodox Christians and Muslims in the Horn of Africa, and the political conflicts, recurring riots and communal violence between Muslims and Christians in Nigeria. The modernization of Africa and the spread of Christianity are likely to enhance the probability of violence along this fault line. Symptomatic of the intensification of this conflict was the Pope John Paul II's speech in Khartoum in February 1993 attacking the actions of the Sudan's Islamist government against the Christian minority there.

On the northern border of Islam, conflict has increasingly erupted between Orthodox and Muslim peoples, including the carnage of Bosnia and Sarajevo, the simmering violence between Serb and Albanian, the tenuous relations between Bulgarians and their Turkish minority, the violence between Ossetians and Ingush, the unremitting slaughter of each other by Armenians and Azeris, the tense relations between Russians and Muslims in Central Asia, and the deployment of Russian troops to protect Russian interests in the Caucasus and Central Asia. Religion reinforces the revival of ethnic identities and restimulates Russian fears about the security of their southern borders. This concern is well captured by Archie Roosevelt:

> Much of Russian history concerns the struggle between the Slavs and the Turkic peoples on their borders, which dates back to the foundation of the Russian state more than a thousand years ago. In the Slavs' millennium-long confrontation with their eastern neighbors lies the key to an understanding not only of Russian history, but Russian character. To understand Russian realities today one has to have a concept of the great Turkic ethnic group that has preoccupied Russians through the centuries.[3]

The conflict of civilizations is deeply rooted elsewhere in Asia. The historic clash between Muslim and Hindu in the subcontinent manifests itself now not only in the rivalry between Pakistan and India but also in intensifying religious strife within India between increasingly militant Hindu groups and India's substantial Muslim minority. The destruction of the Ayodhya mosque in December 1992 brought to the fore the issue of whether India will remain a secular democratic state or become a Hindu one. In East Asia, China has outstanding territorial disputes with most of its neighbors. It has pursued a ruthless policy toward the Buddhist people of Tibet, and it is pursuing an increasingly ruthless policy toward its Turkic-Muslim minority. With the Cold War over, the underlying differences between China and the United States have reasserted themselves in areas such as human rights, trade and weapons proliferation. These differences are unlikely to moderate. A "new cold war," Deng Xiaoping reportedly asserted in 1991, is under way between China and America. . . .

Civilization Rallying: The Kin-Country Syndrome

Groups or states belonging to one civilization that become involved in war with people from a different civilization naturally try to rally support from other members of their own civilization. As the post–Cold War world evolves, civilization commonality, what H. D. S. Greenway has termed the "kin-country" syndrome, is replacing political ideology and traditional balance of power considerations as the principal basis for cooperation and coalitions. It can be seen gradually emerging in the post–Cold War conflicts in the Persian Gulf, the Caucasus and Bosnia. None of these was a full-scale war between civilizations, but each involved some elements of civilizational rallying, which seemed to become more important as the conflict continued and which may provide a foretaste of the future.

First, in the Gulf War one Arab state invaded another and then fought a coalition of Arab, Western and other states. While only a few Muslim governments overtly supported Saddam Hussein, many Arab elites privately cheered him on, and he was highly popular among large sections of the Arab publics. Islamic fundamentalist movements universally supported Iraq rather than the Western-backed governments of Kuwait and Saudi Arabia. Forswearing Arab nationalism, Saddam Hussein explicitly invoked an Islamic appeal. He and his supporters attempted to define the war as a war between civilizations. "It is not the world against Iraq," as Safar Al-Hawali, dean of Islamic Studies at the Umm Al-Qura University in Mecca, put it in a widely circulated tape. "It is the West against Islam." Ignoring the rivalry between Iran and Iraq, the chief Iranian religious leader, Ayatollah Ali Khamenei, called for a holy war against the West: "The struggle against American aggression, greed, plans and policies will be counted as a jihad, and anybody who is killed on that path is a martyr." "This is a war," King Hussein of Jordan argued, "against all Arabs and all Muslims and not against Iraq alone."

The rallying of substantial sections of Arab elites and publics behind Saddam Hussein caused those Arab governments in the anti-Iraq coalition to moderate their activities and temper their public statements. Arab governments opposed or distanced themselves from subsequent Western efforts to apply pressure on Iraq, including enforcement of a no-fly zone in the summer of 1992 and the bombing of Iraq in January 1993. The Western-Soviet-Turkish-Arab anti-Iraq coalition of 1990 had by 1993 become a coalition of almost only the West and Kuwait against Iraq.

Muslims contrasted Western actions against Iraq with the West's failure to protect Bosnians against Serbs and to impose sanctions on Israel for violating U.N. resolutions. The West, they alleged, was using a double standard. A world of clashing civilizations, however, is inevitably a world of double standards: people apply one standard to their kin-countries and a different standard to others.

Second, the kin-country syndrome also appeared in conflicts in the former Soviet Union. Armenian military successes in 1992 and 1993 stimulated Turkey to become increasingly supportive of its religious, ethnic and linguistic brethren in Azerbaijan. "We have a Turkish nation feeling the same sentiments as the Azerbaijanis," said one Turkish official in 1992. "We are under pressure. Our newspapers are full of the photos of atrocities and are asking us if we are still serious about pursuing our neutral policy. Maybe we should show Armenia that there's a big Turkey in the region." President Turgut Özal agreed, remarking that Turkey should at least "scare the Armenians a little bit." Turkey, Özal threatened again in 1993, would "show its fangs." Turkish Air Force jets flew reconnaissance flights along the Armenian border; Turkey suspended food shipments and air flights to Armenia; and Turkey and Iran announced they would not accept dismemberment of Azerbaijan. In the last years of its existence, the Soviet government supported Azerbaijan because its government was dominated by former communists. With the end of the Soviet Union, however, political considerations gave way to religious ones. Russian troops fought on the side of the Armenians, and Azerbaijan accused the "Russian government of turning 180 degrees" toward support for Christian Armenia.

Third, with respect to the fighting in the former Yugoslavia, Western publics manifested sympathy and support for the Bosnian Muslims and the horrors they suffered at the hands of the Serbs. Relatively little concern was expressed, however, over Croatian attacks on Muslims and participation of the dismemberment of Bosnia-Herzegovina. In the early stages of the Yugoslav breakup, Germany, in an unusual display of diplomatic initiative and muscle, induced the other 11 members of the European Community to follow its lead in recognizing Slovenia and Crotia. As a result of the pope's determination to provide strong backing to the two Catholic countries, the Vatican extended recognition even before the Community did. The United States followed the European lead. Thus the leading actors in Western civilization rallied behind their coreligionists. Subsequently Croatia was reported to be receiving substantial quantities of arms from Central European and other Western countries. Boris Yeltsin's government, on the other hand, attempted to pursue a middle course that would be sympathetic to the Orthodox Serbs but not alienate Russia from the West. Russian conservative and nationalist groups, however, including many legislators, attacked the government for not being more forthcoming in its support for the Serbs. By early 1993 several hundred Russians apparently were serving with the Serbian forces, and reports circulated of Russian arms being supplied to Serbia.

Islamic governments and groups, on the other hand, castigated the West for not coming to the defense of the Bosnians. Iranian leaders urged Muslims from all countries to provide help to Bosnia; in violation of the U.N. arms embargo, Iran supplied weapons and men for the Bosnians; Iranian-supported Lebanese groups sent guerrillas to train and organize the Bosnian forces. In 1993 up to 4,000 Muslims from over two dozen Islamic countries were reported to be fighting in Bosnia. The governments of Saudi Arabia and other countries felt under increasing pressure from fundamentalist groups in their own societies to provide more vigorous support for the Bosnians. By the end of 1992, Saudi Arabia had reportedly supplied substantial funding for weapons and supplies for the Bosnians, which significantly increased their military capabilities vis-à-vis the Serbs.

In the 1930s the Spanish Civil War provoked intervention from countries that politically were fascist, communist and democratic. In the 1990s the Yugoslav conflict is provoking intervention from countries that are Muslim, Orthodox and Western Christian. The parallel has not gone unnoticed. "The war in Bosnia-Herzegovina has become the emotional equivalent of the fight against fascism in the Spanish Civil War," one Saudi editor observed. "Those who died there are regarded as martyrs who tried to save their fellow Muslims."

Conflicts and violence will also occur between states and groups within the same civilization. Such conflicts, however, are likely to be less intense and less likely to expand than conflicts between civilizations. Common membership in a civilization reduces the probability of violence in situations where it might otherwise occur. In 1991 and 1992 many people were alarmed by the possibility of violent conflict between Russia and Ukraine over territory, particularly Crimea, the Black Sea fleet, nuclear weapons and economic issues. If civilization is what counts, however, the likelihood of violence between Ukrainians and Russians should be low. They are two Slavic, primarily Orthodox peoples who have had close relationships with each other for centuries. As of early 1993, despite all the reasons for conflict, the leaders of the two countries were effectively negotiating and defusing the issues between the two countries. While there has been serious fighting between Muslims and Christians elsewhere in the former Soviet Union and much tension and some fighting between Western and Orthodox Christians in the Baltic states, there has been virtually no violence between Russians and Ukrainians.

Civilization rallying to date has been limited, but it has been growing, and it clearly has the potential to spread much further. As the conflicts in the Persian Gulf, the Caucasus and Bosnia continued, the positions of nations and the cleavages between them increasingly were along civilizational lines. Populist politicians, religious leaders and the media have found it a potent means of arousing mass support and of pressuring hesitant governments. In the coming years, the local conflicts most likely to escalate into major wars will be those, as in Bosnia and the Caucasus, along the fault lines between civilizations. The next world war, if there is one, will be a war between civilizations.

The West Versus the Rest

The West is now at an extraordinary peak of power in relation to other civilizations. Its superpower opponent has disappeared from the map. Military conflict among Western states is unthinkable, and Western military power is unrivaled. Apart from Japan, the West faces no economic challenge. It dominates international political and security institutions and with Japan international economic institutions. Global political and security issues are effectively settled by a directorate of the United States, Britain and France, world economic issues by a directorate of the United States, Germany and Japan, all of which maintain extraordinarily close relations with each other to the exclusion of lesser and largely non-Western countries. Decisions made at the U.N. Security

Council or in the International Monetary Fund [IMF] that reflect the interests of the West are presented to the world as reflecting the desires of the world community. The very phrase "the world community" has become the euphemistic collective noun (replacing "the Free World") to give global legitimacy to actions reflecting the interests of the United States and other Western powers.[4] Through the IMF and other international economic institutions, the West promotes its economic interests and imposes on other nations the economic policies it thinks appropriate. In any poll of non-Western peoples, the IMF undoubtedly would win the support of finance ministers and a few others, but get an overwhelmingly unfavorable rating from just about everyone else who would agree with Georgy Arbatov's characterization of IMF officials as "neo-Bolsheviks who love expropriating other people's money, imposing undemocratic and alien rules of economic and political conduct and stifling economic freedom."

Western domination of the U.N. Security Council and its decisions, tempered only by occasional abstention by China, produced U.N. legitimation of the West's use of force to drive Iraq out of Kuwait and its elimination of Iraq's sophisticated weapons and capacity to produce such weapons. It also produced the quite unprecedented action by the United States, Britain and France in getting the Security Council to demand that Libya hand over the Pan Am 103 bombing suspects and then to impose sanctions when Libya refused. After defeating the largest Arab army, the West did not hesitate to throw its weight around in the Arab world. The West in effect is using international institutions, military power and economic resources to run the world in ways that will maintain Western predominance, protect Western interests and promote Western political and economic values.

That at least is the way in which non-Westerners see the new world, and there is a significant element of truth in their view. Differences in power and struggles for military, economic and institutional power are thus one source of conflict between the West and other civilizations. Differences in culture, that is basic values and beliefs, are a second source of conflict. V. S. Naipaul has argued that Western civilization is the "universal civilization" that "fits all men." At a superficial level much of Western culture has indeed permeated the rest of the world. At a more basic level, however, Western concepts differ fundamentally from those prevalent in other civilizations. Western ideas of individualism, liberalism, constitutionalism, human rights, equality, liberty, the rule of law, democracy, free markets, the separation of church and state, often have little resonance in Islamic, Confucian, Japanese, Hindu, Buddhist or Orthodox cultures. Western efforts to propagate such ideas produce instead a reaction against "human rights imperialism" and a reaffirmation of indigenous values, as can be seen in the support for religious fundamentalism by the younger generation in non-Western cultures. The very notion that there could be a "universal civilization" is a Western idea, directly at odds with the particularism of most Asian societies and their emphasis on what distinguishes one people from another. Indeed, the author of a review of 100 comparative studies of values in different societies concluded that "the values that are most important in the West are least important worldwide."[5] In the political realm, of course, these differences are most manifest in the efforts of the United States and other Western powers to induce other peoples to adopt

Western ideas concerning democracy and human rights. Modern democratic government originated in the West. When it has developed in non-Western societies it has usually been the product of Western colonialism or imposition.

The central axis of world politics in the future is likely to be, in Kishore Mahbubani's phrase, the conflict between "the West and the Rest" and the responses of non-Western civilizations to Western power and values.[6] Those responses generally take one or a combination of three forms. At one extreme, non-Western states can, like Burma and North Korea, attempt to pursue a course of isolation, to insulate their societies from penetration or "corruption" by the West, and, in effect, to opt out of participation in the Western-dominated global community. The costs of this course, however, are high, and few states have pursued it exclusively. A second alternative, the equivalent of "band-wagoning" in international relations theory, is to attempt to join the West and accept its values and institutions. The third alternative is to attempt to "balance" the West by developing economic and military power and cooperating with other non-Western societies against the West, while preserving indigenous values and institutions; in short, to modernize but not to Westernize. . . .

The Confucian-Islamic Connection

The obstacles to non-Western countries joining the West vary considerably. They are least for Latin American and East European countries. They are greater for the Orthodox countries of the former Soviet Union. They are still greater for Muslim, Confucian, Hindu and Buddhist societies. Japan has established a unique position for itself as an associate member of the West: it is in the West in some respects but clearly not of the West in important dimensions. Those countries that for reason of culture and power do not wish to, or cannot, join the West compete with the West by developing their own economic, military and political power. They do this by promoting their internal development and by cooperating with other non-Western countries. The most prominent form of this cooperation is the Confucian-Islamic connection that has emerged to challenge Western interests, values and power.

Almost without exception, Western countries are reducing their military power; under Yeltsin's leadership so also is Russia. China, North Korea and several Middle Eastern states, however, are significantly expanding their military capabilities. They are doing this by the import of arms from Western and non-Western sources and by the development of indigenous arms industries. One result is the emergence of what Charles Krauthammer has called "Weapon States," and the Weapon States are not Western states. Another result is the redefinition of arms control, which is a Western concept and a Western goal. During the Cold War the primary purpose of arms control was to establish a stable military balance between the United States and its allies and the Soviet Union and its allies. In the post–Cold War world the primary objective of arms control is to prevent the development by non-Western societies of military capabilities that could threaten Western interests. The West attempts to do this through international agreements, economic pressure and controls on the transfer of arms and weapons technologies.

The conflict between the West and the Confucian-Islamic states focuses largely, although not exclusively, on nuclear, chemical and biological weapons, ballistic missiles and other sophisticated means for delivering them, and the guidance, intelligence and other electronic capabilities for achieving that goal. The West promotes non-proliferation as a universal norm and nonproliferation treaties and inspections as means of realizing that norm. It also threatens a variety of sanctions against those who promote the spread of sophisticated weapons and proposes some benefits for those who do not. The attention of the West focuses, naturally, on nations that are actually or potentially hostile to the West.

The non-Western nations, on the other hand, assert their right to acquire and to deploy whatever weapons they think necessary for their security. They also have absorbed, to the full, the truth of the response of the Indian defense minister when asked what lesson he learned from the Gulf War: "Don't fight the United States unless you have nuclear weapons." Nuclear weapons, chemical weapons and missiles are viewed, probably erroneously, as the potential equalizer of superior Western conventional power. China, of course, already has nuclear weapons; Pakistan and India have the capability to deploy them. North Korea, Iran, Iraq, Libya and Algeria appear to be attempting to acquire them. A top Iranian official has declared that all Muslim states should acquire nuclear weapons, and in 1988 the president of Iran reportedly issued a directive calling for development of "offensive and defensive chemical, biological and radiological weapons."

Centrally important to the development of counter-West military capabilities is the sustained expansion of China's military power and its means to create military power. Buoyed by spectacular economic development, China is rapidly increasing its military spending and vigorously moving forward with the modernization of its armed forces. It is purchasing weapons from the former Soviet states; it is developing long-range missiles; in 1992 it tested a one-megaton nuclear device. It is developing power-projection capabilities, acquiring aerial refueling technology, and trying to purchase an aircraft carrier. Its military buildup and assertion of sovereignty over the South China Sea are provoking a multilateral regional arms race in East Asia. China is also a major exporter of arms and weapons technology. It has exported materials to Libya and Iraq that could be used to manufacture nuclear weapons and nerve gas. It has helped Algeria build a reactor suitable for nuclear weapons research and production. China has sold to Iran nuclear technology that American officials believe could only be used to create weapons and apparently has shipped components of 300-mile-range missiles to Pakistan. North Korea has had a nuclear weapons program under way for some while and has sold advanced missiles and missile technology to Syria and Iran. The flow of weapons and weapons technology is generally from East Asia to the Middle East. There is, however, some movement in the reverse direction; China has received Stinger missiles from Pakistan.

A Confucian-Islamic military connection has thus come into being, designed to promote acquisition by its members of the weapons and weapons technologies needed to counter the military power of the West. It may or may not last. At present, however, it is, as Dave McCurdy has said, "a renegades' mutual support pact, run by the proliferators and their backers." A new form of arms competition is thus occurring between Islamic-Confucian states and the

West. In an old-fashioned arms race, each side developed its own arms to balance or to achieve superiority against the other side. In this new form of arms competition, one side is developing its arms and the other side is attempting not to balance but to limit and prevent that arms build-up while at the same time reducing its own military capabilities.

Implications for the West

This article does not argue that civilization identities will replace all other identities, that nation states will disappear, that each civilization will become a single coherent political entity, that groups within a civilization will not conflict with and even fight each other. This paper does set forth the hypotheses that differences between civilizations are real and important; civilization-consciousness is increasing; conflict between civilizations will supplant ideological and other forms of conflict as the dominant global form of conflict; international relations, historically a game played out within Western civilization, will increasingly be de-Westernized and become a game in which non-Western civilizations are actors and not simply objects; successful political, security and economic international institutions are more likely to develop within civilizations than across civilizations; conflicts between groups in different civilizations will be more frequent, more sustained and more violent than conflicts between groups in the same civilization; violent conflicts between groups in different civilizations are the most likely and most dangerous source of escalation that could lead to global wars; the paramount axis of world politics will be the relations between "the West and the Rest;" the elites in some torn non-Western countries will try to make their countries part of the West, but in most cases face major obstacles to accomplishing this; a central focus of conflict for the immediate future will be between the West and several Islamic-Confucian states.

This is not to advocate the desirability of conflicts between civilizations. It is to set forth descriptive hypotheses as to what the future may be like. If these are plausible hypotheses, however, it is necessary to consider their implications for Western policy. These implications should be divided between short-term advantage and long-term accommodation. In the short term it is clearly in the interest of the West to promote greater cooperation and unity within its own civilization, particularly between its European and North American components; to incorporate into the West societies in Eastern Europe and Latin America whose cultures are close to those of the West; to promote and maintain cooperative relations with Russia and Japan; to prevent escalation of local inter-civilization conflicts into major inter-civilization wars; to limit the expansion of the military strength of Confucian and Islamic states; to moderate the reduction of Western military capabilities and maintain military superiority in East and Southwest Asia; to exploit differences and conflicts among Confucian and Islamic states; to support in other civilizations groups sympathetic to Western values and interests; to strengthen international institutions that reflect and legitimate Western interests and values and to promote the involvement of non-Western states in those institutions.

In the longer term other measures would be called for. Western civilization is both Western and modern. Non-Western civilizations have attempted to become modern without becoming Western. To date only Japan has fully succeeded in this quest. Non-Western civilizations will continue to attempt to acquire the wealth, technology, skills, machines and weapons that are part of being modern. They will also attempt to reconcile this modernity with their traditional culture and values. Their economic and military strength relative to the West will increase. Hence the West will increasingly have to accommodate these non-Western modern civilizations whose power approaches that of the West but whose values and interests differ significantly from those of the West. This will require the West to maintain the economic and military power necessary to protect its interests in relation to these civilizations. It will also, however, require the West to develop a more profound understanding of the basic religious and philosophical assumptions underlying other civilizations and the ways in which people in those civilizations see their interests. It will require an effort to identify elements of commonality between Western and other civilizations. For the relevant future, there will be no universal civilization, but instead a world of different civilizations, each of which will have to learn to coexist with the others.

Notes

1. Murray Weidenbaum, *Greater China: The Next Economic Superpower?*, St. Louis: Washington University Center for the Study of American Business, Contemporary Issues, Series 57, February 1993, pp. 2–3.

2. Bernard Lewis, "The Roots of Muslim Rage," *The Atlantic Monthly*, vol. 266, September 1990, p. 60; *Time*, June 15, 1992, pp. 24–28.

3. Archie Roosevelt, *For Lust of Knowing*, Boston; Little, Brown, 1988, pp. 332–333.

4. Almost invariably Western leaders claim they are acting on behalf of "the world community." One minor lapse occurred during the run-up to the Gulf War. In an interview on "Good Morning America," Dec. 21, 1990, British Prime Minister John Major referred to the actions "the West" was taking against Saddam Hussein. He quickly corrected himself and subsequently referred to "the world community." He was, however, right when he erred.

5. Harry C. Triandis, *The New York Times*, Dec. 25, 1990, p. 41, and "Cross-Cultural Studies of Individualism and Collectivism," Nebraska Symposium on Motivation, vol. 37, 1989, pp. 41–133.

6. Kishore Mahbubani, "The West and the Rest," *The National Interest*, Summer 1992, pp. 3–13.

Amartya Sen   ← **NO**

A World Not Neatly Divided

When people talk about clashing civilizations, as so many politicians and academics do now, they can sometimes miss the central issue. The inadequacy of this thesis begins well before we get to the question of whether civilizations must clash. The basic weakness of the theory lies in its program of categorizing people of the world according to a unique, allegedly commanding system of classification. This is problematic because civilizational categories are crude and inconsistent and also because there are other ways of seeing people (linked to politics, language, literature, class, occupation or other affiliations).

The befuddling influence of a singular classification also traps those who dispute the thesis of a clash: To talk about "the Islamic world" or "the Western world" is already to adopt an impoverished vision of humanity as unalterably divided. In fact, civilizations are hard to partition in this way, given the diversities within each society as well as the linkages among different countries and cultures. For example, describing India as a "Hindu civilization" misses the fact that India has more Muslims than any other country except Indonesia and possibly Pakistan. It is futile to try to understand Indian art, literature, music, food or politics without seeing the extensive interactions across barriers of religious communities. These include Hindus and Muslims, Buddhists, Jains, Sikhs, Parsees, Christians (who have been in India since at least the fourth century, well before England's conversion to Christianity), Jews (present since the fall of Jerusalem), and even atheists and agnostics. Sanskrit has a larger atheistic literature than exists in any other classical language. Speaking of India as a Hindu civilization may be comforting to the Hindu fundamentalist, but it is an odd reading of India.

A similar coarseness can be seen in the other categories invoked, like "the Islamic world." Consider Akbar and Aurangzeb, two Muslim emperors of the Mogul dynasty in India. Aurangzeb tried hard to convert Hindus into Muslims and instituted various policies in that direction, of which taxing the non-Muslims was only one example. In contrast, Akbar reveled in his multiethnic court and pluralist laws, and issued official proclamations insisting that no one "should be interfered with on account of religion" and that "anyone is to be allowed to go over to a religion that pleases him."

If a homogeneous view of Islam were to be taken, then only one of these emperors could count as a true Muslim. The Islamic fundamentalist would have no time for Akbar; Prime Minister Tony Blair, given his insistence that tolerance is a defining characteristic of Islam, would have to consider excommunicating Aurangzeb. I expect both Akbar and Aurangzeb would protest, and so would I. A similar crudity is present in the characterization of what is called "Western civilization." Tolerance and individual freedom have certainly been present in European history. But there is no dearth of diversity here, either. When Akbar was making his pronouncements on religious tolerance in Agra, in the 1590's, the Inquisitions were still going on; in 1600, Giordano Bruno was burned at the stake, for heresy, in Campo dei Fiori in Rome.

Dividing the world into discrete civilizations is not just crude. It propels us into the absurd belief that this partitioning is natural and necessary and must overwhelm all other ways of identifying people. That imperious view goes not only against the sentiment that "we human beings are all much the same," but also against the more plausible understanding that we are diversely different. For example, Bangladesh's split from Pakistan was not connected with religion, but with language and politics.

Each of us has many features in our self-conception. Our religion, important as it may be, cannot be an all-engulfing identity. Even a shared poverty can be a source of solidarity across the borders. The kind of division highlighted by, say, the so-called "anti-globalization" protesters—whose movement is, incidentally, one of the most globalized in the world—tries to unite the underdogs of the world economy and goes firmly against religious, national or "civilizational" lines of division.

The main hope of harmony lies not in any imagined uniformity, but in the plurality of our identities, which cut across each other and work against sharp divisions into impenetrable civilizational camps. Political leaders who think and act in terms of sectioning off humanity into various "worlds" stand to make the world more flammable—even when their intentions are very different. They also end up, in the case of civilizations defined by religion, lending authority to religious leaders seen as spokesmen for their "worlds." In the process, other voices are muffled and other concerns silenced. The robbing of our plural identities not only reduces us; it impoverishes the world.

POSTSCRIPT

Are Cultural and Ethnic Rivalries the Defining Dimensions of Twenty-First Century Conflict?

Examining the causes and trends in conflict is a difficult enterprise. Perhaps no area in world politics has been studied and analyzed more than conflict. Arguing that certain forms of conflict are on the upswing or downswing is difficult because we do not have the benefit of historical context. Looking back on the twentieth century, for example, we can clearly see the increasing scope and impact of war; deaths and weapon destructiveness increased at virtually every stage. Determining the likelihood that ethnic conflicts will dominate the security agenda in the coming years, however, is difficult.

The sheer ferocity of ethnic conflicts in the past decade in places like Rwanda, Bosnia, Kosovo, Chechnya, Angola, Afghanistan, and Israel and Palestine seems to support Huntington's thesis that ethnicity and culture will define the battle lines between groups. Yet is culture at work or simply political power? The passion of the combatants may mask the more Machiavellian goals of leaders and individuals.

One element is certain. As the centralized power and influence of large imperial states like the USSR and others diminish, fault lines of conflict that may exist between people in various regions are more readily exploitable, like those in Yugoslavia by former Serb president Slobodan Milosovich. This means that such conflicts are indeed more likely, as leaders take advantage of paranoia, animosity, and fear to create an "other" against which a people may rally, at least for a short time. Whether this is part of a major global trend in ethnic conflict and cultural divides or merely political opportunism is still open to question.

Literature on this topic includes Farhang Rajaee, *Globalization on Trial: The Human Condition and the Information Civilization* (Kumarian Press, 2000); Daniel Patrick Moynihan, *Pandemonium: Ethnicity in International Politics* (Oxford University Press, 1994); and Stephen M. Walt, "Building Up a New Bogeyman," *Foreign Policy* (Spring 1997).

Contributors to This Volume

EDITORS

JAMES E. HARF is a professor of government and world affairs at the University of Tampa where he also serves as director of the Office of International Programs. He spent most of his career at The Ohio State University, where he holds the title of professor emeritus. He is coeditor of *The Unfolding Legacy of 9/11* (University Press of America, 2004) and coauthor of *World Politics and You: A Student Companion to International Politics on the World Stage*, 5th ed. (Brown & Benchmark, 1995) and *The Politic of Global Resources* (Duke University Press, 1986). He recently published his first novel, *Memories of Ivy* (Ivy House Publishing Group, 2005) about life as a university professor. He also coedited a four-book series on the global issues of population, food, energy, and environment, as well as three other book series on national security education, international studies, and international business. His current research interests include the world population problem and tools for addressing international conflict. As a staff member of the Presidential Commission on Foreign Language and International Studies, he was responsible for undergraduate education. He also served 15 years as executive director of the Consortium for International Studies Education.

MARK OWEN LOMBARDI is provost and chief academic officer at the College of Santa Fe. He is coeditor and author of *The Unfolding Legacy of 9/11* (University Press of America, 2004) and coeditor of *Perspectives on Third World Sovereignty: The Post Modern Paradox* (Macmillan, 1996). Dr. Lombardi has authored numerous articles on such topics as African political economy, U.S. foreign policy, the politics of the cold war, and collegiate curriculum reform. Dr. Lombardi is a former executive director of the U.S.-Africa Education Foundation, and he is currently a member of the International Studies Association, the African Studies Association, and the National Committee of International Studies Programs Administrators. His teaching interests include U.S. foreign policy, African politics, third world development, and globalization.

STAFF

Larry Loeppke	Managing Editor
Jill Peter	Senior Developmental Editor
Nichole Altman	Developmental Editor
Beth Kundert	Production Manager
Jane Mohr	Project Manager
Tara McDermott	Design Coordinator
Bonnie Coakley	Editorial Assistant
Lori Church	Permissions Coordinator

AUTHORS

GRAHAM ALLISON is the Douglas Dillon Professor of Government and director of the Belfer Center for Science and International Affairs at Harvard University. During the first term of the Clinton administration, he served as assistant secretary of defense for policy and plans. His publications include *Realizing Human Rights: From Inspiration to Impact* (St. Martin's Press, 2000).

RONALD BAILEY is science correspondent for *Reason* magazine and author of *ECOSCAM: The False Prophets of Ecological Apocalypse*, as well as an adjunct scholar with the Competitive Enterprise Institute.

JAGDISH BHAGWATI is a Columbia University economist and one of the world's foremost authorities on international trade. His latest book is *In defense of Globalization*.

JÉRÔME BINDÉ is director of UNESCO's Analysis and Forecasting Office.

JOSE BOVE is an internationally known French anti-globalization activist and unionist who travels the globe for his cause. He is a sheep farmer and producer of Roquefort cheese.

LESTER R. BROWN is founder, senior researcher, and chairman of the board of directors of the Worldwatch Institute, a research institute devoted to the analysis of global environmental issues. He earned his M.S. in agricultural economics from the University of Maryland in 1959 and his M.P.A. in public administration from Harvard University in 1962. His many books include *Eco-Economy: Building a New Economy for the Environmental Age* (W. W. Norton, 2001).

SYLVIE BRUNEL, a geographer and a specialist on development issues, is a former president of Action Contre la Faim (Action Against Hunger).

XIMING CAI is a research fellow in the Environment and Production Technology Division of the International Food Policy Research Institute. He earned his M.S. in hydrology and water resources at Tsinghua University in Beijing, China, and his Ph.D. in environmental and water resources at the University of Texas at Austin.

SARAH A. CLINE is a senior research assistant in the Environment and Production Technology Division of the International Food Policy Research Institute. She has also served as research assistant at Resources for the Future. She earned her M.S. in agricultural and resource economics from West Virginia University.

DAVID COLE is a professor of law at Georgetown University and author of *No Equal Justice: Race and Class in the American Criminal Justice System* (New Press, 1999).

BENJAMIN COMPAINE is a research consultant at MIT's Program on Internet and Telecoms Convergence, and coauthor of *Who Owns the Media? Competition and Concentration in the Mass Media Industry*, 3rd ed. (Lawrence Erlbaum Associates, 2000).

NETA C. CRAWFORD is an associate professor at Brown University's Watson Institute for International Studies. She is the author of *Argument and Change* *in World Politics: Ethics, Decolonialization, and Humanitarian Intervention* (Cambridge University Press, 2002).

CHRISTOPHER ESSEX is a professor of applied mathematics at the University of Western Ontario. He previously served as a NSERC visiting fellow at the Canadian Climate Centre and an Alexander von Humboldt Research Fellow.

JULIA GALEOTA is the 17-year-old winner of the 2004 Humanist Essay Contest.

KIM R. HOLMES is vice president and director of the Kathryn and Shelby Cullom Davis Institute for International Studies at the Heritage Foundation, where he is the principal spokesman on foreign and defense policy issues. He earned his M.A. and Ph.D. in history from Georgetown University in 1977 and 1982, respectively. His articles have appeared in such journals as *International Security* and *Journal of International Affairs.*

PETER HUBER is a senior fellow at the Manhattan Institute's Center for Legal Policy. His articles on science, technology, environment, and the law have appeared in such journals as the *Harvard Law Review* and the *Yale Law Journal,* and he is a regular columnist for *Forbes* magazine. He is the author of *Hard Green: Saving the Environment From the Environmentalists* (Basic Books, 2000).

SAMUEL P. HUNTINGTON is the Albert J. Weatherhead III University Professor at Harvard University, where he is also director of the John M. Olin Institute for Strategic Studies. He is the author of *The Clash of Civilizations and the Remaking of World Order* (Simon & Schuster, 1996).

INTERGOVERNMENTAL PANEL ON CLIMATE CHANGE (IPCC) was established in 1988 by the World Meteorological Organization (WMO) and the United Nations Environment Programme (UNEP). The role of the IPCC is to assess the scientific, technical, and socioeconomic information relevant to understanding the scientific basis of risk of human-induced climate change, its potential impacts, and options for adaptation and mitigation.

JERRY KANG is an acting professor of law at the University of California, Los Angeles. His teaching and research interests include civil procedure, Asian American jurisprudence, and cyberspace. He earned his J.D. from Harvard Law School in 1993, and he has also worked for the National Telecommunications and Information Administration.

HISHAM KHATIB is honorary vice chairman of the World Energy Council and a former Jordanian minister of energy and minister of planning.

ANDREI KOKOSHIN is a member of the Russian State Duma. Formerly, he served as vice president of the Russian Academy of Sciences, and as first minister of defense and secretary of the Security Council of the Russian Federation.

PHILLIPE LEGRAIN is chief economist of *Britain in Europe, the Campaign for Britain to Join the Euro.* He is the author of *Open World: The Truth about Globalization* (Abacus, 2002).

HARRY G. LEVINE is a professor of sociology at Queens College, City University of New York. He is best known as coauthor of *Crack in America: Demon Drugs and Social Justice.*

BJØRN LOMBORG is an associate professor of statistics in the Department of Political Science at the University of Aarhus in Denmark and a frequent participant in topical coverage in the European media. His areas of interest include the use of surveys in public administration and the use of statistics in the environmental arena. In February 2002 Lomborg was named director of Denmark's national Environmental Assessment Institute. He earned his Ph.D. from the University of Copenhagen in 1994.

PHILLIP LONGMAN is a senior fellow at the New American Foundation and author of *The Empty Cradle* (Basic Books, 2004).

FEDERICO MAYOR is former director-general of UNESCO. He has served as Spain's minister of education and science, as president of the University of Granada, and as a member of the European Parliament.

GORDON McGRANAHAM is director of the Urban Settlements Programme at the International Institute for Environment and Development in London.

ROSS McKITRICK is an associate professor of economics at the University of Guelph in Canada and a senior fellow of the Fraser Institute in Vancouver, B.C.

ROBERT S. McNAMARA was president of the World Bank Group of Institutions until his retirement in 1981. A former lieutenant colonel in the U.S. Air Force, he has also taught business administration at Harvard University, where he earned his M.B.A., and he served as secretary of defense from 1961 until 1968. He has received many awards, including the Albert Einstein Peace Prize, and he is the author of *In Retrospect: The Tragedy and Lessons of Vietnam* (Random House, 1995).

EDWIN MEESE III is the Ronald Reagan Distinguished Fellow in Public Policy at the Heritage Foundation. He also served as U.S. attorney general in the Reagan administration, chairman of the Domestic Policy Council and the National Drug Policy Board, and as a member of the National Security Council. He has also been a professor of law at the University of San Diego. He earned his J.D. from the University of California, Berkeley.

STEVEN W. MOSHER is president of the Population Research Institute and has served as director of the Asian Studies Program at the Claremont Institute in California.

DAVID PIMENTEL is a professor of insect ecology and agricultural sciences at Cornell University, and author of numerous related works.

MARK W. ROSEGRANT is a senior fellow at the International Food Policy Research Institute. He has over 24 years of experience in research and policy analysis in agriculture and economic development, with an emphasis on critical water issues as they impact world food security and environmental sustainability. He earned his Ph.D. in public policy from the University of Michigan.

RENATA SALECL is senior researcher at the Institute of Criminology of the Faculty of Law at the University of Ljubljana and author of *(Per)Versions of Love and Hate* (2000).

DAVID SATTERTHWAITE is director of the Human Settlements Programme at the International Institute for Environment and Development in London, and author of numerous books and articles on the subject.

HERBERT I. SCHILLER, who passed away in 2000, was the most influential media analyst of the left for half a century. He spent much of his academic career at the University of California at San Diego.

AMARTYA SEN is a Nobel Prize–winning economist (1998). Among his notable works are *Collective Choice and Social Welfare* (1970), *Poverty and Famines: An Essay on Entitlement and Deprivation* (1981), *On Ethics and Economics* (1981), and *Development and Freedom* (1999).

JESSICA STERN is a lecturer in public policy in the John F. Kennedy School of Government at Harvard University and a faculty affiliate at the Belfer Center for Science and International Affairs. She has also served as director for Russian, Ukrainian, and Eurasian Affairs at the National Security Council, where she was responsible for national security policy toward Russia and the former Soviet states and for policies to reduce the threat of nuclear smuggling and terrorism. She is the author of *The Ultimate Terrorists* (Harvard University Press, 1999).

JOSEPH E. STIGLITZ is a Nobel Prize winner in economics and professor of economics and finance at Columbia University Graduate School of Business. He is the author of numerous books and articles.

TIMOTHY TAYLOR is the managing editor of the *Journal of Economic Perspectives* at Macalester College. View a site called The Teaching Company for a more in-depth bio.

BARBARA BOYLE TORREY is a member of the Population Reference Bureau's board of trustees and a writer/consultant.

UNITED NATIONS ENVIRONMENTAL PROGRAMME (UNEP) was organized to provide leadership and to encourage partnership in caring for the environment by inspiring, informing, and enabling nations and peoples to improve their quality of life without compromising that of future generations. The executive director of UNEP is Klaus Töpfer, undersecretary general of the United Nations.

ENDY ZEMENIDES is a partner at the Chicago law firm Acosta, Kruse, Raines & Zemenides and a member of the National Strategy Forum Review Editorial Board.